25—

Carl Jay Bajema

PSA 1980

VOLUME ONE

PSA 1980

PROCEEDINGS OF THE 1980
BIENNIAL MEETING
OF THE
PHILOSOPHY OF SCIENCE
ASSOCIATION

volume one

Contributed Papers

edited by

PETER D. ASQUITH
&
RONALD N. GIERE

1980
Philosophy of Science Association
East Lansing, Michigan

Library of Congress Catalog Card Number 72-624169

Cloth Edition: ISBN 0-917586-14-X
Paperback Edition: ISBN 0-917586-13-1

ISSN 0270-8647

Manufactured in the United States of America
by Edwards Brothers, Ann Arbor, Michigan

Part IV. Statistics, Probability and Likelihood

Part V. Literature and the Philosophy of Science

Part VI. Reduction in Biology and Psychology

Part VII. Evolutionary Epistemology and the Sociology of Knowledge

Part XII. Economics and Sociobiology

PREFACE

Among the charges of the By-Laws of the Philosophy of Science Association is furthering "free discussion from diverse standpoints in the field of philosophy of science." To achieve this end PSA engages in a variety of activities including the holding of Biennial Meetings and arranging for the publication of the Proceedings of these meetings. In an effort to achieve timely and affordable publication of the Proceedings PSA itself began to publish the Proceedings starting with the 1976 Biennial Meeting. PSA 1976 appeared in two volumes - Volume 1 containing the contributed papers and some special session papers and Volume II containing the symposia papers. Because it was able to publish PSA 1976 in a timely and affordable manner, PSA has continued to publish its own Proceedings. PSA 1980 is also being published in two volumes - Volume I containing the contributed papers and Volume II containing papers from the symposia and other sessions. As was the case with the previous Proceedings PSA 1980 is utilizing "camera-ready typewritten copy supplied by the authors."

Contributed papers are submitted "publication-ready" well in advance of the Meeting, and acceptance decisions are reached six months or more in advance of the Meeting. This allows sufficient lead time to publish contributed papers prior to the Meeting, permitting the presentation of contributed papers at the Meeting by abstract only. This permits increased time for discussion by an audience that has had an opportunity to read the paper. The PSA membership was polled on the prospect of adopting such a publication arrangement - and overwhelmingly favored it. Consequently, Volume I is being published prior to the Meeting while Volume II will be published after the Meeting as symposia papers are not submitted until after the Meeting.

Printing "camera-ready copy supplied by the author" means that, after submission, there is no "proof stage" during which errors can be corrected. Thus, responsibility for accuracy, proof-reading, and the appearance of the copy rests with those producing the camera-ready copy --in this case the authors themselves. The authors were informed that they had this final responsibility and that PSA could not guarantee that any mistakes would be corrected. Errors did slip by, and when time and technical feasibility allowed, the editors attempted to correct deficiencies in the author-supplied copy. Known errors or infelicities do remain which could not be corrected without encountering substantial delays in the production of the volume. There are also undoubtedly undiscovered errors that remain, as well as some errors that might have been introduced in the process of correcting ones already there. Responsibility for errors in the individual contributions rests with the authors. Responsibility for the portions of the book produced by the editors, of course, rests with the editors.

Our debts are numerous. Thanks go to the 1980 Program Committee -

Richard Grandy, Marsha Hanen, Larry Laudan, Michael Ruse, and Larry Sklar and to the contributors to this volume. Lucy Williams had the responsibility for the mechanics of taking a collection of individual papers and organizing them in a format that would enable them to become a book. Valerie Joseph assisted in the typing. David Boersema and Stan Werne assisted in the proofreading and in the necessary library work.

Peter D. Asquith
Department of Philosophy
and Lyman Briggs College
Michigan State University

Ronald N. Giere
Department of History and
Philosophy of Science
Indiana University

July 9, 1980

1980 PSA PROGRAM [1]

FRIDAY, OCTOBER 17

MORNING

Symposium: *Probability and Causality*

Chairperson: Peter Achinstein, Johns Hopkins

Speakers: Nancy Cartwright, Stanford
Paul Humphreys, Virginia
Wesley Salmon, Arizona

Symposium: *Unity of Science - Group Selection and Sociobiology*

Chairperson: Richard Burian, Drexel

Speakers: David Hull, Wisconsin-Milwaukee
Elliott Sober, Wisconsin-Madison
William Wimsatt, Chicago

Colloquium: *Scientific Problems and Research Traditions*

Chairperson: Larry Lauden, Pittsburgh

Thomas Nickles, Nevada-Reno, "Scientific Problems: Three Empiricist Models"
Peter Barker, Memphis State, "Can Scientific History Repeat?"
Henry Frankel, Missouri-Kansas City, "Problem-Solving, Research Traditions and the Development of Scientific Fields"

[1] Program arrangements as of July 10, 1980

EARLY AFTERNOON

Colloquium: *Quantum Logic and the Interpretation of Quantum Mechanics*

Chairperson: Paul Horwich, M.I.T.

Gary Hardegree, Massachusetts-Amherst, "Micro-States in the Interpretation of Quantum Theory"

R.I.G. Hughes, Princeton, "Quantum Logic and the Interpretation of Quantum Mechanics"

Colloquium: *Species and Evolution*

Chairperson: Michael Ruse, Guelph

Arthur Caplan, Hastings Center, "Have Species Become Déclassé?"
Alexander Rosenberg, Syracuse, "Ruse's Treatment of the Evidence for Evolution"

Colloquium: *Statistics, Probability and Likelihood*

Chairperson: Ronald N. Giere, Indiana

Deborah G. Mayo, V.P.I., "The Philosophical Relevance of Statistics"
Ian Hacking, Stanford, "Grounding Probabilities from Below"
Steven Kimbrough, Wisconsin-Madison, "On the Use of Likelihood as a Guide to Truth"

Colloquium: *Literature and the Philosophy of Science*

Chairperson: E. Fred Carlisle, Michigan State

Walter Creed, Hawaii, "Philosophy of Science and Theory of Literary Criticism: Some Common Problems"
Edward Davenport, CUNY - John Jay College, "Progress in Literary Study"

LATE AFTERNOON

Symposium: *Quantum Theory in Historical Perspective*

Co-Sponsored with History of Science Society

Chairperson: Patrick Heelan, Stony Brook

Speakers: Linda Wessels, Indiana
Paul Teller, Illinois-Chicago Circle
Edward MacKinnon, California State-Hayward

Symposium: *Imagery and Representation and Psychology*

Chairperson: David M. Rosenthal, CUNY - Graduate School

Speakers: Steven Kosslyn, Harvard & George E. Smith, Tufts
Paul A. Kolers, Toronto
Robert Schwartz, Rochester

Symposium: *Social Values and Social Science*

Chairperson: Marsha Hanen, Calgary

Speakers: Sandra Harding, Delaware
Nancy Hartsock, Johns Hopkins
Noretta Koertge, Indiana

EARLY EVENING

Presidential Address

Mary B. Hesse, Cambridge, "Philosophy of Science, 1980: The Hunt for Scientific Reason"

LATE EVENING

Reception

SATURDAY, OCTOBER 18

MORNING

Symposium: *Relativity Principles*

Chairperson: John Earman, Minnesota

Speakers: Michael Friedman, Pennsylvania
Roger Jones, Tennessee
David Malament, Chicago

Symposium: *The Structure of Evolutionary Theory*

Chairperson: Michael Ruse, Guelph

Speakers: Mary B. Williams, Delaware
John Beatty, Harvard
Robert Brandon, Duke

Symposium: *Philosophy, Engineering and Society*

Chairperson: Vivian Weil, Illinois Institute of Technology

Speakers: Jerry W. Gravender, Clarkson
Michael P. Hodges, Vanderbilt
C. Thomas Rogers, Montana
Carol Ann Smith, Missouri-Rolla

EARLY AFTERNOON

Colloquium: *Reduction in Biology and Psychology*

Chairperson: Robert Causey, Texas

Monique Levy, Brussels, "The 'Reduction by Synthesis' of Biology to Physical Chemistry"
A. Lindenmayer and N. Simon, Utrecht, "Is Genetics Formalizable and Is Reduction of Classical to Modern Genetics Possible?"
Robert Richardson, Cincinnati, "Reductionist Research Programmes in Psychology"

Colloquium: *Evolutionary Epistemology and the Sociology of Knowledge*

Chairperson: Robert Butts, Western Ontario

Paul Thagard, Michigan-Dearborn, "Against Evolutionary Epistemology"
Edward Manier, Notre Dame, "Levels of Reflexivity: Unnoted Differences Within the 'Strong Programme' in the Sociology of Knowledge"

Special Session: *Team Teaching the Philosophy of Science*

Chairperson: Michael Bradie, Bowling Green

George Duncan, Bowling Green, "Philosophy of Space and Time"
Mark Gromko, Bowling Green, "Philosophy of Evolution"

LATE AFTERNOON

Symposium: *Locality and Hidden Variables*

Chairperson: Jeffrey Bub, Western Ontario

Speakers: Arthur Fine, Illinois-Chicago Circle
Patrick Suppes, Stanford
Abner Shimony, Boston University

Symposium: *The Problem of Data in Linguistics*

Chairperson: Sarah Stebbins, Douglas

Speakers: Guy Cardan, Yale
Arnold M. Zwicky, Ohio State
Richard Grandy, Rice

Symposium: *Scientific Discourse*

Co-Sponsored with Philosophy of the Social Sciences

Chairperson: Roger Krohn, McGill

Speakers: Karin Knorr, Pennsylvania
Kenneth Morrison, York
Dorothy Smith, Ontario Institute for Studies in Education
Gus Brannigan, Alberta

Colloquium: *History and the Metaphilosophy of Science*

Chairperson: Ernan McMullin, Nortre Dame

Stephen Wykstra, Tulsa, "Toward a Historical Meta-Method for Assessing Normative Methodologies"
Gary Merrill, Loyola, "Moderate Historicism and the Empirical Sense of 'Good Science'"
James Brown, Dalhousie, "History and the Norms of Science"

Panel Discussion: *Philosophy of Science and Science Policy*

Sandra Harding, Delaware
Ruth Macklin, Hastings Center
Jaakko Hintikka, Florida State
Alex Michalos, Guelph
Frederick Suppe, Maryland

SUNDAY, OCTOBER 19

MORNING

Symposium: *Scientific Realism*

Chairperson: Ronald N. Giere, Indiana

Speakers: Richard Boyd, Cornell
Clark Glymour, University of Pittsburgh
Bas van Fraassen, Toronto/USC

Symposium: *Philosophy of History*

Co-Sponsored with Philosophy of the Social Sciences

Chairperson: Karl-Dieter Opp, Hamburg

Speakers: P. H. Nowell-Smith, York
Leon Goldstein, SUNY-Binghamton

Symposium: *Values and Rhetoric*

Co-Sponsored with Philosophy of the Social Sciences

Chairperson: Frank Cunningham, Toronto

Speakers: John S. Nelson, Iowa
George J. Graham, Jr., Vanderbilt
Tom Settle, Guelph
William F. Connolly, Massachusetts

Colloquium: *The Cat Paradox and Quantum Logic*

Chairperson: Michael Gardner, South Carolina

James McGrath, Indiana-South Bend, "A Formal Statement of Schrödinger's Cat Paradox"
S. Bugajski, Katowice, "Only if 'Acrobatic Logic' is Non-Boolean"
E.-W. Stachow, Köln, "A Model-Theoretic Semantics for Quantum Logic"

Colloquium: *Time, Causation and Matter*

Chairperson: Larry Sklar, Michigan

Paul Fitzgerald, Fordham, "Is Temporality Mind Dependent?"
David Kline, Iowa State, "Are There Cases of Simultaneous Causation?"
K. S. Shrader-Frechette, Louisville, "Recent Changes in the Concept of Matter"

Colloquium: *Explanation*

Chairperson: Joseph Hanna, Michigan State

Barbara Klein, Yale, "What Should We Expect of a Theory of Explanation?"
David Gruender, Florida State, "Scientific Explanation and Norms in Science"
Ronald Laymon, Ohio State, "Idealization, Explanation and Confirmation"

Colloquium: *Economics and Sociobiology*

Chairperson: Joseph Pitt, V.P.I.

Daniel Hausman, Maryland, "How to do Philosophy of Economics"
R. Paul Thompson, Toronto, "Is Sociobiology a Pseudoscience?"

SYNOPSIS

The following brief summaries provide an introduction to each of the papers in this volume.

1. *Scientific Problems: Three Empiricist Models.* Thomas Nickles. One component of a viable account of scientific inquiry is a defensible conception of scientific problems. This paper specifies some logical and conceptual requirements that an acceptable account of scientific problems must meet as well as indicating some features that a study of scientific inquiry indicates scientific problems have. On the basis of these requirements and features, three standard empiricist models of problems are examined and found wanting. Finally a constraint inclusion-model of scientific problems is proposed.

2. *Can Scientific History Repeat?* Peter Barker. Although Kuhn, Lakatos and Laudan disagree on many points, these three widely accepted accounts of scientific growth do agree on certain key features of scientific revolutions. This minimal agreement is sufficient to place stringent restraints on the historical development of science. In particular it follows from the common features of their accounts that scientific history can never repeat. Using the term 'supertheory' to denote indifferently the large scale historical entitites employed in all three accounts, it is shown that a supertheory cannot succeed itself, or reappear after a number of intervening scientific revolutions. The relation of these arguments to the details of the three accounts is briefly examined.

3. *Problem-Solving, Research Traditions, and the Development of Scientific Fields.* Henry Frankel. The general thesis that science is essentially a problem-solving activity is extended to the development of new fields. Their development represents a research strategy for generating and solving new unsolved problems and solving existing ones in related fields. The pattern of growth of new fields is guided by the central problems within the field and applicable problems in other fields. Proponents of existing research traditions welcome work in new fields, if they believe it will increase the problem-solving effectiveness of their tradition. Correspondingly, researchers in new fields will graft their work onto established traditions, if they believe it will augment the problem-solving effectiveness of their work. The above claims are defended through using the development of paleomagnetism as a case study.

4. *Micro-States in the Interpretation of Quantum Theory.* Gary M. Hardegree. The interpretation of quantum mechanics is discussed from the viewpoint of quantum logic (QL). QL is understood to concern the

possible properties that can be ascribed to a physical system SYS. The micro-state of SYS at any given moment t is identified with the set of all properties actualized by SYS at time t. Minimal adequacy requirements are proposed for all interpretations of micro-states. A strict interpretation is defined to be one according to which the properties ascribable to SYS are individuated by the projection operators on the associated Hilbert space. Two strict interpretations are examined. Kochen's interpretation is also discussed, and it is argued that it is not a strict interpretation.

5. *Quantum Logic and the Interpretation of Quantum Mechanics.* R.I.G. Hughes. One problem with assessing quantum logic is that there are considerable differences between its practitioners. In particular they offer different versions of the set of sentences which the logic governs. On some accounts the sentences involved describe events, on others they are ascriptions of properties. In this paper a framework is offered within which to discuss different quantum logical interpretations of quantum theory, and then the works of Jauch, Putnam, van Fraassen and Kochen are located within it.

6. *Have Species Become Declássé?* Arthur L. Caplan. Traditionally, species have been treated as classes or kinds in philosophical discussions of systematics and evolutionary biology. Recently a number of biologists and philosophers have proposed a drastic revision of this traditional ontological categorization. They have argued that species ought be viewed as individuals rather than as classes or natural kinds. In this paper an attempt is made to show that (a) the reasons advanced in support of this new view of species are not persuasive, (b) a reasonable explication can be given of the treatment of species as classes that is consistent with current theory and practice in evolutionary biology and systematics, and (c) that once certain confusions concerning the species concept have been clarified, there are good theoretical grounds for maintaining that species are best viewed as classes or kinds.

7. *Ruse's Treatment of the Evidence for Evolution: A Reconsideration* Alexander Rosenberg. It is argued that the assessment of the strength of the evidence for the Darwinian theory of evolution by natural selection offered by Michael Ruse in the Philosophy of Biology is in one respect too weak and in the other too strong. His claim that artificial selection provides at best analogical evidence for the theory is shown to rest on a spurious distinction between artificial and natural selection. His argument that Darwinian theory, unlike its competitors, accounts for the cytological and genetic data is demonstrated to be unwarranted and fails to differentiate the actual degrees of evidential support provided by cytology and genetics for parts of differing and competing theories of evolution. The evidentially secure foundations of Darwin's theory are not challenged in this paper, only Ruse's account of their nature.

8. *The Philosophical Relevance of Statistics.* Deborah G. Mayo. While philosophers have studied probability and induction, statistics has not received the kind of philosophical attention mathematics and physics have. Despite increasing use of statistics in science,

statistical advances have been little noted in the philosophy of science literature. This paper shows the relevance of statistics to both theoretical and applied problems of philosophy. It begins by discussing the relevance of statistics to the problem of induction and then discusses the reasoning that leads to causal generalizations and how statistics elucidates the structure of science as it is actually practiced. In addition to being relevant for building an adequate theory of scientific inference, it is argued that statistics provides a link between philosophy, science and public policy.

9. *Grounding Probabilities From Below.* Ian Hacking. Does the frequency distribution in the population derive from probabilistic facts about the individuals that compose it or are there some stable frequencies that pertain to populations, but do not derive from probabilistic facts about members of the population? The author of this paper suggests that some natural phenomena may be accurately described in propensity terms, while others are accurately described only in frequency terms.

10. *On the Use of Likelihood as a Guide to Truth.* Steven Orla Kimbrough. Confirmation functions are generally thought of as probability functions. The well known difficulties associated with the probabilistic confirmation functions proposed to date indicate that functions other than probability functions should be investigated for the purpose of developing an adequate basis for confirmation theory. This paper deals with one such function, the likelihood function. First, it is argued here that likelihood is not a probability function. Second, a proof is given that, in the limit, likelihood can be used to determine which of two observationally distinct hypotheses is true. Finally, a demonstration is given that, in the presence of a finite amount of experimental information, likelihood can serve as a good estimator of truth.

11. *Philosophy of Science and Theory of Literary Criticism: Some Common Problems.* Walter Creed. Structuralism as well as other methods of literary criticism, take positions analogous to ones espoused in some philosophies of science. Examples are: regarding a discipline as self-contained, having no necessary connection with the external world; taking interpretation (or the postulating of theories) as an arbitrary process, valid if it makes sense of the data, thus avoiding questions of truth; diminishing individuality by overemphasizing the learned aspects of a discipline (reading as governed by assimilated rules, research as controlled by shared goals and methods and implicit agreement on fundamental questions). Interdisciplinary exchange should lead to greater awareness both of difficulties of these positions and of proposed solutions to the difficulties they raise.

12. *Progress In Literary Study.* Edward Davenport. Literary study has been thought incapable of progress because its aims and standards were thought to depend too highly on the value assumptions of the culture which fostered it to permit progress across cultures and across centuries. Max Weber proposed a strategy for working toward objective knowledge and progress in the social sciences generally, which can be applied to

literary study. The Weberian strategy is to make the value assumptions behind a given theory explicit. Applying the Weberian strategy to literary study will enable us to make our knowledge of literature and of literary study more objective because we can replace a model of chaotic and incommensurable work with a model of competing research programs in literary study and we can then learn to compare and improve these programs.

13. *The 'Reduction by Synthesis' of Biology to Physical Chemistry.* Monique H. Levy. The reduction of biology to physical chemistry raises various problems. The present debate implicitly rests on a 'classical' conception of reduction inspired principally by Nagel's work. This conception, not being wholly consistent with scientific practice, should be replaced by 'reduction by synthesis'. This approach is based on relations between chemistry and physics, and offers interesting possibilities for its application to the case of biology.

14. *The Formal Structure of Genetics and the Reduction Problem.* A. Lindenmayer and N. Simon. The discussion of theory reduction in genetics threatens to become more and more confused. The position taken is that before one tries to work out complicated reduction principles which might be applicable to broad areas of biology in their relationships to chemistry and physics, it would be better to attempt first to elucidate the internal structure of some limited biological theories in a formal way and to consider simple constructs for reduction between them. This proposal is elaborated with respect to the original Mendelian genetics, linkage genetics and fine-structure genetics, and their relationship to non-formalized molecular genetics.

15. *Reductionist Research Programmes in Psychology.* Robert C. Richardson. Reductionist research programmes in psychology, and elsewhere, are typified by a number of research strategies and methodological assumptions. The current essay isolates and examines some typical reductionist assumptions as they have been embodied in psychological research. Through a brief examination of the use of lesion studies coupled with functional deficit analyses, it is argued that localizationist approaches to the study of brain function incorporate at least four interlocking hypotheses. Two of the hypotheses are examined in detail. It is urged that neither is warranted, and there is reason to think each is suspect.

16. *Against Evolutionary Epistemology.* Paul Thagard. This paper is a critique of Darwinian models of the growth of scientific knowledge. Donald Campbell, Karl Popper, Stephen Toulmin, and others have discussed analogies between the development of biological species and the development of scientific knowledge: in both kinds of development, we find variation, selection, and transmission. It is argued that these similarities are superficial, and that closer examination of biological evolution and of the history of science shows that a non-Darwinian approach to historical epistemology is needed. An adequate model of the growth of knowledge will have to go beyond evolutionary epistemology in discussing the role of intentional, abductive theory construction in scientific discovery, the selection of theories according to general

criteria, the achievement of progress by sustained application of criteria, and the transmission of selected theories in highly organized scientific communities.

17. *Levels of Reflexivity*. Edward Manier. A basic question confronting programs in the sociology of science is: "Can the thesis that cognitive claims are socially determined be interpreted in a way that preserves the credibility of the sociology of science, when that thesis is reflexively applied to the sociology of science?" That question is approached here by means of a critical comparison of two versions of the "strong programme" in the sociology of knowledge. The key difference is the effort in one of the two versions (B. Barnes') to develop a context within which to articulate the distinction of science and ideology.

18. *Toward a Historical Meta-Method for Assessing Normative Methodologies: Rationability, Serendipity, and the Robinson Crusoe Fallacy*. Stephen J. Wykstra. How can the philosopher use history of science to assess normative methodologies? This paper distinguishes the "intuitionist" meta-methodologies from the "rationability" meta-methodology. The rationability approach is defended by showing that it does not lead to anarchistic conclusions drawn by Feyerabend, Lakatos, and Kuhn; rather, these conclusions are the result of auxiliary assumptions about the nature of rational norms. By freeing the rationability meta-method from these assumptions, the specter of anarchism can be exorcised from it.

19. *Moderate Historicism and the Empirical Sense of 'Good Science'*. G.H. Merrill. Unlike the radical historicist and the radical logicist, the moderate historicist in the philosophy of science adopts the position that neither purely a priori (i.e., logical or philosophical) nor purely historical considerations alone determine the acceptability of a philosophical analysis of science. A dilemma arising from the nature of this position is first described and then it is argued that what is perhaps the most plausible way of avoiding this dilemma is doomed to failure. A particular example of this attempt at escaping the dilemma is considered in some detail, and along the way evidence is amassed in support of the view that no non-trivial statement of moderate historicism will be coherent.

20. *History and the Norms of Science*. James Robert Brown. Starting from the assumption that the history of science is, in some significant sense, rational and thus that historical episodes may serve as evidence in choosing between competing normative methodologies of science, the question arises: "Just what is this history-methodology evidential relation?" After examining the proposals of Laudan, a more plausible account is proposed.

21. *A Formal Statement of Schrödinger's Cat Paradox*. James H. McGrath. Using formal techniques, Schrödinger's 1935 cat argument is reproduced. Assumptions of the argument are made explicit as axioms and rules of inference; from these a contradiction is derived. The formal statement is then used to elucidate several crucial distinctions, to reject several commonly proposed resolutions, and to sketch an Einsteinian perspective for the argument.

22. *Only If 'Acrobatic Logic' Is Non-Boolean.* Slawomir Bugajski. The procedure of 'reading off' quantum logic from the standard quantum mechanics is generalized and described in some detail. It is demonstrated that a recent criticism of quantum logic aimed to discredit the 'read off' procedure is unsound.

23. *A Model Theoretic Semantics for Quantum Logic.* E. -W. Stachow. This contribution is concerned with a particular model theoretic semantics of the object language of quantum physics. The object language considered here comprises logically connected propositions, sequentially connected propositions and modal propositions. The model theoretic semantics arises from the already established dialogic semantics, if the pragmatic concept of the dialog-game is replaced by a "metaphysical" concept of the game. The game is determined by a game tree, the branches of which constitute a set, the set of "possible worlds" of an individual quantum physical system. The semantic concepts like truth, falsity and valuation are defined with respect to this set.

24. *Is Temporality Mind-Dependent?* Paul Fitzgerald. A distinction is made between the indexicality theme and the elapsive theme. The first theme is concerned with the question of whether nowness and other irreducibly indexical A-determinations are mind-dependent or not. It is argued that there are no such A-determinations, within or outside of mind. The second, elapsive theme, which is often not distinguished from the first, deals with whether or not non-indexical felt transiency or elapsiveness is mind-dependent. Four arguments for the mind-dependence of "temporal becoming" are assessed as they apply to these two kinds of temporal becoming.

25. *Are There Cases of Simultaneous Causations?* A. David Kline. Alleged cases of simultaneous causation have played a prominent role in the critique of various accounts of explanation/causation and in the formation of alternative accounts. It is argued that none of the stated cases are genuine instances of simultaneous causations, since they all violate the special theory of relativity (STR). The conditions a genuine case would have to meet in light of the restrictions imposed by STR are outlined.

26. *Recent Changes in the Concept of Matter: How Does 'Elementary Particle' Mean?* K.S. Shrader-Frechette. In this paper the author analyzes the recent history of the concept of matter by examining two criteria, in-principle-observability and noncompositeness, for use of the term 'elementary particle'. Arguing that how these criteria are employed sheds light on a change in what matter means, the author draws three conclusions. (1) Since the seventeenth century, in-principle-observability has undergone a progressive devaluation, if not abandonment, in favor of the criterion of theoretical simplicity. As a consequence, (2) the concept of matter has undergone a "third phase" of dematerialization. (This is an extension of the view of Russ Hanson, who described two such phases.) Finally, (3) the current concept of matter reveals a dilemma: if alleged elementary particles are verified through observation, they are composite and hence not elementary; if they are elementary, they are in-principle-unobservable.

27. *What Should We Expect of a Theory of Explanation?* Barbara V. E. Klein. The purpose of this paper is to provide a meta-theoretic characterization of what ought to be expected of a general theory of explanation, a theory of scientific explanation, and a theory of a kind of explanation. The view presented, called "the logico-normative view", is taken to be implicit in the work of a number of influential writers on the subject, including Hempel. The paper falls into three parts. First, a number of pre-theoretic assumptions are articulated. Second, the two concerns of a general theory of explanation are discussed: (1) to answer a variety of questions pertaining to the nature of explanation and (2) to solve the normative problem for explanation. Finally, the three kinds of theories of explanation are distinguished in terms of the sorts of questions which they address.

28. *Scientific Explanation and Norms in Science.* David Gruender. The paper discusses theories of scientific explanation from the point of view of the norms or ideals of science they exemplify. The relationship of the adoption of these norms to metaphysical positions on determinism is explored, and means for reducing the conflict between methodological and metaphysical issues are suggested.

29. *Idealization, Explanation, and Confirmation.* Ronald Laymon. The use of idealizations and approximations in scientific explanations poses a problem for traditional philosophical theories of confirmation since, strictly speaking, these sorts of statements are false. Furthermore, in several central cases in the history of science, theoretical predictions seen as confirmatory are not, in any usual sense, even approximately true. As a means of eliminating the puzzling nature of these cases, two theses are proposed. First, explanations consist of idealized deductive-nomological sketches plus what are called modal auxiliaries, i.e., arguments showing that if the idealizations used in the initial conditions are improved, then there will be an improvement in the prediction. Second, a theory is confirmed if it can be shown that its idealized sketches can be improved; similarly, a theory is disconfirmed if its idealized sketches cannot be improved. Several examples are given to illustrate both confirmation and disconfirmation achieved by means of the modal auxiliary. These cases are compared with Scriven's bridge example.

30. *How to do Philosophy of Economics.* Daniel M. Hausman. This paper sketches the contemporary turn in philosophy of science and discusses its practical implications for doing philosophy of economics. This turn consists basically of regarding philosophy of science as itself an empirical (social) science. It thus embodies a naturalized epistemology. Some of the circularities inherent in such an epistemology are examined, and it is argued that they are not vicious. Although an empirical approach to the philosophy of science is defended, it is pointed out that there are practical difficulties employing it when studying a discipline like economics in which dispute and controversy are so pervasive. It is argued that the implications of the empirical approach to philosophy of science for day-to-day philosophical practice are undramatic.

31. *Is Sociobiology A Pseudoscience?* R. Paul Thompson. Among the numerous criticisms of sociobiology is the criticism that it is not genuine science. This paper defends sociobiology against this criticism. There are three aspects to the defense. First, it is argued that the testability criterion of pseudoscience is generally problematic as a criterion and that even if accepted it fails to mark sociobiology as a pseudoscience. Second, it is argued that Thagard's more comprehensive and sophisticated criterion of pseudoscience fails to mark sociobiology as a pseudoscience. Third, a positive case for accepting sociobiology as genuine science is presented.

Part I

Scientific Problems and Research Traditions

Scientific Problems: Three Empiricist Models[1]

Thomas Nickles

University of Nevada, Reno

1. The "Problem" Problem

One component of a viable account of scientific inquiry is a defensible conception of scientific problems, i.e., a conception which meets logical demands and which also fits or explains basic "data" arising from the history of that inquiry, including an account of how problems arise. In this paper I shall argue that three standard empiricist models of problems--essentially the logical positivist and Popperian views--satisfy neither the historical requirements nor even the logical conditions (except, weakly, in the case of Popper). These failures are instructive, however. They represent steps toward the solution of the "problem" problem--the problem of developing a model of problems rich enough to account for the data. In the weakest sense (which is all I need), 'accounting for' a datum means explaining how that aspect of scientific inquiry is possible.[2] I shall not have space here to consider promising recent work on problems by Laudan (1977) and others.

I now list a few obvious logical requirements on problems and some uncontroversial data from the history of inquiry, with parenthetical indication of some of the 'how possibly?' questions that they raise. While each item alone may seem so basic as to be platitudinous, we do not have a conception of problems which does justice to them all.

A. Logical and Conceptual Requirements

(1) Problems exist, and some are known. (How is that possible?)

(2) Problems are sometimes solved, i.e., their solutions are discovered. Thus inquiry is possible. (How is inquiry possible?)

(3) Problems are identical only if their solutions (or their classes of admissible solutions) are identical.

PSA 1980, Volume 1, pp. 3-19

(4) Theories (problem solutions) are identical only if the problems they solve are identical.

(5) Two distinct problems may be solved by the same theory and in that sense may have the same solution and, *a fortiori*, the same *range* of admissible solutions (else one theory could solve only a single problem).

(6) A problem may be solved by two or more distinct theories (else there could be no competitive solutions to the same problem).

(7) Problems exist only in relation to goals which have not been achieved.

(8) Problems have objective existence within historical bodies of theory and practice and their goals. Some problems are discovered, some remain unknown or only partially known. The discovery process may be gradual.

(9) Two scientists can have the same problem without knowing the same things about it, and can approach the problem from different directions, even from different fields.

(10) Theories are (putative) problem solutions (but not all problem solutions are theories).

(11) Some problems are subproblems of larger problems.

B. *Evidence that Problems Have Conceptual Depth (see also C)*

(1) Problems can be very puzzling. (E.g., we cannot see how a phenomenon is possible, given what we know; or two powerful and intuitively obvious theories or principles clash.)

(2) Many scientific problems are ill structured, and for substantive (*vs*. purely formal, methodological) reasons.

(3) Despite (2), scientists make reliable judgments about when a problem has been solved. There frequently is near unanimity of agreement. (Kuhn 1962). (How possible?)

(4) Scientists can sometimes make reliable judgments about the solvability of still unsolved problems, including solvability in terms of a given body of theory and the amount of time and effort required to obtain a solution--and thus to evaluate the likely success of alternative research proposals and programs.[3]

(5) Scientists make reliable judgments (and agree) about the cognitive weighting of problems (their intellectual fruitfulness, importance, generality, and centrality). Scientists *know* that some problems are more interesting than others. (Kuhn 1962; Laudan 1977, p. 32). (How possible?)

(6) There exist overdetermined problems, i.e., problems whose constraints cannot all be satisfied (inconsistent constraint sets); but not all problems are overdetermined. (Lugg 1978).

(7) The discovery process typically is structured in time rather than being a momentary psychological experience of the solution popping into someone's head. (Kuhn 1977, Ch. 7).

(8) Complex reasoning typically occurs in problem-solving contexts. (How is such reasoning possible?)

(9) This reasoning falls into many different patterns and is not all inductive. Enumerative induction from the data to a solution is rare. (How is noninductive reasoning to problem solutions possible?)

(10) Problems and problem solutions (e.g., theories) in modern science frequently are highly esoteric (Kuhn 1962; 1977, p. 236) or positively weird (Shapere 1980).

(11) Historically, tradition-bound research in science has been more rapidly and continuously innovative and progressive than problem-solving behaviors not linked to a definite tradition or research program. (Lakatos 1970; Kuhn 1977, p. 234; Laudan 1977).

C. Evidence that Conceptual Constraints Belong to the Problem Itself and Cannot Be Removed to the Background

(1) Some problems are deeper than others.

(2) There are many and diverse types of intellectual problems besides explanation-prediction problems, e.g., determination problems, clarification problems, problems of inconsistency and incoherence (either internally or with other principles, theories, or world views),[4] and others in natural science alone, not to mention problems of pure mathematics, philosophy, etc. Some of these problem types do not have empirical data, at least not as an important component. Many problems can be understood apart from the data, and one may be familiar with all the relevant data without seeing the problem.

(3) Data sometimes do not constitute the problem but serve chiefly as evidence that a problem (or at least a deeper problem) exists. (E.g., the null experiments and relativity.)

(4) Some problems remain unsolved, and some of those are unsolvable, even though the relevant data, if any, have been explained or are explainable.[5]

(5) Problems can be reformulated in significantly (conceptually) different ways, formulated more or less completely, transformed, and reduced to other problems--all without essential change in the

presentation of the data to be explained.[6]

(6) Problems can be modeled on other problems, even when the data or subject matter is dissimilar. The modeling is more substantial than the data alone would permit.[7]

(7) Recognizing and adequately formulating a scientific problem can be very challenging tasks. (Bantz 1980).

(8) The more constraints on the problem solution we know, and the more sharply they are formulated, the more sharply and completely we can formulate the problem, and the better we understand it.

(9) Formulating a good problem is an important theoretical scientific achievement, frequently different from the discovery or production of data for explanation.

The divisions A through C are for convenience of organization and should not be taken to imply a sharp difference between logical and historical requirements. Datum A-1 is an oblique form of the Meno paradox, which is raised explicitly by A-2. The paradox (found in Plato's Meno, 80d-e) raises the most basic question about inquiry: How is inquiry possible? As several methodologists have noted,[8] it would be a mistake to dismiss the paradox as a curiosity of ancient philosophy, of no relevance to 20th-century philosophy of science. The paradox is commonly formulated as a dilemma: Either you know what you are searching for (in trying to solve a problem or acquire new knowledge) or you do not. If you do know, you already have it, whence inquiry is not possible. And if you do not know, you would not recognize it even if you stumbled on it accidentally; hence, again, inquiry is impossible, pointless. The way out of the paradox is to show that the second statement is false: you can know what you are looking for without already having it.

The question, "How is inquiry possible?" has a strong and a weak form. The weak form requests only conditions on what would terminate inquiry, on what would count as obtaining the goal of inquiry. The strong form demands, in addition, guidance as to how to search for the goal state, not simply how to recognize it if you happen to stumble upon it. (Here we may further distinguish degrees of strength - from the provision of algorithms down to the provision of helpful hints and heuristic remarks. One can address the strong problem of inquiry without presupposing the existence of an algorithmic discovery procedure or even a general methodology.) 20th-century empiricists have so totally neglected the strong form of the question as to deny its importance for scientific inquiry. Surely, however, a methodology or lesser account of inquiry which can answer both the weak and the strong forms (insofar as possible) is superior to one which addresses only the weak form.

My task in this paper is a reflexive one. I want to inquire into what a problem itself is. While I cannot hope in this short space to fully and explicitly display this object, I want at least to indicate what sort of thing we are looking for. If successful, I shall at least have clarified the "problem" problem.

2. Three Empiricist Models of Scientific Problems

A. The minimal empiricist model. The traditional empiricist conception of problems is simple, even ascetic. The conceptual cupboard is bare. A problem on this view is an empirical fact in search of explanation and/or prediction--or a process in search of a method of humanly controlling it (depending on the variety of empiricism in question). It is difficult to find passages explicitly expounding this conception, since problems, as such, almost never are (were) discussed by strong empiricists. Nevertheless, this view of problems is clearly implied in discussions of the goals of science as explanatory and predictive systematization (or prediction and control) of observational data as well as in the accounts of explanation, of prediction, and of theories (which are standardly treated as large-scale explanations of data).[9] For as Toulmin (1972) observes, problems are what is left when we subtract current capabilities from disciplinary goals or ideals (see A-7). A problem for an empiricist is a hitherto untreated datum plus the demand that it be explained (predicted, controlled). Derivatively, a theory faces problems insofar as it has not been applied to, or is unable to, explain or predict data in its domain. In short, a problem has very little content or structure.

How well does this minimal empiricist model handle our data on problems? If a problem were only a datum plus the demand that it be explained (predicted, etc.) and no restrictions at all were placed on what counts as an explanation, then we hardly could have a definite problem, much less a definite, general conception of one. We should experience only an unstructured sense of puzzlement on certain occasions, I suppose, without having the least idea what would count as an answer or where to search for one. It would make little sense to speak of the existence of anything so ill defined (A-1). In short, this model cannot resolve the Meno paradox by accounting for the existence of well-defined problems and the possibility of inquiry, whether directed or not.

There is no need to discuss each item on our list, one by one, for it is evident that the minimal empiricist model can explain scarcely any of our data. In order to avoid the accusation of attacking a straw man, I hasten to flesh the model out a bit.

B. The positivist model. Let us therefore suppose that logical empiricist methodological conceptions of explanation, law, theory, confirmation, and the like are built into the model. A problem thus becomes a more complex thing. It now possesses a methodological component--a set of formal, methodological restrictions or constraints on a solution--in addition to the substantive empirical component, the datum to be explained.[10]

To take a concrete example, if we adopt Carl G. Hempel's familiar views (1965, 1966) on the structure of explanation, theories, etc., then the problem of explaining a phenomenon becomes the problem of finding a well-confirmed, general law or theory plus initial and boundary condition statements such that the explanandum statement, describing the phenomenon to be explained, can be logically deduced from the former. I shall call this "souped up" version of the minimal empiricist model the positivist model. Unlike the former, the positivist model includes methodological constraints in addition to the purely observational constraints on the problem solution.

This addition brings us much nearer a resolution of Meno's paradox. For on the positivist model, a reasonably full statement of the problem itself tells us what to look for in a solution. If you stumble on something, your problem itself provides methodological rules for deciding whether or not that something is a solution to the problem. The model therefore solves the problem of inquiry in its weak form. Is it rich enough to handle the strong form?

No, as Hempel would agree. Indeed, the positivists and Karl Popper[11] (at least in his older, classical writings) are one in insisting that there can be no methodology of discovery--only a methodology of justification (corroboration for Popper). In his well known book Hempel (1966, p.12) states that a problem (as he conceives it) is too vague to constitute an efficient starting point for inquiry (and, hence, for methodology). Having a problem, says Hempel, does not even tell us which facts are relevant to our subject, let alone where to look for, or how to construct, an explanatory theory. For Hempel and the positivists, oddly enough, directed inquiry cannot begin from scientific problems; it can only begin once a theory is already in hand. The inquiry consists in testing the theory. There exists an algorithm or quasi-algorithmic procedure for justification but none for discovery on the positivist view. Strong inquiry is impossible in science. Meno was right about strong inquiry.

This positivist solution to Meno's paradox is both too weak and (in another sense) too strong. It is too strong in claiming the existence of a logically rigorous procedure for testing a theory, i.e., an algorithm for determining whether a theory really is a problem solution, whether it really does explain the data. For this implies that scientific problems are, in general, better structured than they in fact are (B-2). Walter Reitman (1964) and Herbert Simon (1973) have pointed out, correctly, that the typical scientific problem is very ill structured indeed, an observation which extends to most intellectually interesting problems. In general, there exists no algorithmic procedure for deciding whether theory T solves problem P.[12] As usual, the positivists attempt to model scientific inquiry too closely on certain areas of logic and mathematics.

The positivist solution to the problem of inquiry is also too weak, since it ignores the presence of complex reasoning--highly directed inquiry--in the context of discovery; and by now numerous historical case studies have verified the existence of such reasoning in the generation phase of scientific ideas as well as in the testing phase. Indeed, the positivist model is inadequate to explain even the possibility of noninductive reasoning _to_ a problem solution, e.g., to a new theory. According to the model, a problem is mostly logical form (the prescribed logical structure of theories, explanations, etc.) and very little content or substance (the data to be explained). These empirical materials are so slight that the only[13] substantive reasoning to a problem solution _possible_ on the positivist model is enumerative inductive reasoning from data to hypothesis (problem solution). However, not even positivistic writers today place such high stock in simple induction. Hempel himself (1966) points out that induction from the facts is a particularly unsatisfactory method of generating deep new theories; and it is precisely where deep problems and solutions are explored that the reasoning is most evident and most subtle. The positivist model is left with no means to explain such reasoning, since a problem (on this model) contains no substantive materials, besides the data, to serve as premises.

Given the fact that the empiricist-positivist conception of problems has constituted the received view, the standard empiricist conclusion that there can be no epistemology of discovery is hardly surprising. On the standard view, discovery is an irrational process, a conclusion further supported by the _mis_identification of rational thought with algorithmic procedures and of methodology with logic.

Finally (to mention only one of several remaining difficulties), the empiricist view in all its variations makes the mistake of casting all problems into the same mold, _viz_.: Explain (predict, control) phenomenon ∅! It is just another instance of the common philosophical mistake of generalizing from a limited variety of examples. As datum B-6 and several of the items under heading C indicate, some problems involve the explanation of phenomena only incidentally, if at all. (Strong empiricists must treat empirical problems as different _sorts_ of things than pure mathematical problems--and what of philosophical problems? They also must assume an "absolute" observational/theoretical distinction as the basis of their model.) Deep problems resulting from the unexpected conflict of two well established theories, such as classical mechanics and Maxwellian electromagnetic theory, cannot even be represented by simple empiricist models of problems. Since they have no way at all to represent conceptual depth of a substantive nature, the empiricist models cannot discriminate deep, central problems from shallow, peripheral ones (and in this respect, too, fail to guide inquiry). The reply that an anomalous datum generates a more serious problem than a routine datum is not really available to the empiricist, since anomaly is defined only against a theoretical background, and the empiricist model makes no allowance for theoretical elements in the problem situation.

It may be objected that the positivists did assume the presence of a theoretical background to problems, and that this device does permit one to handle those data concerned with the conceptual depth of problems. This is probably the case, but since it is Popper who has explicitly advanced this view, I shall call it the Popperian model of problems. It is to this model that I now turn.

C. The Popperian model. Popper, who has stressed the importance of problems to methodology more than any major philosopher of science, analyzes a "problem situation" into a "problem", a "framework", and a "theoretical background" (his terms). In the example Popper (1972, p. 172) provides, one of Galileo's problems was simply to explain the tides, but the Pisan's problem situation was more complex, since he set the problem against the theoretical background of the Copernican viewpoint, to which he was firmly committed, and attempted to solve it in terms of his own conjectural hypothesis (framework) of circular inertia. In short, for Popper a problem proper may be formulated as a simple question or request concerning the explanation of a phenomenon--just the positivist model--but the total problem situation is a complex thing with conceptual depth. Elsewhere, Popper and his followers place more emphasis on problem situations as inconsistencies between theory and data or between two theories.

How well does the Popper model account for our data on problems? By including the theoretical background as well as a more conjectural framework or program in the total problem situation, Popper obtains a much richer view of problems (or rather, problem situations) than the positivist model, a view which enables him to account, more or less adequately, for each item in data sets A and B.[14] The Meno paradox (A-1 and 2) poses no difficulty for the Popperian model. Like the positivist model, Popper's methodology specifies what would count as a problem solution. (I assume that Popper, too, wishes to include methodological constraints in the problem situation--presumably in the theoretical background or as a separate, metatheoretical component--but Popper is silent on the matter.)

Popper's model therefore solves the weak form of the problem of inquiry. Moreover, Popper's model surpasses the positivist model in providing additional, substantive constraints on the problem solution, constraints which can give real direction to inquiry. In an interesting passage, Popper tells us how to come to understand a deep problem situation. First, propose a solution, even an obviously poor one, and then proceed to criticize it. The criticism by which we reject the solution will provide additional conditions on an adequate problem solution or will at least make fully explicit a known condition. By continuing this process of conjecture and critical refutation (which again is not algorithmic but requires imagination), we gain a much clearer understanding of the problem, because (to put it my way), this process makes explicit many additional constraints on the problem solution. The process eventually may cut down the problem space tremendously, permitting the conjectural solutions to be more and more controlled as to focus, however bold and imaginative they may be as hypotheses. Further inquiry can now be very focused indeed.[15]

Thus Popper's model does at least address the strong form of the problem of inquiry, despite his rejection of a methodology of discovery. Popper can at least explain the possibility of detailed, noninductive reasoning in the context of discovery.

Similarly, Popper's model, unlike the minimal empiricist and positivist models, at least addresses condition A-3 (that identical problems have identical solutions), although inadequately. For, as long as we view scientific problems as nothing more than observable phenomena in search of explanation or prediction, we must see Eudoxus, Ptolemy, Copernicus, Newton, and Einstein all as working on the same problem--the problem of the planets. Although there are strong ancestral relationships among (some of) the problems these investigators tackled, historical research shows that there were important differences even in the Eudoxus-Ptolemy and Ptolemy-Copernicus cases. Moreover, it is obvious that Copernican theory solved none of Einstein's problems, and only by the grossest historical stretching could general relativity be said to solve any of Copernicus's problems. Yet if we accept A-3, problems X and Y cannot be the same problem if they have essentially different solutions. Now Popper is at least able to say that Eudoxus, Ptolemy, *et alia* worked in different problem *situations*. To that degree, he can individuate their respective tasks. But I argue (unpub. man.) that it is preferable to incorporate relevant background constraints into the very structure of the problems themselves and (therefore) to say that the problems are *not* identical, albeit related in various ways.[16]

By locating problem situations in his "third world,"--the world of the products of research and scholarship and their relations--rather than in his "second world"--the world of consciousness and belief--Popper (1972, Ch. 4) can easily satisfy conditions A-8 and A-9.

In addition to its relatively high success with group A requirements, Popper's model is the only one of those examined which can handle all of the data in set B. It is now time to test it against data set C. How well can Popper's view discriminate deep problems from shallow ones? (See C-1.) It may seem that Popper has a ready answer. A shallow problem is a datum whose explanation is relatively unimportant to the testing of a theory or to the competition between theories. A deep problem is represented by a datum upon whose explanation the fate of a deep theory turns.

The trouble with this view is that what makes a problem deep is not the problem itself (since Popper continues to conceive a problem proper as simply an empirical datum in search of explanation) but rather its background or setting. This consequence might be tolerable if the only deep *problems* were represented by anomalous data which threaten important *theories*, but that is not so. Deep problems, such as that of resolving the clash between classical mechanics and electromagnetic theory, and the problems of clarifying Newton's theory of gravitation (Finocchiaro 1980) and Planck's quantum theory, need not directly involve additional data to explain at all. In such cases, the explanation and prediction of data serves, rather, to confirm or to disconfirm a problem solution or perhaps to indicate the existence of a conceptual problem. If a theory contains a physical incoherence or an internal inconsistency, this is a serious problem which cannot be expressed simply as a demand for explanation of some data. For conceptually deep problems, the task of explaining or predicting empirical data is at best the surface of the problem, the tip of the iceberg.

Although I cannot argue each item separately, I claim that Popper's model cannot adequately explain any of the data under heading C and is therefore inadequate.[17] The reader will notice that data C-2 through C-8 all presuppose that problems may have conceptual depth.

3. Conclusion: The Constraint-Inclusion Model

It is fair to conclude that a conception of scientific problems which does any justice at all to the data under C (and much greater justice to A-3, 4, and 5 and the items under B) must be still richer than the Popper model. The most obvious way in which to enrich the model is to drop the empiricist restriction that observational data constitute the only constraints on a problem solution which truly help to define the problem itself. This means that we must break down the distinction between problems *per se* and conditions arising out of their historical-theoretical background. We must recognize that at least some theoretical conditions also help to set the problem. But once this distinction is breached, I can see no general grounds for excluding any constraint on a problem solution from the definition of the problem. Can any division between constraints which do and those which do not "belong" to the problem itself be found? Detecting a scientific problem and also giving it an adequate formulation can be very challenging tasks (C-7).[18]

What, then, are problems? My current, short answer is that a problem consists of *all* the conditions or constraints on the solution plus the demand that the solution (an object satisfying the constraints) be found.[19] The demand arises from disciplinary and programmatic goals and is modulated by the domain of information (in Shapere's sense, 1974) produced by the discipline and research program. Specific types of problems will, of course, possess special features. But what else could a problem in general include than the constraints plus the demand? I can think of nothing that this *constraint-inclusion model* leaves out.

Does it include too much? It is agreed on all hands that problems do not arise apart from goals and the demands which derive from them. Furthermore, a problem must include *at least* one constraint on the solution in order to be a well-defined problem at all and in order for inquiry into its solution to be possible. And in this paper I have both argued and suggested (where concise argument is difficult) that there is no basis for drawing an invidious distinction among those constraints which do belong to the problem proper and those which do not. Thus I include all constraints in the problem. For practical purposes, only the more familiar or more important constraints special to a particular problem usually are mentioned in scientific communication and in everyday life; but every single constraint, by definition, rules out some conceivable solutions as inadmissible and thereby (I claim) helps to define the problem. On my view, nearly every problem arising within a discipline or within a human society will have numerous constraints in common; but that fact presents no difficulty of individuation, for the constraint-inclusion model, as its name implies, also contains all constraints which individuate problems on the other models I have reviewed. My model can handle any individuation based on constraints at all. And by including all constraints in the definition of a problem, the model affords the means of saying all that can be said about strong inquiry--reasoning to a problem solution. In my view the old saying that stating the problem is half the solution is literally true! A problem, in a sense, describes its solution.

The role of our "data" on problems in this discussion can now be better appreciated. For these data represent many (though not all) of the constraints we should like to impose on the problem concept itself, on the solution to the "problem" problem. An answer to the question "What is a problem?" must satisfy these constraints (and others which I omitted as less than crucial for the present discussion). The constraints collectively afford a sort of description of a problem. You now see what I meant in saying that our task would be reflexive.

Notes

[1]I am indebted to the National Science Foundation (Grant SOC-7907078) for research support. A longer version of this paper was read at University of Nevada, Las Vegas.

[2]As Dudley Shapere (1980, p. 83) has remarked, "An adequate philosophy of science must show how it is possible that we might know, without guaranteeing that we must know."

[3]Recall that for Kuhn (1962) the paradigm "guarantees" the solvability of the fairly routine problems that he terms "puzzles".

[4]For discussions of these three kinds of problems, see, respectively, Monk (1980); Finocchiaro (1980); and Shapere (1969), Laudan (1977), and Leplin (1980).

[5]E.g., Planck's problem of finding a non-quantum interpretation of the black-body radiation law and his problem of reconciling statistical mechanics with the classical entropy law. See Klein (1963), (1966), and Kuhn (1978).

[6]E.g., the black-body radiation problem was reformulated as a problem concerning cavity radiation. It was treated as a question concerning ideal material oscillators by Planck (1900), as a problem of counting normal modes of vibration by Rayleigh (1900), Ehrenfest

(1906), Debye (1910) et al., as a problem in the theory of electrons by Lorentz (1903), as a problem of atomic state transitions and radiation by Einstein (1916), as a problem of counting occupation numbers by Bose (1924).

[7]For example, see Kuhn's discussion of the pendulum-efflux problem in his postscript to the second edition of (1962).

[8]E.g., Polanyi (1966, p. 22); Simon (1976) and (1977, pp. 160, 338); Hattiangadi (1978); Meyer (1979); Schaffner (1980).

[9]See, e.g., Reichenbach (1938, p. 382) and (1951, p. 255); Hempel (1965, pp. 245, 333); Popper (1957) and (1934, p. 59ff). In (1972, p. 263) Popper writes: "All problems of pure knowledge [i.e., pure research] are problems of explanation" (Popper's emphasis). Toulmin himself states A-7 more narrowly than I do: "Scientific Problems = Explanatory Ideals - Current Capacities" (my emphasis).

[10]Most empiricists seem to treat problems as purely empirical entities with no methodological component. They would distinguish methodology as formal metatheory from both the problem proper and its theoretical background. I claim below and argue in my (unpub. man.) that both moves are mistaken. There can be no sharp separation of methodology from substantive results and interests, nor can problems per se be cleanly separated from their theoretical backgrounds or historical settings. By including methodological constraints in the problem itself, in my discussion of the positivist model, I am making that position stronger than it really is.

[11]At least in his older, classical writings, such as the oft-quoted (1959, pp. 31-32).

[12]Compare remarks like Putnam's in Suppe (1974, p. 362), that in any strict sense there is no logic of justification any more than a logic of discovery. Simon lists several additional conditions for a problem to be well structured and notes that the well/ill-structured distinction is itself ill structured, since there exists no decision procedure for well structuredness.

[13]Positivists can, of course, account for reasoning from a purely conjectural hypothesis, i.e., justificatory reasoning to test whether the hypothesis does solve the given problem. But I am here interested in generative reasoning to the hypothesis.

Nor does the familiar Peirce-Hanson schema for abductive inference constitute an objection to my claim, since the schema pertains to preliminary evaluation more than to generation of new ideas. The "discovered" hypothesis appears in the premises of the abductive argument, not in the conclusion. Although Hanson (1958) stressed the theory-ladenness of observation, he ignored the theory-ladenness of problems and retained a form of the empiricist model of problems. See Schaffner (1980) and my (1980a, Section IV).

[14]Of course, Popper can give no better account than the positivist model if he sticks to problems strictly so called.

[15]Popper, however, makes no attempt to parlay this insight into a positive account of scientific discovery. Apparently, he still locates all reasoning in context of corroboration.

[16]Here I turn the tables on Lugg's (1978) criticism of Brown (1975). Lugg defends a position similar to Popper's. We must (he argues) distinguish problems from their settings, else we end up multiplying problems without necessity. But I claim that Brown is correct: a problem cannot be divorced from its historical-theoretical setting.

[17]The same can be said for the similar model of Lugg (1978).

[18]See, e.g., Bantz's (1980) discussion of the Heitler-London problem of chemical bonding and also Maull's comment on Schaffner (1980), in the same volume. I shall consider elsewhere the objection that theoretical constraints are really only as constraining as the empirical data which support them, that all justification in science, including the determination that a problem solution is correct, ultimately reduces to agreement with the data. This objection, which derives from a foundational conception of epistemology, has the effect of reducing _all_ problems, post-analytically, to _empirical_ problems.

[19]My longer answer attributes definite conceptual structures to problems. For a somewhat crude statement of this view, see my (1978). See also Reitman (1964).

References

Bantz, David. (1980). "The Structure of Discovery: Evolution of Structural Accounts of Chemical Bonding." In Nickles (1980c). Pages 291-329.

Bose, S. N. (1924). "Planck's Gesetz und Lichtquantenhypothese." Zeitschrift für Physik 26: 178-181.

Brown, Harold I. (1975). "Problem Changes in Science and Philosophy." Metaphilosophy 6: 177-192.

Debye, Peter. (1910). "Der Wahrscheinlichkeitsbegriff in der Theorie der Strahlung." Annalen der Physik 33: 1427-1434.

Ehrenfest, Paul. (1906). "Zur Planckschen Strahlungstheorie." Physikalische Zeitschrift 7: 528-532. (As reprinted in Collected Scientific Papers. (ed.) Martin J. Klein. Amsterdam: North Holland, 19 Pages 120-124.)

Einstein, Albert. (1916). "Strahlungs-Emission und -Absorption nach der Quantentheorie." Verhandlungen der Deutschen Physikalischen Gesellschaft 18: 318-323. (Translation of later version reprinted in ter Haar (1967). Pages 167-183.)

Finocchiaro, Maurice. (1980). "Scientific Discoveries as Growth of Understanding: The Case of Newton's Gravitation." In Nickles (1980b). Pages 235-255.

Hanson, N. R. (1958). Patterns of Discovery. Cambridge: Cambridge University Press.

Hattiangadi, J. N. (1978). "The Structure of Problems." Philosophy of the Social Sciences 8: 345-365 and 9: 49-76.

Hempel, Carl G. (1965). Aspects of Scientific Explanation. New York: Free Press.

--------------- (1966). Philosophy of Natural Science. Englewood Cliffs: Prentice-Hall.

Klein, M. J. (1963). "Planck, Entropy, and Quanta, 1901-1906." The Natural Philosopher 1: 83-108.

------------ (1966). "Thermodynamics and Quanta in Planck's Work." Physics Today 19: 23-32.

Kuhn, Thomas. (1962). The Structure of Scientific Revolutions. Chicago: University of Chicago Press. (2nd edition, enlarged, 1970.)

------------- (1977). The Essential Tension. Chicago: University of Chicago Press.

------------- (1978). The Black-Body Problem and the Quantum Discontinuity, 1894-1912. Oxford: Oxford University Press.

Lakatos, Imre. (1970). "Falsification and the Methodology of Scientific Research Programmes." In Criticism and the Growth of Knowledge. Edited by I. Lakatos and A. Musgrave. Cambridge: Cambridge University Press. Pages 91-196.

Laudan, Larry. (1977). Progress and Its Problems. Berkeley: University of California Press.

Leplin, Jarrett. (1980). "The Role of Models in Theory Construction." In Nickles (1980b). Pages 276-283.

Lorentz, H. A. (1903). "On the Emission and Absorption by Metals of Rays of Heat of Great Wave-Lengths." Proceedings of the Amsterdam Academy 5: 666-685. (As reprinted in Collected Papers, Vol. III. The Hague: Martinus Nijhoff, 1936. Pages 155-176.)

Lugg, Andrew. (1978). "Overdetermined Problems in Science." Studies in History and Philosophy of Science 9: 1-18.

Meyer, Michel. (1979). Découverte et Justification en Science. Paris: Editions Klincksieck.

Monk, Robert. (1980). "Productive Reasoning and the Structure of Scientific Research." In Nickles (1980b). Pages 337-354.

Nickles, Thomas. (1978). "Scientific Problems and Constraints." In PSA 1978, Vol. I. Edited by Peter Asquith and Ian Hacking. East Lansing, Michigan: Philosophy of Science Association. Pages 134-148.

---------------- (1980a). "Scientific Discovery and the Future of Philosophy of Science." In Nickles (1980b). Pages 1-59.

--------------- (ed.). (1980b). Scientific Discovery, Logic, and Rationality. (Boston Studies in the Philosophy of Science, Vol. 56). Dordrecht: D. Reidel.

--------------- (ed.). (1980c). Scientific Discovery: Case Studies. (Boston Studies in the Philosophy of Science, Vol. 60). Dordrecht: D. Reidel.

--------------- "What is a Problem that We May Solve It?" Unpublished manuscript.

Planck, Max. (1900). "Zur Theorie des Gesetzes der Energieverteilung im Normalspektrum." Verhandlungen der Deutschen Physikalischen Gesellschaft 2: 237-245. (As reprinted in English translation in ter Haar (1967). Pages 82-90).

Polanyi, Michael. (1966). The Tacit Dimension. Garden City: Doubleday.

Popper, Karl. (1934). Logik der Forschung. Wien: Springer. (Translated as The Logic of Scientific Discovery. New York: Basic Books, 1959).

------------. (1957). "The Aim of Science." Ratio 1: 24-35. (As reprinted in Popper (1972). Pages 191-205.

------------- (1972). Objective Knowledge. Oxford: Oxford University Press.

Rayleigh [Lord]. (1900). "Remarks upon the Law of Complete Radiation." Philosophical Magazine 49: 539-540.

Reichenbach, Hans. (1938). Experience and Prediction. Chicago: University of Chicago Press.

------------------ (1951). The Rise of Scientific Philosophy. Berkeley: University of California Press.

Reitman, Walter. (1964). "Heuristic Decision Procedures, Open Constraints, and the Structure of Ill-defined Problems." In Human Judgments and Optimality. Edited by M. W. Shelly II and G. L. Bryan. New York: John Wiley. Pages 282-315.

Schaffner, Kenneth. (1980). "Discovery in the Biomedical Sciences: Logic or Irrational Intuition?" In Nickles (1980c). Pages 175-205.

Shapere, Dudley. (1969). "Notes Toward a Post-positivistic Interpretation of Science." In The Legacy of Logical Positivism. Edited by P. Achinstein and S. Barker. Baltimore: Johns Hopkins University Press. Pages 115-160.

--------------- (1974). "Scientific Theories and Their Domains." In Suppe (1974). Pages 518-570.

--------------- (1980). "The Character of Scientific Change." In Nickles (1980b). Pages 61-116.

Simon, Herbert. (1973). "The Structure of Ill Structured Problems." Artificial Intelligence 4: 181-201. (As reprinted in Simon (1977). Pages 304-325.)

--------------- (1976). "Discussion: The Meno Paradox." Philosophy of Science 43: 147-151. (As reprinted in Simon (1977). Pages 338-341.).

--------------- (1977). Models of Discovery and Other Topics in the Methods of Science. Dordrecht: D. Reidel.

Suppe, Frederick (ed.). (1974). The Structure of Scientific Theories. Urbana: University of Illinois Press.

ter Haar, Dirk (ed.). (1967). The Old Quantum Theory. Oxford: Pergamon Press.

Toulmin, Stephen. (1972). Human Understanding. Princeton: Princeton University Press.

Can Scientific History Repeat?[1]

Peter Barker

Memphis State University

Kuhn (1962), Lakatos (1978) and Laudan (1977) all subscribe to a view of science which admits the existence of large scale entities differing from theories. Kuhn called these entities 'paradigms', Lakatos called them 'research programs' and Laudan calls them 'research traditions'. These entities are distinguished from theories by their historical durability and the unique manner in which they succeed one another. The history of science may be described as a succession of eras during each of which one (or more) of these entities was dominant. During its era of dominance one of these entities may support several incompatible theories which replace one another successively but share certain common features. These features may be (part of) a common ontology, a common set of epistemological standards employed in gathering observational evidence, or common methodological or conceptual directives employed in solving new scientific problems. Although theories may change, these features (or some of these features) do not, instead they are the province of the large scale entity. In the absence of a single neutral term to describe these entities I shall call them all, indifferently, *supertheories*.

The models of science adopted by Kuhn, Lakatos and Laudan suggest that, in the simplest case, the history of science will consist of a chain of supertheories. Two obvious questions about such chains are:
--What is the connection between the links in the chain, that is: How does one supertheory replace another? and
--Can the chain contain a loop? or Can scientific history repeat?

All three authors agree on one key aspect of the answer to the first of these questions. A supertheory cannot be overturned by empirical evidence alone. Empirical evidence is impotent without a new supertheory to back it. More specifically, in the technical vocabulary of Kuhn (1962, p. 52) and Lakatos (1970, p. 48 fn. 2), empirical evidence, for example in the form of experimental failures, which counts *prima facie* against a supertheory, can, during the era of dominance of that

PSA 1980, Volume 1, pp. 20-28

supertheory, only be classified as an anomaly. Such evidence is never, on its own, sufficient to overturn the dominant supertheory, although it may lead to the replacement of some theory which articulates the supertheory. The connection between supertheories and theories is not, therefore, deductive. Theories may be falsified by empirical evidence (given the background of the supertheory) but a supertheory can be overthrown only by a combination of an anomaly and a new supertheory (and perhaps the satisfaction of various other conditions). From the viewpoint of the new supertheory the empirical evidence which had been classified as an anomaly can be reclassified as a counterexample to the supertheory dominant when the anomaly was recognized. I will consider the consequences of treating the existence of an anomaly for the predecessor supertheory, which is converted into a counterexample by the successor supertheory, as a necessary condition for such replacement. The process of supertheory replacement corresponds, of course, to the phenomenon of the scientific revolution.[2]

Although Kuhn, Lakatos and Laudan disagree on the many other conditions which must be met when one supertheory replaces another, the minimal agreement sketched above is enough to place quite stringent restraints on the historical development of science. In particular, as I intend to show, it follows from this account that scientific history can never repeat. Once discarded a supertheory can never reappear.

Consider first the question of whether a supertheory can succeed itself. If we accept the account of scientific revolutions which requires the conversion of an anomaly to a counterexample, could one and the same supertheory both precede and succeed a given revolution? The answer is clearly negative. One and the same empirical phenomenon is to stand as an anomaly to the supertheory which precedes the revolution and as a counterexample to that same supertheory in the light of the supertheory which follows the revolution. Clearly no (non-self-contradictory) supertheory can stand in both of these relations to the same empirical phenomenon.

It is immaterial which account we choose of the relation between the empirical phenomenon and the supertheories. The relation may be explanatory, in which case we can make the point by saying that a single supertheory cannot at one time both fail to explain and explain the same phenomenon, or it may be one of problem solving, in which case the point can be made by saying that a given empirical phenomenon cannot, at one and the same time, count as an unsolved problem and a solved problem for a single supertheory. Regardless of what we pick, and this seems to be the point of Kuhn's original choice of the vocabulary 'anomaly' and 'counterinstance' as well as Lakatos' use of 'anomaly' and 'counterexample', it is clear that being an anomaly with respect to a given supertheory is logically incompatible with being a counterexample with respect to that same supertheory.

The question remains whether a given supertheory can succeed itself, not immediately, but after a period during which various other supertheories intervene. For the purposes of argument let us consider that the history of a given science can be divided into n successive eras corresponding to the periods of dominance of supertheories S_1. . . S_n, each of which succeeds its predecessors in the manner already described. The n-1 scientific revolutions which led, successively, to the replacement of S_1 by S_2. . . S_n were each preceded by an anomaly, call them A_1 . . . A_{n-1}, which in the light of the succeeding supertheory was recognized as a counterexample to the supertheory whose suffix it bears. Can S_1 succeed S_n?

According to the model under discussion any supertheory which succeeds S_n can do so only if some empirical phenomenon, call it A_n, is an anomaly for S_n and is converted into a counterexample by the supertheory which succeeds it. As noted before this is a necessary but perhaps not sufficient condition for the replacement of S_n. Simplifying this, for the case of S_1 replacing S_n:

(I) S_1 succeeds S_n only if A_n is an anomaly (and is converted into a counterexample. . . etc.).

The classification, or if you prefer, the recognition, of A_n as an anomaly depends upon previous scientific history in the following way. A_n counts as an anomaly only against the background of S_n which was itself established by means of a scientific revolution. As each scientific revolution involved an anomaly which was converted to a counterexample, A_n is an anomaly for S_n only if A_{n-1}, an anomaly for S_{n-1}, was converted to a counterexample to S_{n-1} by S_n, and so on all the way back through the history of the subject, to the point where A_1 was converted from an anomaly to a counterexample to S_1 by S_2. Condensing this chain of reasoning we can conclude:

(II) A_n is an anomaly only if A_1 is a counterexample to S_1 in the light of S_2.

As noted in the preceding argument, which established that a supertheory cannot immediately succeed itself, a given phenomenon cannot be both an anomaly and a counterexample with respect to the same supertheory. Further, its classification depends primarily on the dominant supertheory at the time the classification is made. During the era of S_1, A_1 was an anomaly. During the era of S_2, A_1 was classified as a counterexample to S_1 in the light of S_2. However, if S_1 replaces S_n, establishing a circle in the historical development of the subject, the previous evaluation of A_1, in the light of S_2, must be reversed restoring it to the status of an anomaly.

Notice that this re-re-classification of A_1 will not occur in the general case. In the case of three successive non-identical supertheories S_p, S_q and S_r, the transition from S_q to S_r will have no effect on the classification of A_p as a counterexample with respect to S_p in the light of S_q. Given a non-cumulative account of supertheory replacement (scientific revolution) S_r may not even deal with the phenomenon A_p.

Further, to anticipate my main argument slightly, if from the viewpoint of S_r there was some doubt that A_p was indeed a counterexample to S_p, this would equally cast doubt on the legitimacy of S_q, and cast doubt on the enterprise in which we were then engaged, that is the replacement of S_q by S_r.[3]

Only in the case of the historical recurrence of a supertheory will a revision take place, and then only in the status of the counterexample which led to the demise of the recurring supertheory. All intervening reclassifications of anomalies as counterexamples will remain unchanged, by the preceding argument. Thus: If S_1 succeeds S_n then A_1 is not a counterexample (it goes back to being an anomaly). Which, by contraposition, becomes:

(III) If A_1 is a counterexample then it is not the case that S_1 succeeds S_n.

We have now accumulated the following premises:

(I) S_1 succeeds S_n only if A_n is an anomaly.

(II) A_n is an anomaly only if A_1 is a counterexample.

(III) If A_1 is a counterexample then it is not the case that S_1 succeeds S_n.

From which it follows that:

(IV) S_1 succeeds S_n only if it is not the case that S_1 succeeds S_n.

From (IV) it follows by propositional logic, that:

(V) It is not the case that S_1 succeeds S_n.

The argument may be summarized by considering a possible objection to the formulation above. The original statement of (II) concluded ". . . A_1 is a counterexample to S_1 in the light of S_2." The argument preceding proposition (III), however, established only that if A_1 is evaluated after S_1 has replaced S_n, A_1 may be re-evaluated as an anomaly, and not a counterexample, in the light of the newly re-introduced S_1. A_1 may indeed now be regarded as a mere anomaly in the light of S_1, but this does not alter the fact that in the light of S_2 the phenomenon designated A_1 is a counterexample to S_1. The argument therefore collapses, as the second constituent of (II) is not the first constituent of (III).

In dealing with this objection, recall, first, that S_1 is supposed to replace S_n by means of a scientific revolution which fits the prescribed pattern. This pattern requires that during the era of S_n an anomaly A_n emerged, and that S_1 succeeded S_n by, among other things, converting this anomaly to a counterexample. The replacement of S_n by S_1 in this manner simultaneously endorses the replacement of S_{n-1} by S_n. Without the prior endorsement of S_n it would hardly make sense to claim the superiority of S_1 on the basis that it converts A_n, the anomaly of S_n, into a counterexample while itself suffering no similar defect. The endorsement of S_n as a replacement for S_{n-1} requires that we endorse S_{n-1} as a replacement for S_{n-2}, and so on all the way back to S_2. S_2, however, replaced S_1 (in part) by converting A_1 to a counterexample against S_1.

Accepting A_1 as a counterexample in the light of S_2 is required if we are to replace S_n by S_1 in accordance with this model of scientific revolutions. But if we now accept S_1 again, it seems that we must admit that we were mistaken in our original abandonment of S_1 and that A_1 was and is not a counterexample but an anomaly. We must therefore abandon our classification of A_1 in the light of S_2. But this undermines the whole chain of revolutions leading to S_n, and hence our reason for reintroducing S_1 in the first place. So the situation seems to be this: either we cut the ground out from under our own feet when we replace S_n by S_1, because this requires that we repudiate the basis for introducing S_2 and all the intervening history up to and including S_n, in which case we no longer have the necessary pieces to replace S_n by S_1 according to the model, or else we abandon the entire history of the discipline from the point where S_1 was rejected, an option which may be consistent with dogmatism. It is, of course, always open to us to retain or return to a supertheory by refusing to consider alternatives, in which case our anomalies will remain permanently anomalies. The remaining possibility, which has already been ruled out, is to regard the present case as one where S_1 succeeds itself. It seems, then, that there are no circumstances in which a supertheory can succeed itself by revolution.

There is nothing in the argument I have just presented to prevent two supertheories replacing a single predecessor. If no decision is reached between two such supertheories, their persistence introduces a branch structure into the linear chain of supertheories considered so far. Each in turn may be superceded by two, (or more) supertheories, leading to an elaborate branch structure. The irreversibility argument works on any branch, tracing it backwards to the supertheory which began the whole process.

Returning to the two supertheories which replaced a single predecessor, if one later supercedes the other, or generally if a single supertheory replaces predecessors on more than one branch, a loop will appear in the otherwise linear chain of supertheories. It is not the sort of loop I have excluded by means of the argument for irreversibility--no supertheory appears more than once. The argument for irreversibility can be applied to any chain of supertheories traced backwards through any member of such a loop.

I constructed the preceding arguments in terms of 'supertheories' because I did not wish to be limited by the details of any specific account, however I must now consider some of these details.

The argument for the irreversibility of science clearly applies to the Kuhnian model. By the Kuhnian model I mean the model of science on the large scale which is to be found in The Structure of Scientific Revolutions. According to this model a supertheory is a paradigm. Supertheory replacement is preceded historically by a period of normal science during which an anomaly arises which leads to a crisis state (Kuhn 1962, p. 82). During the crisis state brought on by the anomaly, abnormal research is permitted and a new paradigm emerges (Kuhn 1962, p. 84). From the viewpoint of the new paradigm the anomaly may be seen as a counterinstance to the old paradigm (Kuhn 1962, p. 153). All the elements required to construct the argument for irreversibility are therefore present in the Kuhnian model. Kuhn's reservations about this reading of his book are now well known. However, many people have adopted the Kuhnian model and for them the irreversibility argument provides a partial defense against the claim that the model denies scientific progress.

Lakatos' methodology of scientific research programs was intended before all else to give an account of science on the large scale (Lakatos 1970, p. 69). Its version of a supertheory is a string of connected theories called a research program. Although Lakatos clearly intended his account to supercede Kuhn's, and especially to counter the idea that there are no objective criteria for choosing one supertheory over another, it is not clear that one supertheory can ever finally supercede another on Lakatos' model.

Supertheory appraisal is based on the existence of an heuristic for each supertheory which directs a series of 'problemshifts'. Problemshifts are responses to empirical anomalies and may be classified as progressive or degenerating (Lakatos 1970, p. 33-4). Supertheories are said to progress (Lakatos 1970, p. 48-9), degenerate (Lakatos 1970, p. 68) and stagnate (Lakatos 1971, p. 112). One would expect that a progressive supertheory is preferable to a stagnating or degenerating one. Not so. However decayed a supertheory becomes it may always experience a burst of heuristic power (Lakatos 1970, p. 77 text to fn.'s 2, 3), resolve the anomalies which led to the degeneration in a progressive (or 'content increasing') fashion and again deserve attention.

The situation is exacerbated by Lakatos' treatment of crucial experiments. An experiment can only be seen to be crucial with hindsight (Lakatos 1970, p. 72). From the vantage point of a new supertheory an anomaly of a rival may be reclassified as a counterexample to the supertheory which produced it. This exactly parallels Kuhn's treatment of anomalies and counterinstances, to which Lakatos refers (Lakatos 1970, p. 72 fn. 3). However, Lakatos denies that crucial experiments are either necessary or sufficient conditions for supertheory replacement. The deciding factor in supertheory replacement is always appraisal of progressiveness or degeneration, and this is always reversible. All talk about crucial experiments is to be treated as harmless propaganda by the adherents of the current supertheory.

I take exception to two aspects of this account. First, Lakatos insists on the importance of history to the philosophy of science but denies the significance of crucial experiments despite the importance attached to them, historically, by working scientists. A position which does greater justice to history is desirable. Second, I cannot accept an account of science that holds out the permanent possibility of a comeback for Aristotle's physics or Ptolemaic astronomy.

Both defects may be remedied by admitting crucial experiments as one feature of supertheory replacement. Appraisal of supertheories as progressive or degenerating may remain the central feature of supertheory replacement, but included in the account will be crucial experiments understood as anomalies converted to counterexamples in the light of a new supertheory. If crucial experiments are made at least a necessary condition for supertheory replacement, the argument for the irreversibility of science can be made in terms of Lakatos' model and Aristotle and Ptolemy can be laid to rest.

In conclusion, let me consider the most recent and, to me, the most interesting account of science on the large scale, that of Laudan in Progress and Its Problems. Like Lakatos, Laudan argues against the alleged consequence of the Kuhnian model that one cannot give objective criteria for selecting supertheories. However Laudan denies that Lakatos' criteria find any application in history (Laudan 1977, p. 77 text to fn. 16).

In his own model, Laudan introduces a new class of problems to be considered in supertheory appraisal, calling them 'conceptual problems' (Laudan 1977, p. 45 ff.). These differ from the 'empirical problems', typified by Kuhn's anomalies. Similar problems are the basis of problem shifts for Lakatos. For Laudan, supertheory replacement rests on a minimax strategy, minimizing conceptual problems and maximizing solved empirical problems (Laudan 1977, p. 68, 119). These judgments are necessarily made in the context of several competing supertheories, called by Laudan 'research traditions'. Laudan does not hold that crucial experiments or the conversion of anomalies to counterexamples are necessary features of scientific revolutions. He also deprecates excessive emphasis on scientific revolutions (Laudan 1977, p. 183) and he can, I think, establish the irreversibility of science on grounds other than those presented in this paper. However the details of his account suggest an interesting interpretation of my main argument.

Laudan points out a new class of anomalies called by him 'nonrefuting anomalies' (Laudan 1977, p. 29). Nonrefuting anomalies are not predicted by one supertheory but are satisfactorily accommodated by a rival. I think these nonrefuting anomalies will serve just as well as the original variety for constructing my argument against science repeating itself. The main points are that, although not logically incompatible with a predecessor supertheory while compatible with its successor, such anomalies are held against the predecessor and counted in favor of the successor. A given anomaly cannot stand in both relations to a single supertheory. That such anomalies can only be recognized from the viewpoint of the supertheory they support is a common feature of Kuhn's, Lakatos' and Laudan's models. If the argument for scientific irreversibility can be made using Laudan's nonrefuting anomalies, then the use of 'anomaly' and 'counterexample' in the original argument is seen to be in-essential. The 'anomalies', 'counterinstances' and 'counterexamples' of Kuhn and Lakatos are just particular choices from a wider class. More generally one might say that the argument for scientific irreversibility, if correct, shows that it is the existence of phenomena asymmetrically connected to rival supertheories that establishes the direction of scientific change.

Notes

[1]My thanks to Larry Laudan, Edward Madden and Sheldon Reaven for criticisms of earlier drafts of this paper.

[2]These features of scientific revolutions are described in Kuhn (1962, Chapters VI-IX), Lakatos (1970, pp. 68-90, especially p. 86 ff.), and Laudan (1977, especially p. 29 and pp. 118-120).

[3]This is to say S_r may supercede both S_q and S_p by:

(1) converting an anomaly of S_q to a counterexample to S_q in the light of S_r.

(2) converting A_p into a counterexample to S_p from the viewpoint of S_r.

In this case we would probably say that S_q should never have succeeded S_p and regard S_p as being directly superceded by S_r. Notice that in this case A_p still ends up a counterexample.

References

Kuhn, Thomas S. (1962). The Structure of Scientific Revolutions. Chicago: University of Chicago Press.

Lakatos, I. (1970). "Falsification and the Methodology of Scientific Research Programmes." In Criticism and the Growth of Knowledge. Edited by I. Lakatos and A. Musgrave. Cambridge: Cambridge University Press. Pages 91-196. (As reprinted in Lakatos (1978). Pages 8-101.)

----------. (1971). "History of Science and Its Rational Reconstructions." In PSA 1970. (Boston Studies in the Philosophy of Science, Vol. 8.) Edited by R.C. Buck and R.S. Cohen. Dordrecht: Reidel. Pages 91-135. (As reprinted in Lakatos (1978). Pages 102-138.)

----------. (1978). The Methodology of Scientific Research Programmes. (Philosophical Papers, Vol. 1.) Edited by J. Worrall and G. Currie. Cambridge: Cambridge University Press.

Laudan, Larry (1977). Progress and Its Problems. Berkeley: University of California Press.

Problem-solving, Research Traditions, and the Development of Scientific Fields[1]

Henry Frankel

University of Missouri, Kansas City

1. Introduction

If science is essentially a problem-solving activity, it stands to reason that a good part of the rationale behind the development and growth of new scientific fields (or subfields) is the generation of problem-solutions. I should like to argue that the origin and, especially, the development of new scientific fields is primarily a research procedure for solving existing unsolved problems and generating and solving new unsolved problems.[2] By a scientific field I have in mind the following:

> A scientific field (F) has the following characteristics: (a) a given subject matter and (b) a data base about the subject matter, at least some of which is puzzling or problematic, (c) a set of problems arising from the puzzling subject matter (internal problems), (d) a set of problems arising from data in another field (applied or external problems), but thought solvable through appeal to the data base of F, (e) a set of techniques for expanding the data base, and (f) a set of theories or solutions to the internal problems.

If a field is relatively new it might not contain any of its own theories or solutions. Those fields which primarily deal with external or applied problems are applied fields. Moreover, I am not sure whether strict identity conditions can be developed for clearly distinguishing one field from another, or telling whether a given technique, set of data or problems falls within the scope of one field or another without resorting to stipulation. However, I don't think the issue is too important because fields aren't neatly isolated, discrete entities. Investigators in related fields often deal with the same subject matter, share data and techniques, attempt to solve the same problems and endorse similar solutions, because it often proves useful to formulate and attack problems in a variety of ways.

PSA 1980, Volume 1, pp. 29-40

I shall defend this general claim that scientific fields develop as a means for solving existing unsolved problems and generating new unsolved ones by outlining the overall development of paleomagnetism during the 1950's. Specifically, I shall argue the following:

Claim A The development of a scientific field (subfield) is controlled primarily by its ability to solve and generate new problems, and its pattern of development is as follows:

A1 A new sort of discovery is made. The discovery could have been accidental, uncovered through an investigation undertaken for some other purpose, or predicted on the basis of a solution to an existing problem in an established field. The discovery is thought to be puzzling in its own right and becomes data for a new unsolved problem within the framework of the developing field (internal problem) or is taken to be relevant for solving an existing problem in an established field (applied problem).

A2 An overall research strategy ensues which is designed to develop and improve solutions to existing problems. This strategy involves improvement of the data base, grounding of any controversial background assumptions employed in the proposed solution or in gathering and interpreting the data, and comparative evaluation of the proposed solutions. All of these activities are intertwined. Improvement of the data base consists in obtaining useful data; that is, data which can be employed in testing and developing solutions. Here researchers design new instruments and devise techniques for increasing the size and reliability of the data, construct various techniques for displaying and interpreting the data, and augment the data base for further testing and construction of solutions. Grounding of assumptions employed in the suggested solutions or methods of data collection and interpretation involves appeal to solutions to other problems and data either from the new field or from another field, depending upon the nature of the assumptions. Evaluation of the proposed solutions consists of a comparative assessment of the solutions. Researchers consider such factors as the reasonableness of the utilized background assumptions, reliability of the supportive data, relative incompleteness of the various solutions and special difficulties encountered by the proposed solutions.

A3 The field remains active as long as it continues to generate unsolved problems or has data believed to be relevant for solving problems in other fields. Of course, it may be rejuvenated through application of its data, techniques and solutions to new problems in related fields, realization that other problems arise from its existing data base, or development of some way to improve its data base so that it again becomes useful for solving existing problems.

In addition, if problem-solving effectiveness is the mark by which theories are evaluated, then workers in new fields will graft their problem-solutions onto existing research traditions if they believe the graft will increase the problem-solving effectiveness of their own work. Moreover, proponents of established research traditions will endorse the results of new field work insofar as it increases the problem-solving effectiveness of their tradition. Thus, I also shall defend the following two claims:

Claim B Researchers in a new field will graft their work onto an established research tradition if they believe that their work and the tradition mutually support each other and that such a graft will increase the problem-solving effectiveness of their own work.

Claim C Proponents of existing research traditions will endorse and promote the development of new fields which improve or promise to improve the problem-solving effectiveness of their tradition. Proponents of research traditions whose problem-solving effectiveness is lessened through work in a new field, often attempt to discredit work in the new field.

2. Case Study: Part I, the Development of Paleomagnetism -- A Defense of Claim A

2.1. The Early Development of Paleomagnetism

Although paleomagnetic studies extended back into the nineteenth century, they failed to attract much attention until around 1950. Nevertheless, the existence of remanent magnetism (the general subject matter of paleomagnetism) had been substantiated, investigators offered solutions to several problems and suggested how paleomagnetism could be applied to problems in geomagnetism (the study of the earth's magnetic field) as well as the debate over continental drift. The most notable problem they posed arose out of the discovery of lava samples that possessed a remanent magnetism roughly opposite to the present geomagnetic field. Several researchers suggested that these lava samples had formed when the geomagnetic field had had an opposite polarity since these rocks had probably formed when the field had been reversed. Embedded in this solution was the thesis that lava, when formed, acquires a magnetism which it keeps (its remanent magnetism) which is parallel to the then existing field, and that this remanent magnetism is sufficiently stable to remain within the rock sample. To substantiate part of this thesis, they heated samples of lava in the laboratory and noted that it acquired a magnetism parallel to the existing field upon solidification. The basis behind their suggestion that paleomagnetic studies could prove useful to the field of geomagnetism was that remanent magnetism appeared to be stable. Thus, if rocks acquire their remanent magnetism in a direction parallel to the earth's magnetic field, they could learn about the history of the field through studying remanence. The suggestion that paleomagnetic studies might prove useful for the continental drift debate depended upon the stability of remanent magnetism and the thesis that, if the present correlation between the earth's rotational and magnetic axes held throughout geological time, researchers could determine the previous positions of the continents or shifts in the geomagnetic poles.

2.2. The Expansion of Paleomagnetism

The major reasons for the growth of paleomagnetism were as follows. a) Geomagnetists were having trouble developing a solution to the problem of the origin of the earth's magnetic field. This led to the recognition of the importance of paleomagnetic studies for solving the problem and subsequent application of such studies to the problem. Consider, for example, what happened to P.M.S. Blackett, who, along with S.K. Runcorn, led a number of British scientists into paleomagnetism. Blackett (1952), who became interested in the problem of the origin of the geomagnetic field through his work on cosmic rays, adopted and expanded an hypothesis of H.A. Wilson's that any rotating body produces a magnetic field simply because of its rotating. In order to test it, Blackett decided to measure the magnetic field produced by a 10 x 10 cm. gold cylinder at rest and thereby rotating with the earth. If his hypothesis were correct, he reasoned that the produced field should be of an order of 10^{-3} -- 10^{-4} gammas. Because this was so small, he had to design an extremely sensitive magnetometer. Although he overcame the technological problems, the experiment proved to be negative, and Blackett dismissed the hypothesis. But he recognized the importance of his magnetometer and the value of paleomagnetism for offering useful data to solve the problem of the geomagnetic field's origin. "When I had completed the magnetometer, I realized that it was in many ways admirably suited to the measurement of the magnetic properties of very weakly magnetized geological specimens, for instance sedimentary rocks." (Blackett 1952, p. 311). Blackett had read a 1948 paper entitled "Pre-history of the Earth's Magnetic Field," (Blackett 1956, p. 5) and this led him to review extensively the paleomagnetic literature because: "Such information would not only be of importance for its own sake but would be of immense value in an attempt to understand the physical mechanism giving rise to the field. For one of the main difficulties of finding a plausible theory of the origin of the earth's field lies in the fact that we have direct measurements of the earth's magnetism by means of a compass needle only over the last 400 years, compared with the more than 500 million years of reliable geological history." (Blackett 1956, p. 5). Heeding his own advice, Blackett set up a research team at Imperial College, and his group was responsible for some of the major paleomagnetic studies in the fifties and early sixties. Indeed, Blackett's Lectures on Rock Magnetism which he delivered in 1954 and subsequently published in 1956, outlined much of paleomagnetism's ensuing course of development. Nor was Blackett an exception; other geomagnetists turned to paleomagnetism or advocated such a move. Runcorn, a former graduate student of Blackett's, moved easily from his earlier work on the origin of the geomagnetic field to paleomagnetic studies. In 1948 Runcorn (1948) calculated how the strength of the geomagnetic field should change with depth in accordance with Blackett's hypothesis, and in 1950 he found negative evidence for Blackett's theory by measuring the strength of the field in mines. Then in the years to come, Runcorn became one of the leaders in paleomagnetism, used it to understand the geological history of the earth's magnetic field, helped develop techniques for collection and interpretation of the data, and applied his work to the problem of reversals and the debate over continental drift. Nor was the situation too much different with Sir Edward Bullard. Bullard (1949), following W.M. Elsasser (1946), developed the self-exciting dynamo theory, which became the dominant solution in the 1950's for the origin of the earth's magnetic field, and both of them (1955) revised their models in light of the paleomagnetic studies. In particular, they developed variations so as to account for the continued uncovering of samples with reversed remanent magnetism. Moreover, Bullard (1968) encouraged the pursuit of paleomagnetic work at Cambridge and its application to the debate over continental drift.

b) Developments in rock magnetism which led to a better understanding of how different kinds of rocks acquire their magnetism. Only after researchers had a better understanding of this central problem in rock magnetism (the study of how rocks acquire their magnetism) could they get a better idea about remanent magnetism. This need to develop a more complete understanding of rock magnetism was particularly acute since the topic proved to be quite complex, and the depth of understanding was shallow and limited almost entirely to magnetization of igneous rocks until near the end of the forties. Perhaps, those most responsible for the growth of rock magnetism were T. Nagata (1953) of Japan and L. Néel (1951) of France. Both laid out much of the theoretical basis for how rocks became magnetized, and extended their analysis to include work on sedimentary and metamorphic rocks. Their works served as standard references for paleomagnetists throughout the fifties. The major reasons for why rock magnetism presented an intractable problem were as follows: It possessed difficulties far beyond the study of magnetic elements. Rocks are a heterogeneous lot, and their magnetization occurs in a variety of ways. Moreover, the processes are quite often different than exhibited by magnetic elements, and consequently, researchers simply couldn't extrapolate from classical studies on magnetic elements. The intensity of magnetization of rocks is, in general, much less than acquired by magnetic elements. Futhermore, there were additional complications with the extension of their work to remanent magnetism. The intensity of the earth's field is much less than fields manufactured to study magnetic elements, and they had to consider the additional factor of what happens to remanent magnetism over long periods of time.

c) The development of highly sensitive magnetometers. These were initially developed as a means for testing a solution to the problem of the origin of the earth's magnetic field, and were required so as to expand the study of remanent magnetism to non-igneous rock which possess a remanent magnetism of much less intensity than igneous rock. Blackett's astatic magnetometer (1952) was only the first among many new magnetometers that were eventually designed to further paleomagnetic investigations. What is of interest is that investigators

(Collinson, 1967) constructed their new instruments with particular kinds of studies in mind, and changed their designs to fit the requirements of their particular study.

d) Reconsideration of the problem of reversals with the ensuing development of an alternative solution to the problem. Some researchers suggested that reversals were caused by self-reversals of the found sample rather than as a response to a global reversal of the magnetic field. Subsequently, theoretical mechanisms for self-reversals were proposed and tested. Indeed, some of the predictions were first thought to be confirmed. This led to increased interest in the problem of reversals, laying out of the connections between the various solutions to the problems of the origin of the geomagnetic field and how rocks acquire their magnetism, the undertaking of various experiments to determine which, if any, of the various solutions was correct, and finally, using paleomagnetic reversal data to construct a highly successful technique for dating geological events.

In 1949 J. Graham (1949), an American rock magnetist, wondered if reversals could be caused by internal or mineralogical changes in the given sample. He wrote Néel about the idea. Néel (1951) predicted four theoretical mechanisms for self-reversals, and in 1952 Nagata (1953) and his co-workers found a natural rock sample that underwent self-reversal in the laboratory. This led to a massive research effort by the paleomagnetic community to solve the reversal problem. Numerous researchers explicitly mapped out the overall strategy, wherein they spelled out how both hypotheses related to other problems and devised experimental programs to test the hypotheses. They pointed out that if the global hypothesis were correct, then any account of the origin of the geomagnetic field must at least allow for field reversals. Thus, Bullard and others revised their self-exciting dynamo theories accordingly. Given the assumption of self-reversals, Néel and other theorists in rock magnetism devised alternative models for how it could occur. The experimental part of the program concentrated on devising "crucial" tests for choosing one of the hypotheses. Strategies were devised to determine if rocks of the same age have the same polarity regardless of rock type or location. Here the idea was to see if there were world-wide stratigraphic successions of reversals. They also concentrated on finding beds of one rock variety which were cut across by reversed sequences. This work led to the use of potassium-argon techniques for dating reversals and development of the geomagnetic reversal time scale with the vindication of the global hypothesis in the middle sixties. Others sought to determine whether there were any systematic chemical and mineralogical differences between rocks of positive and negative polarity. Besides engaging in field studies, workers attempted to produce self-reversals in the laboratory by using found and synthetic samples and putting them through changes suggested by the self-reversal models.

e) The application of paleomagnetism as a means for settling the debate about continental drift. Continental drift theory, as noted by Frankel (1979), had reached a low point in the 1940's. Most of the initial rise of interest enjoyed in the 1950's is attributable to paleomagnetism. It offered a seemingly independent and quantitative approach for determining the past positions of landmasses. But the application of paleomagnetism to the debate and the overall problem of the position of continents throughout geological time proved to depend upon a key theoretical assumption relating to the origin of the geomagnetic field, and was plagued by experimental difficulties. Nevertheless, despite the difficulties -- indeed, because of the difficulties -- it is clear that these paleogeographic directional studies were undertaken explicitly to answer questions about continental drift and polar wandering. Their efforts were directed to solving this major problem; they were determined to obtain useful but reliable data, explicate any theoretical assumptions they employed, remove unclarities from the proposed solutions, and design studies which would allow for elimination of alternative solutions (Irving 1964).

The basic idea behind using paleomagnetic data for fixing the previous position of a landmass was that the ancient _geographic_ positions could be extrapolated from the ancient _geomagnetic_ ones. Thus, researchers had to have (i) reliable paleomagnetic information, (ii) a general procedure for interpreting the paleomagnetic data in terms of paleomagnetic positions and (iii) a means for determining specifically what could be concluded from the paleomagnetic information about ancient positions of the landmasses. There were serious problems with all three aspects. With respect to (i) they had to determine such things as the following: the age of the tested sample, whether the remanent magnetism was stable, whether the remanent magnetism was original or acquired at some time after the sample's formation, and whether the given sample had undergone subsequent regional re-orientation after its formation. Graham, Blackett and his co-workers, Runcorn, E. Irving and many others worked out a whole list of procedures that should be followed in the treatment of a paleomagnetic sample. The basic assumption involved in (ii) was that the earth's magnetic field, when averaged over periods of several thousand years, is that which would be maintained by a geocentric axially directed dipole, i.e., that the geomagnetic and geographic poles have, on the average, coincided. In defense of this assumption they appealed to the paleomagnetic studies and compass readings which indicated that the inclination of the magnetic field from the geocentric poles averaged out to 0° within periods of several thousand years, paleoclimatological data which proved to be supportive of the assumption, and, more importantly, to the Elsasser-Bullard self-exciting dynamo theory for the origin of the geomagnetic field since it required a coincidence of the magnetic and rotational (geographic) poles. These difficulties surrounding (iii) were over precisely what could be concluded about past latitudes and longitudes of given samples and the appropriate landmasses, even if the sample's ancient magnetic dip (inclination) and declination were known. Early on, Blackett and his group (Clegg, _et al_. 1954) argued that Indian samples indicated that India had drifted northward across the equator to its present position -- in support of continental drift -- on the basis of the paleomagnetic inclination of the analyzed samples. In other words, they supposed that ancient latitudes could be determined on the basis of paleomagnetic inclinations. But Runcorn (1955) and others argued that the inclinational data showed only the position of the landmass with respect to the ancient geomagnetic poles, and therefore, one could assume from a change in magnetic inclination only that the landmass had drifted (continental drift), the geomagnetic poles had wandered (polar wandering), or a combination of the two. Then, Runcorn (1956), after investigating roughly contemporaneous samples from Europe and North America, argued in favor of longitudinal shifts of North America or Europe and subsequent opening of the Atlantic. He based his claim on the fact that paleomagnetic declinational data from both continents gave polar-wandering curves which did not coincide unless the two continents originally had been united, as well as on Irving's work on Australian samples (Irving 1964) which indicated that the magnetic pole would have had to have been in more than one place at the same time, if continental drift had not occurred. Thus, among other things, Runcorn assumed that shifts in ancient geographical longitudes could be determined from ancient magnetic declinations. However, Blackett, Irving and others pointed out that derivation of longitudinal shifts only from declinational data was illicit since declinational differences also could be explained in terms of rotational shifts. Because of these difficulties, paleomagnetists, during the early sixties, took extreme care in presenting their results; undertook studies designed to maximize the amount of information they could infer from paleomagnetic inclinational and declinational data; and augmented their paleomagnetic studies with paleoclimatological ones in an attempt to settle the debate over continental drift.

3. Case Study: Part II, Paleomagnetism and Continental Drift -- A Defense of Claims B and C

The general features which characterize the relation between the rise of paleomagnetism and the debate over continental drift were as follows.

a) Except for Blackett, none of the new paleomagnetists had been supporters of continental drift theory, and even Blackett had not been an active supporter of the tradition but a physicist-geophysicist who expressed sympathy in the tradition. Runcorn adopted drift theory after concluding that polar wandering was not sufficient to account for the paleomagnetic data. Irving endorsed drift theory after and because of his paleomagnetic studies. Bullard became a proponent of drift theory primarily because he thought the paleomagnetic studies were based upon well-founded principles and that drift theory was the only tradition which could account for the paleomagnetic results. These paleomagnetists became drifters because the initial paleomagnetic results supported drift, and they continued their work to find more support.

b) Active proponents of drift theory quickly viewed the paleomagnetic studies as excellent support for their tradition, and argued that it was the only overall view which could account for the new findings. This is evidenced by the endorsement of paleomagnetism by the major proponents of drift theory in the early fifties, namely, Arthur Holmes, W. Carey and L. King. They all were favorably disposed toward the directional studies and incorporated them into their discussion of drift theory.

c) Opponents of the drift tradition generally ignored the advances in paleomagnetism or attempted to undermine the findings by launching broadside attacks against the methodology employed by paleomagnetists. For example, Sir Harold Jeffreys (1959) considered any paleomagnetic study risky business, and H. Meyerhoff (1972) attacked their assumption that the earth's magnetic field has always been a geocentric dipole.

The support for claims B and C is straightforward. Fact (a) supports B, while (b) and (c) are supportive of claim C.

4. Conclusion

Each of the three claims supported by this case study are tied together by the more general thesis that science is mainly a problem-solving activity. New fields are generated not by just a novel discovery, but by one which gives rise to a problem or is thought to be relevant for solving one in some other field. Research activities are directed toward solving problems, and workers within the field will often graft their solutions onto existing traditions, if such a graft increases the problem-solving effectiveness of their own work. Proponents of existing traditions will quickly endorse the work in new fields when it fits in with their tradition, for such a tactic allows them to increase the number of problems their traditions can solve.

There are two other claims suggested by this case study which are related to the view that science is essentially a problem-solving activity, and both are relevant to what counts as an important problem.[3]

Claim D: The epistemic weight or importance of a solved problem is a function of the reliability of its data base. The more reliable the data base, the more important the problem.

Claim E: Two sets of an equal number of solved problems may have different epistemic weights, even if their members individually have the same weight; for it is more to the credit of a research tradition to solve problems from diverse fields rather than from a single field. By solving problems in diverse fields, a research tradition demonstrates its ability to unify work from originally unrelated fields and provides independent evidence for itself.

Of course the proponents of continental drift were pleased with the paleomagnetic studies because they had a solution for the puzzling data, while the fixist opponents could only attack the methodology of the new science. But, I think, their pleasure was derived from more than their having gained another solved problem which their opponents were unable to solve. For they were particularly pleased by their success in paleomagnetism since (a) the paleomagnetic data base appeared, at least initially, to be much more reliable than the data bases in, say, paleoclimatology and paleogeography, and (b) they were able to solve a problem in a completely different field. Drifters had been repeatedly shot down for offering solutions based upon somewhat unreliable data. But paleomagnetism, at least, seemed to offer a much firmer data base. Old drifters and the new paleomagnetic converts all spoke about the comparative greater reliability of the paleomagnetic data compared to other branches of earth science. Indeed, I don't think it was an accident that the major thrust of the non-drifters was to question the reliability of paleomagnetism's data base. Finally, by solving a problem in a new field, they argued that they were able to provide independent evidence for their position and offer a more unified overall view than non-drifters.

Notes

[1]This is based on research supported by the National Science Foundation's History and Philosophy of Science Program. I should also like to thank Rachel Laudan and Nanette Biersmith for their aid. An earlier version of this paper was read at the 1980 Joint Atlantic Seminar in the History of the Physical Sciences

[2]The overall thesis that science is essentially a problem-solving activity has been most recently developed and defended by Larry Laudan (1977). Laudan offers a preliminary account of theory choice in terms of problem-solving effectiveness, and relates it to his notion of 'research traditions' -- his surrogate for Kuhn's 'paradigms' and Lakatos' 'research programmes'. Part of the aim of this paper is to extend some of Laudan's ideas to the question of the development of scientific fields and how proponents of established research traditions react to the development of new fields. In addition, my account of the development of new fields is similar to one proposed by Lindley Darden (1978). There are, however, differences between our accounts. I place more stress on the role of relevant problems in guiding how new fields develop, allow for the development of new fields through the generation of <u>new</u> problems, and consider the general way in which work in new fields is received by proponents of established research traditions.

[3]Laudan (1977) correctly argues that problem-solving effectiveness is more than a function of the number of solved, unsolved, anomalous and conceptual problems possessed by a theory or tradition, for he argues that the relative epistemic weight or importance of a problem is relevant. But, Laudan's "calculus of problem weighting" must be further developed. Claims D and E are two preliminary suggestions for buttressing his account. Claim D offers one factor for ranking solved problems, and claim E suggests that two sets of solved problems with the same number of solved problems, may have different overall epistemic weights, even if the individual problems have equal epistemic weights. This is a possibility that Laudan doesn't consider.

References

Blackett, P.M.S. (1952). "A Negative Experiment Relating to Magnetism and the Earth's Rotation." Philosophical Transactions of the Royal Society of London A245: 309-370.

---------------. (1956). Lectures on Rock Magnetism. Jerusalem: The Weizmann Science Press of Israel.

---------------. (1961). "Comparison of Ancient Climates with the Ancient Latitudes Deduced from Rock Magnetic Measurements." Proceedings of the Royal Society of London A263: 1-30.

Bullard, E.C. (1949). "The Magnetic Field within the Earth." Proceedings of the Royal Society of London A197: 433-455.

------------. (1955). "The Stability of a Homopolar Dynamo." Proceedings of the Cambridge Philosophical Society 51: 744-760.

------------. (1968). "The Barderian Lecture, 1967: Reversals of the Earth's Magnetic Field." Philosophical Transactions of the Royal Society of London A263: 481-524.

Clegg, J.A., Almond, M. and Stubbs, P.H.S. (1954). "The Remanent Magnetism of Some Sedimentary Rocks in Britain." Philosophical Magazine 45: 583-598.

Collinson, D.W., Creer, K.M. and Runcorn, S.K. (eds.). (1967). Methods in Paleomagnetism. Amsterdam: Elsevier.

Darden, Lindley. (1978). "Discoveries and the Emergence of New Fields in Science." In PSA 1978, Volume One. Edited by P.D. Asquith and I. Hacking. East Lansing: Philosophy of Science Association. Pages 149-160.

Elsasser, W.M. (1946). "Induction Effects in Terrestrial Magnetism: Part I. Theory." Physics Review: 69: 106-116.

-------------. (1955). "Hydromagnetism I & II: A Review." American Journal of Physics 23 & 24: 590-609 and 85-110.

Frankel, H. (1979). "The Career of Continental Drift Theory: An Application of Imre Lakatos's Analysis of Scientific Growth to the Rise of Drift Theory." Studies in History and Philosophy of Science 10: 21-66.

Graham, J.W. (1949). "The Stability and Significance of Magnetism in Sedimentary Rocks." Journal of Geophysical Research 59: 131-167.

Irving, E. (1964). Paleomagnetism and Its Application to Geological and Geophysical Problems. New York: Wiley.

Jeffreys, H. (1959). The Earth (4th ed.). Cambridge: Cambridge University Press.

Laudan, Larry. (1977). Progress and Its Problems. Berkeley: University of California Press.

Meyerhoff, A.A. and Meyerhoff, H.A. (1972). "The New Global Tectonics: Major Inconsistencies." Bulletin of the American Association of Petroleum Geologists 56: 269-336.

Nagata, T. (1953). Rock Magnetism. Tokyo: Maruzen.

Néel, L. (1951). "L'inversion de l'aimantation permanente des roches." Annales de Geophysique 7: 90-102.

Runcorn, S.K. (1948). "The Radial Variation of the Earth's Magnetic Field." Proceedings of the Physical Society of London 61: 373-381.

------------. (1955). "Rock Magnetism -- Geophysical Aspects." Advances in Physics 4: 244-291.

------------. (1956). "Paleomagnetic Comparisons Between Europe and North America." Proceedings of the Geological Association of Canada 8: 301-316.

Runcorn, S.K., Benson, A.C., Moore, A.F. and Griffiths, D.H. (1951). "Measurements with Depth of the Main Geomagnetic Field." Philosophical Transactions of the Royal Society of London A244: 113-151.

Part II

Quantum Logic and the Interpretation of Quantum Mechanics

Micro-States in the Interpretation of Quantum Theory

Gary M. Hardegree

University of Massachusetts, Amherst

1. Introduction

In the present work, I discuss the interpretation of quantum mechanics (QM) from the viewpoint of quantum logic (QL). I regard the objects of QL to be the possible (accidental) properties that can be ascribed to a quantum system SYS. The basic idea is that, at any given moment t, SYS actualizes some properties, but not others. The *micro-state* of SYS at time t is identified with the set of all properties that SYS actualizes at time t.

One thing an interpretation of QM is supposed to do, I believe, is delineate the admissible quantum micro-states. Since the characterization of quantum micro-states is intimately related to the characterization of quantum logical consistency, the interpretation of QM is intimately tied to the interpretation of QL.

Two kinds of interpretations are discussed. *Strict interpretations* are based on the assumption that the properties of a system are individuated by the projection operators on the associated Hilbert space. *Non-strict interpretations* deny this assumption, and allow the possibility that distinct properties are associated with the same projection operator.

I examine two strict interpretations. One, which I refer to as the *lattice interpretation*, is associated with the interpretations of Jauch, Piron, and van Fraassen. The other, which I refer to as the *minimal strict interpretation*, has not been proposed before.

I also examine Kochen's interpretation, which I argue is *not* a strict interpretation. I conclude by hypothesizing a principle of individuation for quantum properties that is consistent with Kochen's definition of quantum micro-states. I remark that the resulting "logic" of properties has the structure of a *semi-Boolean algebra*.

PSA 1980, Volume 1, pp. 43-54

2. Quantum Logic

QL traces to the classic work on von Neumann (1932), in which the theories of Heisenberg and Schroedinger were unified into a common mathematical formalism based on Hilbert space. According to von Neumann's formalism, every physical system SYS is described by a Hilbert space H. The magnitudes ("observables") pertaining to SYS are represented by the self-adjoint operators on H, and the statistical states pertaining to SYS are represented by the density operators on H. For each magnitude m, and each statistical state w, the formalism prescribes a probability measure p_m^w on the set B(R) of Borel subsets of the real line R; $p_m^w(X)$ is customarily interpreted as the probability (in state w) that a measurement of m will/would yield a value in X.

The QL approach to QM is based on singling out the projection operators (projections) on H for special consideration, an idea suggested by von Neumann (1932) and pursued by Birkhoff and von Neumann (1936). The basic idea underlying the QL approach is that the projections on H represent the possible (accidental) properties that can be ascribed to SYS.

The properties ascribable to a system group into natural families, each associated with a magnitude. In accordance with the general theory of measurement, every magnitude m may be decomposed into a family [m] of properties, where [m] is a Boolean sigma-algebra $[m(X): X \in B(R)]$. Intuitively, m(X) corresponds to the property that a system has when a measurement of m would yield a value in X. As a simple example, let m be length in centimeters, and let X be [1,2]; then m(X) corresponds to the property of being at least 1 cm. long but no more than 2 cm. long.

The quantum theoretic version of the general theory of measurement is known as the Spectral Theorem, according to which every self-adjoint operator may be decomposed into a Boolean sigma-algebra of projections. In addition to the Spectral Theorem, other major theorems in QL include Gleason's Theorem (1957), which explicates quantum statistical states, and Wigner's Theorem (1931), which explicates quantum symmetries.

3. Micro-States

Although QL provides an explication of magnitudes, symmetries, and statistical states in QM, it does not provide a straightforward explication of quantum <u>micro-states</u>. The chief difficulty is that the logical structure of quantum properties cannot be deduced from the extant quantum formalism without tacitly assuming an interpretation of QL and QM.

By "logical structure" I have in mind two notions — <u>forcing</u> and <u>exclusion</u>, which are defined as follows. Here, Q is any property, and Γ is any set of properties.

(D1) To say that Γ forces Q is to say that a system cannot actualize every property in Γ without also actualizing Q.

(D2) To say that Γ excludes Q is to say that a system cannot actualize every property in Γ and also actualize Q.

In the case that Γ is a singleton $\{P\}$, we say that P forces (resp., excludes) Q.

The notions of forcing and exclusion enable us to define the notion of pre-micro-state as follows.

(D3) A collection S of properties is said to be a pre-micro-state if it satisfies the following restrictions.

(a) For every property P, if S forces P, then $P \in S$.

(b) For every property P, if S excludes P, then $P \notin S$.

The set of all pre-micro-states is partially ordered by set inclusion; the maximal elements of this set are referred to as micro-states.

(D4) A pre-micro-state S is said to be a micro-state if for any pre-micro-state S*, $S \subseteq S^*$ only if $S = S^*$.

The following theorems provide a characterization of forcing and exclusion in terms of micro-states.

(T1) Γ forces Q iff for every micro-state S, if $\Gamma \subseteq S$, then $Q \in S$.

(T2) Γ excludes Q iff for every micro-state S, if $\Gamma \subseteq S$, then $Q \notin S$.

Having defined forcing, exclusion, and micro-states, I now wish to propose various principles to serve as minimal adequacy requirements on any construal of micro-states. The first principle formally states that every possible property is actualizable.[1]

(P1) For every possible property P, there is a micro-state S such that $P \in S$.

The remaining principles concern the property family [m] of an arbitrary magnitude m. The first such principle formally states our assumption that each [m] is a Boolean sigma-algebra. The others impose various logical restraints on admissible micro-states.

(P2) For every magnitude m, the family $[m(X): X \in B(R)]$ is a Boolean sigma-algebra.

(P3) If $X \subseteq Y$, then m(X) forces m(Y).

(P4) If $X \cap Y = \emptyset$, then m(X) excludes m(Y).

(P5) $\{m(X_i): i \in I\}$ forces $m(\bigcap\{X_i: i \in I\})$.

Although Principle (P5) is plausible, it involves technical problems in the case of magnitudes that do not have pure discrete spectra. In order to circumvent these problems, we concentrate exclusively on discrete-valued magnitudes, and we treat continuous-valued magnitudes as infinite families of discrete-valued magnitudes, rather than as autonomous entities. Restricting our attention to discrete-valued magnitudes allows us to replace Principle (P2) by the following.

(P2*) For every (discrete-valued) magnitude m, the family $[m(X): X \in B(R)]$ is a complete atomic Boolean algebra.

Complete atomic Boolean algebras are precisely the Boolean algebras that are isomorphic to power set Boolean algebras. In the case of [m], the atoms are the properties $m(\{r\})$, where r is in the spectrum of m. For simplicity, we write m(r) instead of $m(\{r\})$.

The set of m-properties actualized by a micro-state S is the set $S \cap [m]$, which we denote S[m]. The following theorem states the basic restraint of S[m].

(T3) For every magnitude m and micro-state S, S[m] is a principal filter on [m], provided that S[m] is not empty.

A <u>principal filter</u> on a lattice L is any subset F of L for which there is an element $a \in L$ such that $F = \{x \in L: a \leq x\}$. According to (T3), if any m-property is actualized by S, then there is a <u>principal</u> m-property actualized by S. In particular, by (P5), if any m-property is actualized by S, then there is a smallest Borel subset X_o such that $m(X_o)$ is actualized by S; $m(X_o)$ is the principal m-property actualized by S.

This enables us to define a <u>valuation function</u> V_S (for each micro-state S), which is a partial function on the class of magnitudes, defined as follows.

(D5) If $S[m] \neq \emptyset$, then $V_S(m)$ is the smallest Borel subset X_o such that $m(X_o) \in S$.

If $S[m] = \emptyset$, then $V_S(m)$ is not defined.

If it exists, the set $V_S(m)$ is called the "value" assigned to m by S. There are three cases of interest.

(C1) If $V_S(m)$ does not exist, then we say that S <u>assigns no value</u> to m.

(C2) If $V_S(m) = \{r\}$ for some $r \in R$, then we say that S <u>assigns an exact value</u> to m.

(C3) If $V_S(m) = X \neq \{r\}$ for any $r \in R$, then we say that S <u>assigns an inexact value</u> to m.

4. Strict Interpretations

In the previous section, I have proposed principles that I regard as minimal constraints on any characterization of micro-states. In addition to these minimal principles, there are further (non-minimal) principles that might be imposed on micro-states. In order to identify these principles, as well as the theorems that depend upon them, I mark them with various letters.

The first non-minimal principle that I consider is customarily associated with QL. According to this principle, not only is every quantum property associated with a projection on the relevant Hilbert space H, the projections on H *individuate* the properties in the sense that distinct properties are associated with distinct projections. This principle may be formally stated as follows; here, a and b are magnitudes, and X and Y are Borel subsets of R.

(P6S) If properties a(X) and b(Y) are associated with the same projection, according to the spectral decompositions of a and b, then a(X) = b(Y).

It is worthwhile to note that (P6S) is equivalent to the following.

(P6S*) If a(X) and b(Y) are statistically equivalent, then a(X) = b(Y).

Statistical equivalence is defined as follows.

(D6) a(X) and b(Y) are said to be *statistically equivalent* if for every statistical state w, $p^w_a(X) = p^w_b(Y)$.

In saying that properties P and Q are identical, I have *at least* the following in mind.

(I) If P = Q, then for every micro-state S, P ∈ S iff Q ∈ S.

By a *strict interpretation*, I mean any interpretation that conforms with Principle (P6S). According to strict interpretations, there is a strict (1-1) correspondence between the properties ascribable to a system and the projections on the associated Hilbert space. This allows us to read the logic of properties from the "logic" of projections. This yields the following theorems; here, $\hat{P}$ is the projection associated with property P.

(T4S) If $\hat{P} \leq \hat{Q}$, then P forces Q.

(T5S) If $\hat{P} \perp \hat{Q}$, then P excludes Q.

(T6S) If $\{\hat{P}_i: i \in I\}$ are jointly compatible, and $\inf\{\hat{P}_i: i \in I\} = \hat{R}$, then $\{P_i: i \in I\}$ forces R.

Granted the correspondence between properties and projections, for convenience, we can simply treat properties and their corresponding projections as interchangeable. For example, we can regard micro-states in strict interpretations as collections of projections. We can then understand Theorems (T4S)-(T6S) as follows. (T4S) claims that every micro-state is closed under the partial order relation $\leq$, which is consistent with the usual interpretation of QL. (T5S) claims that orthogonal projections cannot be simultaneously actualized, which is also consistent with the usual interpretation of QL. Finally, (T6S) claims that every micro-state is closed under the formation of compatible infima (of arbitrary cardinality).

Note very carefully that, in virtue of (T4S), every micro-state is closed under the formation of some non-compatible infima. For example, suppose that S contains an atomic projection $\hat{P}$. Then, in virtue of (T4S), S contains every projection $\hat{Q}$ such that $\hat{P} \leq \hat{Q}$. Now, for the most part, the projections $\hat{Q}$ such that $\hat{P} \leq \hat{Q}$ are incompatible. At the same time, however, if $\hat{P} \leq \hat{Q}$ and $\hat{P} \leq \hat{R}$, then $\hat{P} \leq \inf\{\hat{Q},\hat{R}\}$, irrespective of whether $\hat{Q}$ and $\hat{R}$ are compatible. Thus, (T4S) requires that every micro-state is closed under some non-compatible infima.

5. The Lattice Interpretation

In a strict interpretation, although every micro-state is required to be closed under compatible infima, it is not required to be closed under non-compatible infima, except as required by (T4S). Now, the projections on a Hilbert space H form a complete atomic lattice, which we denote L(H), so every subset of L(H) has an infimum. Nevertheless, a strict micro-state is not required by the principles so far proposed to be closed under arbitrary non-compatible infima, only compatible infima. If we wish to impose this further restriction on micro-states, then it must be added as a further principle, which may be stated as follows.

(P7L) If $\inf\{\hat{P}_i : i \in I\} = \hat{R}$, then $\{P_i : i \in I\}$ forces R.

Principle (P7L) may be understood as claiming that the lattice structure of L(H) is physically significant, so I refer to (strict) interpretations that conform with (P7L) as lattice interpretations. Examples include the interpretations proposed by Jauch (1968), Piron (1976), and van Fraassen (1973).

By adding (P7L) to our list of principles, we can show the following theorems.

(T7L) Every pre-micro-state corresponds to a principal filter on L(H).

(T8L) Every micro-state corresponds to a maximal principal filter on L(H).

In an atomic lattice L, the maximal principal filters are biuniquely associated with the atoms on L. In the case of L(H), the atomic projections are biuniquely associated with the rays (one-dimensional subspaces) on H, so we have the following theorem.

(T9L) Every micro-state corresponds to a ray on H.

Note that (T7L)-(T9L) characterize all lattice interpretations. This means that all lattice interpretations agree concerning the characterization of quantum micro-states, although they need not agree on other aspects of the interpretation of QM. Since we are primarily interested in micro-states, let us simply refer to all lattice interpretations as "the" lattice interpretation.

According to the lattice interpretation, every micro-state has a minimal, or principal, element. In other words, not only is there a principal a-property, a principal b-property, etc., there is a principal property simpliciter. This is analogous to classical mechanics (CM), according to which the principal property of a system at any given moment of time is its "location" in phase space at that moment. In QM, according to the lattice interpretation, we might plausibly think of the principal property of a quantum system as its "location" in Hilbert space.

The disanalogy between QM and CM concerns the associated valuation functions. In CM every micro-state assigns an exact value to every magnitude. According to the lattice interpretation of QM, no micro-state assigns an exact value to every magnitude, although every micro-state assigns a (possibly inexact) value to every magnitude. A lattice micro-state assigns an exact value to a magnitude just in case the corresponding ray is an eigen-ray of the corresponding self-adjoint operator.

6. The Minimal Strict Interpretation

Recall that, according to Principle (P1), every possible property is actualized by at least one micro-state. In the case of a strict interpretation, the possible properties correspond to the non-zero projections on the relevant Hilbert space H. By Theorem (T4S), every micro-state is closed under the partial order relation on L(H). Putting these ideas together, we obtain the following theorem.

(T10S) Every maximal principal filter on L(H) is contained in at least one micro-state.

Here, we think of a micro-state as a collection of projections on L(H).

Now, within the limits imposed by (T10S), there are more or less conservative/liberal construals of micro-states, according to the size of the admissible micro-states. The smaller/larger the micro-states, the more conservative/liberal the interpretation.

Evidently, the lattice interpretation is the most conservative strict interpretation, since it claims that the micro-states coincide with the maximal principal filters on L(H).

In the present section, we consider the opposite extreme among strict interpretations. The interpretation we examine is referred to as the minimal strict interpretation, since it involves the most liberal construal of quantum micro-states that is consistent with Principles (P1)-(P6S).

To characterize the minimal strict construal of micro-states, we look at all the restrictions forced by the strictness principle, as well as the minimal principles, and we take these restrictions as defining the micro-states. This yields the following theorem, which characterizes the minimal strict interpretation.

(T11M) Every pre-micro-state corresponds to a principal ortho-filter on L(H).

By a principal ortho-filter on L(H) I mean a subset F of L(H) that is closed under the partial order relation and under the formation of arbitrary compatible infima. Every such filter has principal (minimal) elements; each one is the infimum of a maximal compatible subset of F. Note that in the case of Boolean lattices, principal ortho-filters coincide with ordinary principal filters.

Applying the definition of micro-state, we have the following theorem.

(T12M) Every micro-state corresponds to a maximal principal ortho-filter on L(H).

In a maximal principal ortho-filter on L(H), the principal elements are all atomic projections, no two of which are orthogonal. Indeed, any collection of non-orthogonal rays (atomic projections) on H determines a pre-micro-state S, which is a micro-state if and only if every ray not in S is orthogonal to at least one ray in S. Thus, we have the following theorem.

(T13M) Every micro-state corresponds to a maximal collection of non-orthogonal rays on H.

It is interesting to compare minimal micro-states with lattice micro-states. Whereas a lattice micro-state S corresponds to a single ray R_S on H, a minimal micro-state S corresponds to a collection $\mathbb{R}_S$ of rays on H. Consequently, minimal micro-states assign exact values to more magnitudes than lattice micro-states. In particular, a minimal micro-state S assigns an exact value to magnitude m if and only if at least one ray in $\mathbb{R}_S$ is an eigen-ray of m.

Although minimal micro-states assign exact values to many more magnitudes than lattice micro-states, they do not assign exact values to every magnitude, unless d(H) = 2. If d(H) $\geq$ 3, then in virtue of the theorems of Gleason (1957) and Kochen and Specker (1967), no strict micro-state assigns an exact value to every magnitude.

7. Non-Strict Interpretations

Having examined strict interpretations, let us now briefly consider non-strict interpretations, that is, interpretations that deny the strict (1-1) correspondence between the properties ascribable to a system and the projections on the associated Hilbert space. According to these interpretations, although every property is associated with a projection, the correspondence is not 1-1; in particular, distinct properties may be associated with the same projection.

Non-strict interpretations fall into two natural groups: whereas conservative interpretations are more conservative than the lattice interpretation, liberal interpretations are more liberal than the minimal strict interpretation. Examples of conservative interpretations include those proposed by Bohr, Reichenbach (1944), and Kochen (1980). Examples of liberal interpretations include Fine's interpretation (1974) and van Fraassen's anti-Copenhagen variant of the modal interpretation (1973).

For the sake of brevity, I concentrate on Kochen's interpretation, and I concentrate on showing that it is not a strict interpretation. The fundamental principles of Kochen's construal of quantum micro-states may be stated as follows.

(K1) Every property m(X) is associated with a projection on H, which we denote $\hat{m}(X)$. The set of all m-projections is denoted $[\hat{m}]$.

(K2) Two properties P,Q are simultaneously actualizable only if the associated projections $\hat{P},\hat{Q}$ commute.

(K3) Every micro-state corresponds to a maximal compatible principal filter on L(H).

(K4) Every micro-state corresponds to an ordered pair $<\hat{P},B>$, where $\hat{P}$ is an atomic projection on H, and B is a maximal Boolean sublattice of L(H) that contains $\hat{P}$.

(K4*) Every micro-state corresponds to an ordered pair $<R,B>$, where R is an ray on H, and B is a maximal Boolean sublattice of L(H) that contains the projection onto R.

By a (maximal) compatible principal filter on L(H), I mean a (maximal) principal filter on any maximal Boolean sublattice of L(H). Each such filter is obtained by selecting a maximal Boolean sublattice of L(H) and then selecting an atomic element, as well as all the elements above it, in this Boolean sublattice.

In order to see that Kochen's construal of micro-states is more conservative than the lattice construal, we need merely observe that compatible filters are generally smaller than ordinary filters. For example, let H3 be a 3-dimensional Hilbert space; then every maximal Boolean sublattice of L(H3) has 8 elements, so every maximal compatible principal filter on L(H3) has 4 elements. By contrast, an ordinary maximal principal filter on L(H3) contains infinitely many elements, and is in fact isomorphic to L(H2), where H2 is a 2-dimensional Hilbert space. Thinking geometrically (in terms of subspaces, rather than projections), a maximal principal filter on L(H3) consists of a line ℓ together with all the planes that pass through ℓ, as well as the whole space H3. On the other hand, a maximal compatible principal filter on L(H3) consists of a line ℓ together with a particular pair of orthogonal planes through ℓ, as well as H3.

The conservative nature of Kochen's micro-states is reflected in the associated valuation functions. Let S be a micro-state $\langle \hat{P}, B\rangle$, in accordance with (K4). Suppose that the projection $\hat{m}(X)$ is in B, and suppose that $\hat{P} \leq \hat{m}(X)$. Does it follow that m(X) is actualized by S? Well, if S actualizes m(X), then in virtue of Principle (P3), S actualizes m(Y) for every Y such that $X \subseteq Y$. It follows that every projection $\hat{m}(Y)$ such that $X \subseteq Y$ is in B, as are all the orthocomplements of these projections.

For example, suppose that X = {r}, where r is in the spectrum of m. Then $\hat{P} \leq \hat{m}(r)$ if and only if $\hat{P}$ is an eigen-projection of m belonging to eigenvalue r. Now, S (= $\langle P, B\rangle$) actualizes m(r) only if it actualizes m(X) for every X such that $r \in X$. So if S actualizes m(r), then every projection $\hat{m}(X)$ such that $r \in X$ is in B, as are all their orthocomplements. But this means that every m-projection is in B. We can also argue in the converse direction, so we have the following theorem about Kochen's micro-states.

(K5) A micro-state S (= $\langle \hat{P}, B\rangle$) actualizes m(r) if and only if $\hat{P} \leq \hat{m}(r)$ and $[\hat{m}] \subseteq B$.

Note that in the special case that m is a maximal (non-degenerate) magnitude, $[\hat{m}]$ is a maximal Boolean sublattice of L(H), so $[\hat{m}] \subseteq B$ only if $[\hat{m}] = B$.

(K5) should be carefully compared with the lattice interpretation. According to the lattice interpretaion, every micro-state S corresponds to a ray R_S on H, and S assigns an exact value to m if and only if R_S is an eigen-ray of m. By contrast, according to Kochen's interpretation, a micro-state corresponds to an ordered pair $\langle R_S, B_S\rangle$ – in accordance with (K4*) – and S assigns an exact value to m if and only if R_S is an eigen-ray of m and every m-projection $\hat{m}(X)$ is contained in B_S.

Now, any given ray R on H is an eigen-ray of infinitely many non-compatible maximal magnitudes, each one associated with a different maximal Boolean sublattice of L(H) containing the projection onto R. In the lattice interpretation, the micro-state associated with R assigns an exact value to every one of the magnitudes. On the other hand, in Kochen's interpretation, a micro-state <R,B> assigns an exact value to exactly one of them, namely, the one associated with B.

We thus conclude that Kochen's interpretation is not strict. For any given atomic projection $\hat{P}$, there are magnitudes a and b, and eigenvalues r and s, such that $a(r) = b(s) = \hat{P}$, but inasmuch as a and b are associated with different Boolean sublattices of L(H), a(r) and b(s) are actualized by different micro-states, so they are not identical.

Since the projections on H do not individuate the properties, in Kochen's interpretation, we must ask exactly how the properties are individuated. One plausible interpretation of Kochen's theory is expressed in the following principle.

(K6) Every quantum property m(X) is characterized by a compatible principal filter on L(H), where the principal element is the projection $\hat{m}(X)$.

Thus, every property is associated with a projection, but distinct properties may be associated with the same projection. A projection P represents as many properties as there are compatible principal filters determined by P. For example, suppose that P is an atomic projection. Then there is a distinct property represented by P for each maximal Boolean sublattice containing P. On the other hand, the orthocomplement of P, I-P, determines exactly one compatible principal filter, so only one property is associated with I-P.

If we examine the logic of properties associated with Kochen's interpretation (in terms of forcing and exclusion, as in Section 3), we find that this logic has the structure of a semi-Boolean algebra (Abbott 1967), not a partial Boolean algebra. One obtains the partial Boolean algebra usually associated with QL by identifying any pair of properties that are statistically equivalent.

At the moment, not much is known about the semi-Boolean approach to QL (cf.,Hardegree and Frazer 1980), and not much is known about the relation between semi-Boolean algebras and the interpretation of QM (cf.,Hardegree 1979). These topics deserve serious attention in future research on the foundations of QM.

Notes

[1]The notion of a possible property is ambiguous. An _abstractly possible_ property is a property that can be ascribed to a system, truly or falsely. QL concerns abstractly possible properties. Principle (P1) concerns _physically possible_ properties, that is, properties that a system might actually possess. Every property m(X) is abstractly possible, but m(X) is physically possible if and only if X contains at least one point in the spectrum of m. Thus, for example, m(∅) is abstractly possible but not physically possible.

References

Abbott, J.C. (1967). "Semi-Boolean Algebra." Mathematicki Vesnik 4: 177-198.

Birkhoff, G. and von Neumann, J. (1936). "The Logic of Quantum Mechanics." Annals of Mathematics 37: 823-43.

Fine, A. (1974). "On the Completeness of Quantum Mechanics." Synthese 29: 257-89.

Gleason, A.M. (1957). "Measures on the Closed Subspaces of Hilbert Space." Journal of Mathematics and Mechanics 6: 885-93.

Hardegree, G.M. (1979). "Reichenbach and the Interpretation of Quantum Mechanics." In Hans Reichenbach: Logical Empiricist. Edited by W.C. Salmon. Dordrecht: D. Reidel. Pages 513-566.

--------------- and Frazer, P.J. (1980). "Charting the Laybrinth of Quantum Logics." In Proceedings of a Workshop on Quantum Logic. Edited by E. Beltrametti. New York: Plenum Press. In press.

Jauch, J.M. (1968). Foundations of Quantum Mechanics. Reading, MA: Addison-Wesley.

Kochen, S. and Specker, E.P. (1967). "The Problem of Hidden Variables in Quantum Mechanics." Journal of Mathematics and Mechanics 17: 59-87.

----------. (1980). The Interpretation of Quantum Mechanics." Unpublished manuscript.

Piron, C. (1976). Foundations of Quantum Physics. Reading, MA: W.A. Benjamin.

Reichenbach, H. (1944). Philosophic Foundations of Quantum Mechanics. Los Angeles and Berkeley: University of California Press.

van Fraassen, B.C. (1973). "Semantic Analysis of Quantum Logic." In Contemporary Research in the Foundations and Philosophy of Quantum Theory. (The University of Western Ontario Series in the Philosophy of Science, vol. 2). Edited by C.A. Hooker. Dordrecht: D. Reidel. Pages 80-113.

von Neumann, J. (1932). Mathematische Grundlagen der Quantenmechanik. Berlin: Springer. (Translated by R.T. Beyer as The Mathematical Foundations of Quantum Mechanics. Princeton: Princeton University Press, 1955).

Wigner, E. (1931). Gruppentheorie und ihre Anwendung. Vieweg: Braunschweig. (Translated by J.J. Griffen as Group Theory. New York: Academic Press, 1959).

Quantum Logic and the Interpretation of Quantum Mechanics [1]

R.I.G. Hughes

Princeton University

1. Introduction

There is no such thing as "The Quantum Logical Interpretation of Quantum Mechanics". Rather, there is a cluster of interpretations, all of which can be described as "quantum logical". Here I provide a general framework for discussing interpretations of this kind, and then locate various suggestions within it. The presentation owes much to van Fraassen (see, in particular, van Fraassen 1974); his "modal interpretation" is one of those I discuss, along with those of Jauch, Putnam and Kochen. I begin by rehearsing some orthodox quantum theory.

2. Quantum Mechanics and Hilbert Space

Within quantum mechanics we deal with a set O of measurable quantities, or observables (position, momentum, components of spin and so on). Experiment can determine the value of an observable for a given system: the values so determined will be real numbers. A maximal amount of information about what the result would be for a given system, whatever experiment we chose to perform on it, is available once we know the state of the system. Given the state, quantum theory allows us to assign a probability to each event describable as, "An experiment to measure A yields a result within Δ," where A is any observable in O, and Δ any Borel subset of the real numbers. (In what follows Δ will always denote a Borel subset of the reals.)

The algorithm which enables us to do this is set out in terms of Hilbert spaces, complex vector spaces on which an inner product has been defined. Quantum theory uses Hilbert spaces of at most denumerable dimension.[2]

Observables are represented by the self-adjoint operators on a Hilbert space $\mathcal{H}$. To each such operator A corresponds a family

PSA 1980, Volume 1, pp. 55-67

$\{P^A_\Delta: \ \Delta \varepsilon B(R)\}$ such that

$$P^A_\Lambda = 0$$

$$P^A_R = I$$

and, if $\Delta \subseteq \Gamma$, then

$$P^A_\Delta \leq P^A_\Gamma$$

The ordering relation among these projection operators exactly corresponds to the relation of set inclusion among the subspaces onto which they project.

In the case when A admits a complete set of eigenvectors, the subspace L^A_Δ onto which P^A_Δ projects is the subspace spanned by those eigenvectors of A with eigenvalues lying within Δ.

States are represented by density operators on H, that is, by certain weighted sums of projection operators: ρ is a density operator on H iff there is an orthonormal basis $\{\psi_i\}$ for H; for each i, $P(\psi_i)$ is the projection operator onto the one-dimensional subspace spanned by ψ_i; and

$$\rho = \sum_i c_i P(\psi_i)$$

where $\sum_i c_i = 1$ and, for each i, $c_i \geq 0$.

In the case when, for some i,

$$\rho = P(\psi_i)$$

(i.e., $c_j = 0$ for all $j \neq i$), the state is known as a pure state, and may be specified by the vector ψ_i. Otherwise the state is termed a mixed state. In general, the subspace $S[\rho]$ of H spanned by the set $\{\psi_i: \ c_i \neq 0\}$ of basis vectors is known as the image space of ρ.

The probability $p(A,\Delta)$ that a measurement of A will yield a result within Δ, if the system is in the state ρ, is given by

$$p(A,\Delta) = Tr(\rho P^A_\Delta)$$

In the special case, when ρ is the pure state ψ, we have

$$p(A,\Delta) = \langle\psi | P^A_\Delta \psi\rangle$$

It follows that

$$p(A,\Delta) = 1 \quad \text{iff} \quad S[\rho] \subseteq L^A_\Delta \quad .$$

Or, in the special case,

$$p(A,\Delta) = 1 \quad \text{iff} \quad \psi \varepsilon L^A_\Delta \quad .$$

3. Non-Boolean Algebras and Quantum Logic

The last two equations point to the importance of the set of subspaces of Hilbert space within quantum theory. Like the power set of a given set, the set of subspaces of $\mathcal{H}$ forms an algebra; however, whereas the former is Boolean, the latter is not. It may be regarded either as an orthocomplemented lattice ($\mathcal{B}(\mathcal{H})_{OM}$) or as a partial Boolean algebra ($\mathcal{B}(\mathcal{H})_{PB}$).[3] In both cases there are two binary operations involved and one singulary. In the case of the orthocomplemented lattice, we have

$$L_1 \wedge L_2 = \text{Inf } \{L_1,L_2\}$$

$$L_1 \vee L_2 = \text{Sup } \{L_1,L_2\}$$

where Inf and Sup are defined with respect to the ordering relation of inclusion among subspaces. The singulary operation is that of orthocomplementation, taking L into $L^\perp$.

On the alternative approach the binary operations are only partial operations; $L_1 \wedge L_2$ and $L_1 \vee L_2$ are only defined where L_1 and L_2 are compatible:

L_1 and L_2 are <u>compatible</u> ($L_1 \updownarrow L_2$)

iff

L_1 and L_2 are orthogonal except (possibly) for an overlap

iff

$$P_1P_2 = P_2P_1$$

(where P_i is the projection operator onto L_i)

Where they are defined, these operations yield the same subspaces as they do on the lattice approach. Again, the singulary operation is that of orthocomplementation, which is everywhere defined.

Note that on both approaches the set of objects we deal with is the same, namely the set of subspaces of a Hilbert space. The structure $\mathcal{B}(\mathcal{H})_{OM}$ conforms to the axioms of an orthocomplemented lattice, and in addition to the orthomodularity axiom:

$$a \wedge (a^{\perp} \vee (a \wedge b)) \leq b$$

The structure $\mathcal{B}(\mathcal{H})_{PB}$, on the other hand, can be thought of as a family of Boolean algebras all of which share the same maximum and minimum elements and which may overlap elsewhere as well. Each maximal Boolean subalgebra of $\mathcal{B}(\mathcal{H})_{PB}$ (and, come to that, of $\mathcal{B}(\mathcal{H})_{OM}$) is the algebra generated by the set of one-dimensional subspaces associated with an orthogonal basis for $\mathcal{H}$, and conversely. If we wish, we may introduce infinitary operations on $\mathcal{B}(\mathcal{H})_{OM}$ and $\mathcal{B}(\mathcal{H})_{PB}$ to make them respectively an orthomodular σ-lattice and a partial Boolean σ-algebra.

Within both $\mathcal{B}(\mathcal{H})_{PB}$ and $\mathcal{B}(\mathcal{H})_{OM}$ each subspace L generates a <u>filter</u>, the set of those subspaces of $\mathcal{H}$ which contain L. If L is a one-dimensional subspace (an <u>atom</u> of the algebra), the filter generated by L is an <u>ultrafilter</u> on $\mathcal{B}(\mathcal{H})_{PB}$ (resp. on $\mathcal{B}(\mathcal{H})_{OM}$). A <u>Boolean ultrafilter</u> on $\mathcal{B}(\mathcal{H})_{PB}$ generated by a one-dimensional subspace L of $\mathcal{H}$ is the intersection of the ultrafilter generated by L with a maximal Boolean subalgebra of $\mathcal{B}(\mathcal{H})_{PB}$ which contains L. A given atom L will generate an infinite number of such Boolean ultrafilters; the union of all these Boolean ultrafilters is the ultrafilter on $\mathcal{B}(\mathcal{H})_{PB}$ generated by L.

Where I want to talk about the algebraic structure of the set of subspaces of H without specifying whether it is to be regarded as a lattice or as a p.B.a, I will designate it by $\mathcal{B}(H)$. It is this structure which plays the key role within quantum logic.

The development of such a logic can be broken down into stages. First the syntax of a formal propositional language $\mathcal{Q}$ is given, consisting of rules for forming well-formed formulae from propositional variables, connectives and so on. Secondly there is a definition of an interpretation of $\mathcal{Q}$ within $\mathcal{B}(H)$. We select which version of $\mathcal{B}(H)$ we wish to use; an interpretation is then a function a* mapping formulae of $\mathcal{Q}$ onto elements of this structure, and conforming to certain constraints, so that for instance,

$$a^*(A \,\&\, B) = a^*(A) \wedge a^*(B)$$

and so forth. To obtain a valuation of $\mathcal{Q}$, (i.e.,a mapping of the set of formulae of $\mathcal{Q}$ onto a set of truth-values) we need a further function t from $\mathcal{B}(H)$ onto the set of truth values, which also obeys various conditions (ensuring, for instance, that the maximum of the algebra always receives the value "True", and the minimum the value "False"). Call such a function a t-function. Then, given an interpretation a* and a t-function t, the composition $t \cdot a^*$ will be a valuation of $\mathcal{Q}$.

Finally, we can try to produce a proof theory for $\mathcal{Q}$ corresponding to the semantics that has emerged.

All this will produce a quantum logic, but it will not on its own yield a quantum-logical interpretation of quantum mechanics. To do that we must (at least) say (a) what the elements of the algebra, $\mathcal{B}(H)_{OM}$ or $\mathcal{B}(H)_{PB}$ as the case may be, are taken to represent, (b) what motivates our particular choice of t-function, (c) what connection exists between a given t-function and the state of a quantum mechanical system, and (d) how our answers to (a) (b) and (c) interlock.

I will give an account of the possible answers to (a), and then see how that and the other questions are answered in four different quantum-logical interpretations of quantum theory.

4. M-Statements, MC-Statements, P-Statements, PS-Statements

Consider the event described by the sentence, "An experiment to measure A will yield a result in the Borel set $\Delta \subseteq R$". This was the event (A,Δ) such that (from Section 2)

$$p(A,\Delta) = 1 \text{ iff } S[\rho] \subseteq L^A_\Delta .$$

Thus to each event (A,Δ) there corresponds a subspace L^A_Δ of $\mathcal{H}$, such that the event is certain if and only if the image space of the state of the system (the state vector in the pure case) is contained in it. Rather more controversially, we can also attach an event (or an equivalence class of events) to each subspace of $\mathcal{H}$. Now for each event we can find a sentence saying that the event occurs; we use the same notation to name both the sentence and the event:

$(A, \Delta) =_{def}$ "An experiment on the system to measure A yields a result within Δ."

Clearly, to each member of $\mathcal{B}(\mathcal{H})$ now corresponds an (equivalence class of) such sentences, which may be called M-statements. Now one way to regard quantum logic is as the logic of M-statements. For the interpretation of $\mathcal{Q}$ within $\mathcal{B}(\mathcal{H})$ can now be thought of, quite literally, as an interpretation: the function a* tells us what each sentence in the formal language is to mean. On this approach, the constraints on the t-functions are dictated by which set of M-statements can be true together.

There is a presupposition that has to be fulfilled before we can think of a truth value being assigned to the M-statement (A,Δ), namely that an A-experiment is performed. (Which is to say that the probability $p(A, \Delta)$ is conditional on the performance of the experiment.) This would be conveyed by regarding (A,Δ) as the sentence, "An experiment on the system to measure A would yield a result within Δ"; I have instead talked in terms of presuppositions to avoid ambiguities of modality, for it is also useful to introduce two modal operators to act on the set of M-statements. We introduce $\Box$ and $\Diamond$, with the semantic requirement that

$\Box(A, \Delta)$ is true iff $p(A,\Delta) = 1$

$\Diamond(A, \Delta)$ is true iff $p(A, \Delta) \neq 0$

Thus these operators carry a reference to the state of the system: we can regard $\Box$ as expressing state-determined necessity. Sentences of the form $\Box(A, \Delta)$ I will call MC-statements. Whether $\Box$ and $\Diamond$ should properly be described as operators is moot: it is not clear what their iteration would signify. For present purposes I will disallow such iteration, and merely allow them to prefix M-statements, or statements built up from M-statements by the connectives corresponding to $\wedge, \vee$ and $^\perp$.

This gives us an alternative conception of quantum logic: we can think of it as the logic of MC-statements. On the other hand, an advocate of quantum logic who is also a realist may want to associate each subspace of $\mathcal{H}$, not only with a possible event (and hence an M-statement and an MC-statement), but also with a possible property of a microsystem. A microsystem may have the <u>property</u> that the value of A for that system lies within Δ, and this property can be associated with the subspace L^A_Δ of $\mathcal{H}$. As interpretations of the formal language the realist will use property ascriptions or <u>P-statements</u>. We abbreviate such statements as follows:

> $\langle A,\Delta\rangle$ names the sentence, "The value of A for the system lies within Δ ."[4]

As we shall see, on certain interpretations, some but not all of these property ascriptions are true by virtue of the state of the system: we need to distinguish such ascriptions and write:

> $[A, \Delta]$ names the sentence "By virtue of the state of the system, the value of A for the system lies within Δ ."

We call such sentences PS-statements. They are, in some sense, modal versions of P-statements, as the MC-statements were of the M-statements.

We have distinguished four kinds of statements, M-statements, MC-statements, P-statements, and PS-statements.[5] Any realist interpretation of quantum theory must tell us how the truth conditions for corresponding statements of the four kinds interlock. In classical mechanics the situation is simple; experiments reveal the properties which systems possess: moreover, they possess those properties if and only if they are in specific states. In the classical case

$$(A, \Delta) \text{ is true iff } \Box(A, \Delta) \text{ is true iff } \langle A,\Delta\rangle \text{ is true}$$

$$\text{iff } [A, \Delta] \text{ is true.}$$

In quantum mechanics matters are not so straightforward.

5. The Logic of MC-Statements

Arguably the least problematic, but also the least useful way to look at quantum logic is to think of it as the logic of MC-statements. On this account the element L^A_Δ of $\mathfrak{B}(\mathcal{H})$ corresponds to the sentence $\Box(A, \Delta)$. We now stipulate (1) that each sentence obtained by linking MC-statements with quantum logical connectives should also be an MC-statement, and (2) that the connectives correspond to operations on $\mathcal{B}(\mathcal{H})$.

Given these two stipulations the differences between quantum logical and classical connectives appear naturally. The effect of a quantum logical connective is that of a classical connective modified by the operator $\Box$: thus the sentence

$$\Box(A,\Delta) \ \& \ \Box(B,\Gamma)$$

is equivalent to the sentence

$$\Box((A,\Delta) \text{ and } (B,\Gamma)) .$$

Within this logic any pure state of a system functions as an admissible valuation, assigning the value "True" to each member of the ultrafilter generated by that state. We can make negation truth functional by assigning "False" to an element of $\mathcal{B}(\mathcal{H})$ only if it is the orthocomplement of a member of the ultrafilter, and leave the truth values of other elements of $\mathcal{B}(\mathcal{H})$ indeterminate, or we may emphasize the modal nature of the logic by assigning "False" to anything lying outside the ultrafilter, as we wish.

But no interpretation occurs when we develop this logic: we merely restate a selection of the theory's predictions in a different way.

6. Jauch

The propositions dealt with by Jauch in his "propositional calculus of physical systems" (see Jauch 1968, p. 72 *et seq.*) are "yes-no experiments", or, in our terms, events. This is also true for, e.g., Bub (see Bub 1974, p. 93); however, Bub regards the set of events as a partial Boolean algebra, whereas Jauch thinks of it as an orthomodular lattice. As we have noted, to each event we can attach an M-statement, and so, effectively, Jauch's quantum logic is the logic of M-statements; in fact it is less confusing to think of (equivalence classes of) M-statements as Jauch's propositions. (That M-statements rather than MC-statements are involved is clear from the discussion on p. 93 of Jauch (1968)).

I have argued elsewhere (Hughes 1980) that the realist stance adopted by Jauch in his discussion of a physical system requires that the M-statements within his logic also be regarded as property ascriptions, and on one occasion (Jauch 1968, p. 107) he makes this identification explicit. However, there are problems generated by so doing. For the property in question cannot be the property of *having passed* the experiment, as this is associated with the corresponding MC-statement, whereas, as we have seen, Jauch's propositions are effectively M-statements. Yet to associate a property with each M-statement would lead Jauch to the same difficulty as Putnam, to whom we now turn.

7. Putnam

In Putnam (1969) it is very clear that we are dealing with a set of P-statements: he writes, "Statements of the form m(S) = r - 'the magnitude m has the value r in the system S' - are the sorts of statements we shall call *basic physical propositions* here." (Putnam 1969, p. 220). Further, these are the kinds of statements dealt with in quantum logic, which attach, in other words, to the elements of $\mathcal{B}(\mathcal{H})$. Thus to supply the semantics of his quantum logic we need to know what properties a system can possess, on Putnam's view, at any one time. But he claims that every observable (or magnitude in his terminology) has a definite value at a given time; even assuming a particle to be in an eigenstate of position (and to simplify the exposition he has assumed that such eigenstates exist) then "it is still true... that 'the particle *has* a momentum'; and if I measure it I shall find it." (Putnam 1969, pp. 229-30). A similar claim was made by Demopoulos (1974, p. 727): "For any property P it is completely determinable whether or not P holds." Now, to the set of P-statements referring to the possible values of a single observable there corresponds a Boolean subalgebra of $\mathcal{B}(\mathcal{H})$. The requirement that, for a system, each observable has a specific (even if not specifiable) value at a given time means that within each Boolean subalgebra of $\mathcal{B}(\mathcal{H})$ just one (Boolean) ultrafilter be assigned the value "True", which in turn means that the whole algebra $\mathcal{B}(\mathcal{H})_{PB}$ be homomorphically mapped onto $\{0,1\}$. But as Kochen and Specker (1967) showed, no such homomorphisms exist for a Hilbert space $\mathcal{H}$ of dimension 3 or greater.[6]

This failure of the Putnam programme was pointed out by Friedman and Glymour (1972). They commented (p. 28): "The central problem of quantum theory remains unmoved."

8. van Fraassen

The "central problem" they referred to was the problem of measurement. It is this problem which, *inter alia*, the rather more subtle analyses of van Fraassen and Kochen aim to resolve. I will not here examine these analyses fully, but merely indicate their place in this general account of quantum-logical interpretations.[7]

Both accounts deal with the properties of a system. For van Fraassen the properties a system possesses at any given time are associated with the members of an ultrafilter on $\mathcal{B}(\mathcal{H})$. If the system is in a pure state, the ultrafilter in question is that generated by the pure state. For a system in such a state <A,Δ> implies [A,Δ] implies □(A, Δ).

But not all states are pure. In particular, after an A-measurement the state may well be the mixed state $\rho = \sum_i c_i P(\psi_i)$ where each of the ψ_i is an eigenvector of A. In fact, if prior to the measurement the system has been in a pure state which is not an eigenstate of A then this will certainly be the case. Van Fraassen now rejects the ignorance interpretation of mixed states, which suggests that the system after the measurement really is in one of the states ψ_i, but replaces it by, as it were, an ignorance interpretation of properties: the system has the properties within one of the ultrafilters of $\mathcal{B}(\mathcal{H})$ which includes S[ρ], but we do not know which. In this case <A,Δ> does not imply [A, Δ]. Nevertheless <A,Δ> always implies (A,Δ) and so the system behaves as though the projection postulate holds. (See van Fraassen (1974)). Also, we always have □(A, Δ) implies [A, Δ] implies <A,Δ> .

Note that when a system is in a state $\rho = \sum_i c_i P(\psi_i)$ not every ultrafilter on $\mathcal{B}(\mathcal{H})$ which includes S[ρ] may qualify as a possible set of properties of the system. (I am reading "possible" here as "having non-zero probability".) If, for a distinct i and j, $c_i = c_j$, then the decomposition of ρ into a weighted sum of projection operators is not uniquely determined: the projection operators $P(\psi_i)$ and $P(\psi_j)$ may be replaced by $P(\psi_i')$ and $P(\psi_j')$, where ψ_i' and ψ_j' are any orthonormal pair of vectors spanning the same subspace as $\{\psi_i, \psi_j\}$. But the only pair which generate possible ultrafilters of properties of the system are those which are eigenvectors of the observable which has just been measured. This is not a criticism of van Fraassen, but an indication of how, on his approach, physical considerations are needed to supplement formal ones.

9. Kochen

For Kochen, as for van Fraassen, a system's properties are determined by the interaction it has just undergone, but with this difference. Associated with an A-experiment is a Boolean subalgebra of possible P-statements involving A. The sets of possible properties of a system after such an interaction are the Boolean ultrafilters on $\mathcal{B}(\mathcal{H})_{PB}$ generated by the atoms within this subalgebra, rather than the ultrafilters on the whole p.b.a. Only the properties within this Boolean subalgebra receive a truth-value. On this approach, the t-functions of quantum logic are partial functions.

The set of properties possessed by a system Kochen refers to as its state, or, sometimes, individual state (Kochen 1979 , p. 28).

What we have hitherto called the state is referred to as the *statistical state*. On this approach PS-statements are misleading, since systems cannot be said to possess properties *by virtue of* the statistical state. The statistical state represents available information: it tells us what can be known about the individual state, i.e., about the properties of a system. However, from a complete specification of a system's properties we can deduce its (statistical) state, namely the atom in $\mathcal{B}(\mathcal{H})$ which generates the Boolean ultrafilter of properties. It follows that $\langle A,\Delta\rangle$ implies $\Box(A,\Delta)$.

However, oddly enough the converse need not be true. Consider two observables A and B which share an eigenvector ψ, but do not commute. Then there may be a subspace containing ψ in the Boolean subalgebra of $\mathcal{B}(\mathcal{H})$ generated by the eigenvectors of A, which is not a member of that generated by the eigenvectors of B. Call this subspace L^A_Δ. Now assume that the system has just emerged from a B-measurement; its properties all lie within the Boolean subalgebra associated with B. Thus $\langle A,\Delta\rangle$ has no truth value. Yet the statistical state of the system may well be ψ, in which case $\Box(A,\Delta)$ will be true. Thus, on this account, the fact that a measurement of A will yield a result within Δ with certainty does not allow us to say that it has the corresponding property; to accept Kochen's interpretation, then, is to reject a principle which Einstein, Podolsky and Rosen (1935) put forward as beyond question.

Notes

[1]I would like to thank Ed Levy and Bas van Fraassen for their comments on a previous draft of this paper.

[2]For a full version of what is sketched out here, see von Neumann (1932) or Fano (1971).

[3]For more on orthomodular lattices, see Holland (1970); for more on partial Boolean algebras see Kochen and Specker (1965).

[4]I leave aside the question raised by Teller (1979), whether this is an appropriate form for a property ascription involving a continuous quantity A.

[5]Compare Heelan (1970).

[6]For more on this see Healey (1979).

[7]Specifically, I talk of the "Copenhagen Variant" of van Fraassen's Modal Interpretation. See van Fraassen (1972 , 1973 , 1974).

References

Bub, J. (1974). The Interpretation of Quantum Mechanics. (University of Western Ontario Series in Philosophy of Science, Vol. 3.) Dordrecht, Holland: Reidel.

Demopoulos, W. (1974). "What is the Logical Interpretation of Quantum Mechanics?" In PSA 1974: Proceedings of the 1974 Biennial Meeting of the Philosophy of Science Association. Edited by R.S. Cohen et al. Boston: Reidel. Pages 721-28.

Einstein, A., Podolsky, B. and Rosen, N. (1935). "Can Quantum Mechanical Description of Physical Reality be Considered Complete?" Physical Review 47: 777-80.

Fano, G. (1971). Mathematical Methods of Quantum Mechanics. New York: McGraw Hill.

Friedman, M. and Glymour, C. (1972). "If Quanta Had Logic." Journal of Philosophical Logic 1: 16-29.

Healey, R. (1979). "Quantum Realism: Naivete is no Excuse." Synthese 42: 121-44.

Heelan, P. (1970). "Quantum and Classical Logic: Their Respective Roles." Synthese 21: 2-33.

Holland, S.S., Jr. (1970). "The Current Interest in Orthomodular Lattices." In Trends in Lattice Theory. Edited by J.C. Abbott. New York: Van Nostrand. Pages 41-116.

Hughes, R.I.G. (1980). "Realism and Quantum Logic." In Proceedings of the Erice Workshop on Quantum Logic. Edited by E. Beltrametti et al. London: Plenum Press. In Press.

Jauch, J.M. (1968). Foundations of Quantum Mechanics. Reading, MA: Addison Wesley.

Kochen, S. and Specker, E.P. (1965). "Logical Structures Arising in Quantum Theory." In Symposium on the Theory of Models. Edited by J. Addison et al. Amsterdam: North Holland. Pages 177-189.

---------. (1967). "The Problem of Hidden Variables in Quantum Mechanics." Journal of Mathematics and Mechanics 17: 59-87.

---------. (1979). "The Interpretation of Quantum Mechanics." Unpublished Manuscript.

Putnam, H. (1969). "Is Logic Empirical?" In Proceedings of the Boston Colloquium for the Philosophy of Science, 1966/1968. (Boston Studies in the Philosophy of Science, Vol. V.) Edited by R.S. Cohen and M. Wartofsky. Boston: Reidel. Pages 216-241.

Teller, P. (1979). "Quantum Mechanics and the Nature of Continuous Physical Quantities." Journal of Philosophy 76: 346-61.

van Fraassen, B.C. (1972). "A Formal Approach to the Philosophy of Science." In Paradigms and Paradoxes: the Philosophical Challenge of the Quantum Domain. (University of Pittsburgh Series in the Philosophy of Science, Vol. 5.) Edited by R. Colodny. Pittsburgh: University of Pittsburgh Press. Pages 303-366.

-----------------. (1973). "Semantic Analysis of Quantum Logic." In Contemporary Research in the Foundations and Philosophy of Quantum Theory. (University of Western Ontario Series in Philosophy of Science, Vol. 2.) Edited by C.A. Hooker. Dordrecht, Holland: Reidel. Pages 80-113.

------------------. (1974). "The Einstein-Pololski-Rosen Paradox." Synthese 29: 291-309.

von Neumann, J. (1932). Mathematische Grundlagen der Quantenmechanik. Berlin: Springer. (Translated by R.T. Beyer as The Mathematical Foundations of Quantum Mechanics. Princeton: Princeton University Press, 1955.)

Part III

Species and Evolution

Have Species Become Déclassé?[1]

Arthur L. Caplan

The Hastings Center and Columbia University

1. Two Views of the Ontology of Species

There is no more popular pastime in the literature of the philosophy of biology than analyzing the concept of species. This is partly due to the fact that the concept is such a prominent one in presentations of evolutionary theory. It is also a result of the fact that philosophers and biologists have struggled without a great deal of success to formulate a definition of species that would be acceptable for all the diverse purposes of the biological sciences.

For some time, philosophers of biology assumed that, whatever problems of definition and explication exist regarding the concept of a species, the ontological status of the concept was certain. Species have long been viewed as classes. Indeed, they have been viewed as paradigmatic examples of classes of a special variety. Since the traits of organisms vary from creature to creature as well as from generation to generation, a special set of properties must be used to group or aggregate organisms into species. Unlike most classes of objects, there often is no single property or trait which is present in all the members of a species. Rather, there are sets of properties some number of which are instantiated in any given organism (Beckner 1968, pp. 60-66; Ruse 1973, pp. 122-139).

Thus, species in biology have been viewed as good examples of cluster concepts (Bambrough 1960-61). The properties used to group individuals into classes are such that, while no single organism possesses all of them, each organism possesses some of the set or cluster. Thus, for example, if the set of properties used to define the class of organisms we know as *Cygnus olor* are white color, long necks, downy feathers, large wings, black eye and beak markings, short legs, and good swimming ability, traditionally it has not been viewed as necessary that every bird possess all of these traits to be included in this species. It is entirely possible that a swan may be found which lacks one or

PSA 1980, Volume 1, pp. 71-82

another of these various properties. Nevertheless, the bird might still be classified as a member of the species Cygnus olor. As long as a given organism possessed a high proportion of the relevant diagnostic properties, the species membership of the organism was thought to be secure (Beckner 1968, pp. 60-72).

Recently, however, there has been a shift in the thinking of a number of philosophers and biologists concerning the ontological status of species. They have argued that if we are to justifiably group creatures with diverse traits into species, we must begin to rethink the ontological status of the species concept.

The proposal which has captured the allegiance of a number of scholars (Ghiselin 1975; Hull 1976, 1978; Mayr 1976; Reed 1979; Van Valen 1976) in the past few years is that, rather than viewing species as classes, they be viewed as spatiotemporally localized individuals. By individuals proponents of this ontological shift mean "spatiotemporally localized, cohesive and continuous entities" (Hull 1978, p. 336). On this view organisms would not be members of the classes of species to which they belong. Instead, they would have the relationship of a part to a whole in the way that cells, tissues, and organs are parts of and not members of individual human beings.

The cohesion and continuity used to individuate species as individuals resides in the fact that organisms can be grouped according to their lineage or phylogeny. Since evolution occurs partly as a consequence of the (imperfect) transmission of genes from one organism to another, it is possible to trace continuous lines of descent among many different organisms. As David Hull has observed "...organisms form lineages. The relevant organismal units in evolution are not sets of organisms defined in terms of structural similarity but lineages formed by the imperfect copying processes of reproduction." (1978, p. 341). It is the continuity and coherence of organic lineages that provide the grounds for thinking of species as individuals.

Those who have advocated this change in the ontological status of species have not done so lightly; they are well aware of the fact that viewing species as individuals represents a drastic break from previous thinking about the species concept (Ghiselin 1975; Hull 1976, 1978). However, they feel that such an ontological shift is necessitated both by current theoretical work in evolutionary biology, and, by the conceptual benefits to be garnered from this maneuver. In the remainder of this paper I shall try to show that (a) the reasons advanced for viewing species as individuals rather than classes are not persuasive, (b) a reasonable explication can be given of species as classes that is consistent with the tenets of the modern synthetic theory of evolution, and, (c) that a cost/benefit computation of the two ontological views of the species concept actually favors the classic rather than the individualistic interpretation.

2. The Rationale for Viewing Species as Individuals

One of the most important reasons underlying the claim that species are best viewed as individuals and not classes or sets of organisms is that current evolutionary theory has as one of its main implications the conclusion that species are best treated as individuals. This implication is drawn from two disparate sources: (a) the specific claims about species and speciation made by evolutionists and (b) the role played by species in evolutionary theorizing.

While there is a good deal of uncertainty as to the exact nature of the tenets of the modern synthetic theory of evolution, there is no dispute concerning the fact that evolutionary change is an expected outcome of this theory. This is a result of the fact that in a world of scarcity containing a number of promiscuous creatures, there are enough sources of variation and selection present to guarantee continuous change in the phenotypic constitution of these creatures over long periods of time. Continuing phenotypic change is to be expected as soon as the mechanisms and makeup of the biological world are even roughly discerned (Caplan 1979).

If this is so then the class view of species is doomed. Essentialism is, quite simply, not compatible with the continuous change model of population evolution dictated by current evolutionary theory. There are, in the long run, simply not going to be any essences which persist long enough to permit systematists to aggregate organisms into groups (Griffiths 1974). Even if one gets clever and tries to use clusters of properties to construct polythetic definitions, the probability is very great that <u>all</u> of the observable traits of creatures will eventually evolve to the point where any proposed cluster of properties would be useless as the criteria for classification.[2]

Not only is continuous change an implication of evolutionary theory, it is also an empirical fact. Few contemporary ornithologists would use the cluster of traits now thought diagnostic of <u>Cygnus olor</u> to identify the distant avian ancestors of this species. There are few essentialists in the trenches of paleontologists.

Not only does evolutionary theory dictate the occurrence of continuous change, it also mandates the need for continuity. As Hull observes, "Evolution is a selection process, and selection processes require continuity" (1976, p. 190). If selection is to act upon variation to produce new phenotypic alterations, it must do so by acting upon variants produced by genotypes which are transmitted from one generation to the next. The reproduction and replication of organisms requires precisely the sort of continuity and spatiotemporal localization appropriate to individuals. Since species ultimately result from selection among genetic lineages, and, since genetic lineages are individuals, species must, to be selected, be individuals as well.

If it is true that species are the basic units of evolutionary change and if species can be seen as equivalent to segments of evolutionary lineages or phylogenies, then the view that species are best understood as individuals appears quite credible. Devotees of this line of reasoning note that if species are really treated as individuals by evolutionary theorists, this fact should be reflected in their language, both in describing and in individuating species. And, indeed, numerous examples can be found of evolutionary theorists referring to species as "super-organisms" or "homeostatic systems". Also biologists do often refer to new species as "splitting off", "budding", "fusing", and, even, as "unique"--all terminology more appropriate to the individuation of individuals than it is to the individuation of classes (Ghiselin 1975, pp. 542-543; Hull 1978, pp. 344-350).

3. Difficulties With These Arguments

It is certainly true that evolutionary theory, when supplemented with the appropriate empirical information for describing the contingencies of the past and present biotic world, provides much support for the view that change is an expected outcome of evolutionary processes. However, the notion that species are classes is quite capable of accommodating this fact.

It is not the case that the ontological interpretation of species as classes implies any commitment to essentialism. No property or properties need be deemed eternal or immutable in trying to generate traits by which organisms can be grouped or aggregated. The idea of family resemblance and the related notion of a cluster concept were specifically introduced into the analysis of the species concept to provide a means of grouping organisms into classes despite the fact of continuous evolutionary change in species membership. Merely noting the implications of evolutionary theory or the facts of phylogeny relative to change does not <u>per se</u> invalidate the utility of these modes of classification.

Nor is it the case that grouping individuals into classes is incompatible with complete and continuous change in the properties used to define a given class. When the vast majority of class defining properties are no longer instantiated by any organisms it is reasonable to assume that the class is either extinct or has evolved into a different class. Thus, at some point Archaeopteryx crossed the class line dividing reptiles and birds. Reptilian properties gradually disappeared while avian characteristics emerged.

Surprisingly, if evolutionary theory mandates continuous and irreversible change in the traits of organisms, the concept of a class is more appropriate to describing this state of affairs than is the concept of individuals. Individuals have clear, datable, unambiguous origins and terminations. Their parts are usually contiguous in space and time. Species, however, have fuzzy beginnings and ambiguous demises. Their members are rarely spatially contiguous. If one views them over time they are likely to either disappear with a whimper or slowly transmute into a new variety. Neither process is clear, datable, or localizable as would be required in individuating individuals. The individual Charles Darwin has a datable beginning and end. The evolution of reptiles into birds is not amenable to similar spatiotemporal demarcations.

It is true that biologists talk about species as budding, fusing, and splitting. However, the criteria they look to more than any other are the capacity for interbreeding and descent from a common ancestor. Biological language can be opaque in such uses because it is not always clear whether in talking about species biologists have reference to the taxonomic category "species" or to the actual groups of organisms (taxa) to be seen, directly, in nature, and, by inference, in the fossil record.

The taxonomic category "species" refers to a particular level of biological organization in a classification. It makes no reference to any actual organism or organisms. Indeed, the category "species" is clearly treated as a class concept in biology. Its defining properties are (a) the ability to produce fertile offspring between members of a group and (b) descent from a common ancestor (Mayr 1970; Simpson 1961). Most of the references to species made by biologists are to the category of species and not the taxa of the actual biological world. Statements to the effect that the "species" is the central unit of evolution, or, about the budding or splitting of species are claims about species as a general category. Certainly no evolutionist believes that claims such as "the species is the primary locus of selection" refer in an oblique way to a finite and denumerable list of extant and extinct species taxa. The dual function of the term species as a category and as a description of actual taxa in nature is misleading. The role played by species in evolutionary theory is, thus, ambiguous as to whether it refers to a particular level of biological organization characterized by descent and gene exchange, or, to the properties of a particular group of creatures which possess this type of biological organization.

4. Are Species Taxa Individuals Or Classes?

Traditionally species taxa have been grouped on the basis of phenotypic properties. Observable traits of morphology, physiology, or behavior have been used to lump or group individual organisms into species. When a systematist attempts to decide whether a given fly is a member of Drosophila melanogaster or Drosophila persimilis the decision is based upon a comparison of the traits of the unidentified fly with the traits of known individuals from these two species. The search for similarities and differences in phenotypic traits is at the heart of most classifications of taxa in biology (Bock 1977; Sokal 1973).

However, the story is more complex than this. In many cases biologists are concerned with establishing the cause of a particular similarity between organisms in classifying them. Biologists are particularly concerned to know whether the similarities they observe among creatures are a consequence of common ancestry at some earlier point in phylogeny. The issue of whether a given trait is homologous, analogous, or simply an artefact of measurement or allometry is one that occupies a good deal of attention in the classification and analysis of taxa (Bock 1977; Mayr 1969).

The focus on similarities and their proper classification as homologous or analogous is illustrative of the fact that the grounds for grouping organisms into species is less a matter of establishing coherence, continuity, or, functional role than it is a matter of similarity among the traits of organisms. Similarity of traits and establishing the causes of these observed similarities seem decisive criteria for grouping organisms into species.

When biologists address themselves to the question of why certain organisms manifest similarities among their traits they tend to offer two major explanations--similar genotypes as a result of common descent (homology), or, similar life histories and environmental circumstances (analogy). The fact that organisms exist with similar and relatively stable properties is thus held to be a result, not of the ontological nature of species, but to similarities and stabilities in the causes and circumstances affecting organisms (Stebbins 1977). Similar causes produce similar effects. So the similarities in the traits of organisms can be attributed to the stability and permanence of certain genotypes and environments. While it is true that genes and environments change, this fact is in no way incompatible with the observation that similar genes in similar environments will produce organisms endowed with similar traits.

It is unnecessary to posit any new ontological classification to explain the unity and coherence of species taxa. The unity and coherence of species is a direct consequence of evolutionary theory which pinpoints the relevant causes of species unity and coherence. The theory also reveals the fact that phenotypic traits are best understood as criteria for establishing commonalities at a more basic level. Commonality of genotype and environment are pivotal for understanding overt similarities among organisms. Species taxa are perhaps best understood as groups of individuals which share similar genotypes and similar environments. The hidden 'essences' of species taxa are the genotypes and environments which produce the similarities of traits we observe among organisms. The problem is that the resolving power of phenotypic properties for discerning similarities of genotype or environment is relatively poor (Lorenz 1970; Mayr 1969). Nevertheless, the characterization of a species taxon as a class aggregated on the basis of environmental and genotypic similarities seems quite adequate for explaining the unity and coherence of species and the kind of references biologists make in describing species taxa.

5. Ontological Costs and Benefits

One of the most important arguments cited in favor of the designation of species as individuals concerns the role played by species in the laws and generalizations of current evolutionary theory, or rather the absence of a role for species in the generalizations of evolutionary theory.

Most philosophical analyses of laws in natural science depict them as describing timeless regularities in nature. Reference to particulars is seen as simply incompatible with the universality required of scientific laws (Hempel 1966, pp. 33-40; Ruse 1970, 1977). Thus, if species are treated as individuals, they cannot appear in the generalizations of evolutionary theory. But since, this argument continues, phrases such as "swans are white" rarely grace the pages of modern textbooks of evolutionary theory, the validity of the species as individuals view is confirmed (Hull 1978, pp. 353-355). Treating species as individuals will free evolutionary biology from the charge that it can have no true laws because lawlike claims about albinotic mice, black crows, and other assorted creatures of zoonomic concern will no longer count as evidence for the provincial nature of evolutionary biology.

There is something quite peculiar about this line of argument. The argument explains away a problem that does not really exist and claims it as a benefit. Few biologists seriously believe that generalizations about the colors of swans or crows are central tenets of evolutionary theory, or, mainstays in evolutionary explanations.

Insofar as there is a need to explain the absence of references to specific taxa in the laws of evolutionary biology, perfectly sound reasons for this absence can be given that require no radical ontological shifts. First, descriptive generalizations about observable empirical phenomena are rarely viewed as plausible candidates for the status of laws in any science. In physics descriptions of the behavior of balls rolling down inclined planes are viewed as explananda--things to be explained and not the explanations of physical science. The fact that no evolutionary biologist posits laws about bird colors or the breeding patterns of mice can be understood as a consequence of the view that observable empirical regularities require explanation.

Second, the distinction between species as a category and species as a means of denoting a particular taxon is relevant to understanding the absence of certain types of generalizations in evolutionary theory. Biologists study taxa they suspect possess enough properties (genotypes and environments) to count as species. They then generalize, on the basis of the observations that they make concerning these taxa, to hypotheses about any group of organisms that might be organized and related in this fashion--the species category. When the term species does in fact appear, in such evolutionary generalizations as "two species can rarely occupy the same niche," "speciation is only the multiplication of species," or, "populations of a species may start to diverge before the appearance of an external barrier," the concept "species" is meant to refer to the category species and not to a finite number of spatiotemporally localized groups of taxa.

Those who object that biology is not a science or is, at best, a provincial science (Munson 1975), conflate the categorical and taxonomic senses of the species concept. The properties used to group taxa into species may be limited in their proper predication to entities which occupy a small segment of space and time. However, the criteria used to characterize the species category--commonality of descent and the ability to exchange genetic information--can, and have (Williams 1970) been described without reference to the particularities of individuals. And it is the latter sense which dominates biological usage in evolutionary generalizations.

There are further twists and turns to the argument about the benefits to be obtained from allowing species to be treated as individuals in scientific theorizing in biology. A key benefit that is held to accrue to this view of species is a broadening of the concept of scientific explanation. Hull argues that:

> Because many scientists, especially those working in historical disciplines...can rarely derive the sequences of events which they investigate from any laws of nature, philosophers tend to dismiss the efforts of these scientists as not being genuinely explanatory... . Historical narratives describing the evolution of mammals, the splitting of Pangea into the various continents, and the rise and fall of the Third Reich certainly seem explanatory. ...Historical narratives can be just as explanatory as derivations from scientific laws even though they concern unique sequences of events. (1976, p. 188).

If species are individuals then their continuity and coherence can provide the "glue" necessary for constructing explanatory narratives in historical sciences.

The problem here is that the supposed benefit of viewing species as individuals is not contingent upon ontology. Debates about the nature and structure of historical explanations are legion. But the fact is that there is sufficient unity and continuity in the common properties of the members of a class to allow class terms to serve as the subjects of narrative statements. Think about the sagas that have been spun about the glorious achievements of the Boston Celtics, the Green Bay Packers, and the graduates of Harvard College. All of these entities are classes, yet they all act as unifying reference points for historical narratives of great complexity and, even, explanatory power.

Not only can classes serve the role of the subjects of historical accounts, but, the use of species in this role is quite compatible with the modes of explanation current in evolutionary theory. Evolutionists often do describe the changes that have taken place in a particular phylogeny or lineage segment. But it would be inaccurate to say that such narratives are the end of explanatory matters. For modern evolutionists (Simpson 1961; Williams 1966; Wilson 1975) believe that all of the trends and patterns exhibited in phylogeny can be explained by means of atemporal laws governing the processes of genetic duplication, gene transmission, development, and selection. The key properties of evolving species are genotype and environment--class categorizing properties that allow for the analysis of historical sequences of events by atemporal causal laws. The explanation of historical sequences of events by subsuming each event under a set of various atemporal nomological generalizations is the most distinctive of all the explanatory patterns found in evolutionary biology. Tracing phylogenies and spinning historical narratives may be legitimate explanatory techniques, but it is only because such techniques are placeholders for the actual application of the atemporal mechanistic laws of population biology, genetics, ecology and demography to events in the history of the characteristic genotypic and environmental properties of the members of species.

Perhaps the strangest of all the benefits that has been claimed in the name of rugged species individualism is the fact that such a view results in the elimination of any science from the roster of science that purports to study only one species. Surely it would be ludicrous to have a discipline devoted to the study of a single individual--no possible laws or generalizations could emerge from such a sample. Thus, much of social science, ethology and various sub-fields in biology such as primatology, parasitology, and ornithology will either be consigned to the scientific dustbin, or, to the unglamorous status of explanatory narration.

The proposition that theories in the social and biological sciences which explain events in single species ought be dismissed as unscientific is likely to receive a rather cool reception from practitioners in these areas. Many scientists appear to be quite content to construct nomic generalization about small numbers of species. The touting of the recategorization of the social sciences as a welcome benefit of a species as individuals view appears especially dubious at a time when many scientists are renewing their efforts to construct systematic theories of human social behavior.

Perhaps the difficulties associated with generalizations in social science are better understood as problems with predicates and not ontology. Attempts to link the phenotypic properties of a class of entities into laws may fail simply because such efforts overlook the causally efficacious role of genotypic and environmental properties in producing such traits. Social scientists and others concerned with single species may need to construct new nomic generalizations that attempt to link the truly significant attributes of such classes. Indeed, many biologists and anthropologists (Alexander 1975; Chagnon and Irons 1979; Wilson 1975) have recently advanced this claim in some of the sociobiological literature. Scope, range, and particular reference may be false leads in understanding the difficulties confronting those who study mice and men.

The costs of treating species as individuals do not appear to be neutralized by corresponding gains in conceptual clarity. Nor do current biological theories or usage seem to demand or support any ontological shift. Without demonstrable need or benefit there would seem to be no reason for declassifying species.

Notes

[1] I would like to thank Walter Bock, David Hull, Janet Caplan, and Carola Mone for their helpful comments on an earlier draft of this paper.

[2] This view overlooks the fact that the properties of cluster concepts can themselves be viewed as classes. One solution to accommodating continuous change with the view of species as classes is to view cluster concepts as classes and attempt to locate "characteristic" properties for particular clusters (See Suppe 1973).

References

Alexander, R. D. (1975). "The Search For a General Theory of Behavior." Behavioral Science 10: 77-100.

Bambrough, R. (1960-61). "Universals and Family Resemblances." Proceedings of the Aristotelian Society 61: 207-222. (As reprinted in Universals and Particulars. Edited by Michael Loux. Garden City, NY: Doubleday & Co., 1970. Pages 109-127.)

Beckner, M. (1968). The Biological Way of Thought. Berkeley: University of California.

Bock, W. (1977). "Foundations and Methods of Evolutionary Classification." In Major Patterns in Vertebrate Evolution. Edited by M. Hecht. New York: Plenum. Pages 851-895.

Caplan, A. L. (1979). "Darwinism and Deductivist Models of Theory Structure." Studies in the History and Philosophy of Science 10: 341-353.

Chagnon, N. A. and Irons, W. (eds.). (1979). Evolutionary Biology and Human Social Behavior: An Anthropological Perspective. North Scituate, Mass.: Duxbury Press.

Ghiselin, M. T. (1975). "A Radical Solution to the Species Problem." Systematic Zoology 23: 536-544.

Griffiths, G. C. D. (1974). "On the Foundations of Biological Systematics." Acta Biotheoretica 23: 85-131.

Hempel, C. G. (1966). Philosophy of Natural Science. New York: Prentice-Hall.

Hull, D. L. (1976). "Are Species Really Individuals?" Systematic Zoology 25: 536-544.

----------. (1978). "A Matter of Individuality." Philosophy of Science 45: 335-360.

Lorenz, K. (1970). Studies in Animal and Human Behavior, Vol. I. Trans. Robert Martin. Cambridge: Harvard University Press. (Originally published as Über Tierisches und Menschliches Verhalten: Gesammelte Abhandlungen, Band I. Munich: R. Piper Verlag, 1970.)

Mayr, E. (1969). Principles of Systematic Zoology. New York: McGraw-Hill.

-------. (1970). Populations, Species, and Evolution. Cambridge: Harvard.

-------. (1976). "Is the Species a Class or an Individual?" Systematic Zoology 25: 192.

Munson, R. (1975). "Is Biology a Provincial Science?" Philosophy of Science 42: 428-447.

Reed, E. S. (1979). "The Role of Symmetry in Ghiselin's Radical Solution to the Species Problem." Systematic Zoology 28: 71-78.

Ruse, M. (1970). "Are There Laws in Biology?" Australasian Journal of Philosophy 48: 234-246.

-------. (1973). The Philosophy of Biology. London: Hutchinson.

-------. (1977). "Is Biology Different From Physics?" In Logic, Laws and Life. (University of Pittsburgh Series in the Philosophy of Science, Vol. 6.) Edited by R.G. Colodny. Pittsburgh: University of Pittsburgh Press. Pages 89-128.

Simpson, G.G. (1950). "Evolutionary Determinism." In This View of Life. New York: Harcourt, Bruce & World. Pages 176-189. (As reprinted in Ma and Nature. Edited by R. Munson. New York: Dell, 1971. Pages 200-212.

------------. (1961). Principles of Animal Taxonomy. New York: Columbia University Press.

Sokal, R. R. (1973). "The Species Problem Reconsidered." Systematic Zoology 22: 360-374.

Stebbins, G. L. (1977). Processes of Organic Evolution. Englewood Cliffs: Prentice-Hall.

Suppe, F. (1973). "Facts and Empirical Truth." Canadian Journal of Philosophy 3: 197-212.

Williams, G. C. (1966). Adaptation and Natural Selection. Princeton: Princeton University Press.

Williams, M. B. (1970). "Deducing the Consequences of Evolution: A Mathematical Model." Journal of Theoretical Biology 29: 343-385.

Wilson, E. O. (1975). Sociobiology: The New Synthesis. Cambridge, Mass.: Harvard University Press.

Van Valen, L. (1976). "Individualistic Classes." Philosophy of Science 43: 539-541.

Ruse's Treatment of the Evidence for Evolution: A Reconsideration

Alexander Rosenberg

Syracuse University

The Darwinian theory of evolution through natural selection is beyond a doubt an extremely well established theoretical edifice that satisfies any reasonable conceptual and empirical criterion of scientific respectability. It is a body of well confirmed nomological generalizations with great explanatory and predictive power. And yet the most accessible account of the relation between this theory, its competitors and the empirical evidence seems seriously defective in its account of these relations. Indeed, it may seem so defective as to leave the question open, not whether the theory of natural selection is well confirmed (this cannot be doubted), but exactly what sort of evidence does confirm it and via what bridge principles. The account of the matter which I have described as most accessible (in the philosophy of science at any rate), and yet seriously bedeviled is to be found in Michael Ruse's otherwise excellent introduction to the *Philosophy of Biology* (Ruse 1973). In the chapter of this book devoted to "The Evidence" Ruse aims to lay to rest the notion that evidence for Darwin's theory and against his competitor's views is anything less than substantial and unequivocal. Although I agree with Ruse's conclusions, I hope to show that the evidence is in important respects both stronger and weaker than Ruse's treatment reveals.

A serious obstacle to finding evidence that bears directly on the hypothesis, that evolution results from differential selection over heritable variations, is itself biological. That is, when the temporal rate of reproduction of organisms under observation is close to our own, and when environmental conditions are subject to changes of geological slowness, the individual Homo sapiens will simply not live long enough to detect many changes that confirm the theory. These facts have two related consequences for defenders of the theory. They permit its exponents to explain away the absence of confirming evidence from the evolution of middle sized mammals, whose evolution interests us most and would provide the psychologically most striking and forceful confirmation of the theory and of its explanatory range. On the other

PSA 1980, Volume 1, 83-93

hand, the evolutionist's temptation to cite these facts as excusing the absence of certain sorts of evidence can be turned to his disadvantage at the hands of an opponent of the theory who wishes mistakenly to find it guilty of unfalsifiability. Nevertheless, there have been detectable chances in wild populations that strongly confirm the limited claim that evolution at least sometimes proceeds in accordance with Darwinian mechanisms. Among examples that Ruse cites are the cases of industrial melanism in moths, and evolved resistance to myomatosis among Australian rabbits. But as Ruse also notes, in these cases "the amount of evolutionary chance involved is very small... . Hence were someone to put forward a theory allowing for the possibility of small change due to selection, but denying that large scale changes have the same cause, evidence [of this] type would be unable to decide between the theory [of natural selection] and it." (Ruse 1973 , p. 101-2). The absence of evidence from wild populations for anything more than the existential claim, that natural selection over heritable variation obtains, is essential. For the contemporary Darwinian theory asserts not the mere occurrence of this sort of evolution, it claims ubiquity for it; it claims that all evolution proceeds by selection over small variations. Because of this feature of the theory, and because of the facts about the biology and environment of the agents that test the theory, recourse to other arenas of confirmation is essential.

For Darwin himself, and for Ruse as well, one of the most crucial of these further arenas of test is to be found in the phenomenon of artificial selection. Ruse raises and answers affirmatively but qualifiedly three questions: "can artificial selection, done purely for practical or aesthetic reasons,...support the claims of...evolutionary theory? ...can one devise experiments using artificial selection which will support...natural selection? ...can one achieve natural selection in artificial surroundings, and if so what support can this give to the synthetic theory ?" (Ruse 1973 ,p. 102). The qualification in all three of Ruse's affirmative answers reflects a misunderstanding of the relation between artificial and natural selection which unduly mitigates the force of the evidence provided by the former for the latter.

According to Ruse the argument from the character of artificial selection to the occurrence of natural selection is an analogical one with a merely modal conclusion. The two sorts of selections bear similarities (both manifest differential reproduction with cumulative effect involving relatively slight intergenerational variation). But, writes Ruse, "the analogy between...artificial and natural selection breaks down. In particular, one cannot argue (from these instances) that *natural selection does ever occur*, or if it occurs that the kinds of variations to be selected will be the same as those selected by breeders. Because the breeders of Afghan hounds select for long coats, it does not necessarily follow that there will be selection for long coats in wild populations of dogs. Consequently, the evidence from these cases of artificial selection is very limited. They show that selection *can* have big effects; *but they do not show that natural selection exists* or that if it does exist, in what direction such selection points." (Ruse 1973, p. 104, *emphasis added*).

Now it is clear that in advancing the theory of natural selection Darwin argued analogically from the existence of artificial selection to the possibility of natural selection. But on the assumption that Darwin's theory is correct, the distinction between these two sorts of selection is not that of exhaustive and exclusive difference, but rather that of general case to special case. Artificial selection is not another form of selection, different from natural selection; it is a species of natural selection. The forces operating in artificial selection are not just similar to those acting in natural selection, they are a special subset of those forces. The operative force in artificial selection is a systematic set of behavior of one species (*Homo sapiens*) that determines the reproductive success of members of other species (e.g., *canus familiaris*, *felus domesticus*, or any other domesticated species). Regularities in the behavior of a predatory species determines the differential reproductive successes of members of its prey species, and this behavior is rightly deemed a natural selective force. It differs from other selective forces, like climate, geological stability, available food supply for its prey, etc., but it is undeniably a selective force. How are we to distinguish the literal selection for fitness by the plant and animal breeder from the metaphorical selection for fitness by the predator? Surely the fact that in the former case the causally relevant variables include the conscious agricultural intentions of the *Homo sapiens*, and in the latter case the causally relevant variable do not include such conscious purposes is no reason to distinguish artificial and natural selection as mutally exclusive mechanisms of evolution. To suppose otherwise is tantamount to erecting that barrier between *Homo sapiens* and the other species that Darwin's theory did the most to bring down.

But if artificial selection is a type of natural selection then Ruse's conclusions about its bearing on Darwin's theory are either too weak or false. It will clearly turn out to be false that the existence of artificial selection does not imply the existence of natural selection. For if the former is a species of the latter whose existence is obviously assumed by all parties in the present dispute natural selection occurs if artificial selection does. Similarly, Ruse is wrong to say that the selection of traits which breeders make reveals nothing about the selection for traits that nature makes. Since artificial selection is a form of natural selection the selection for traits that nature makes will include the ones that humans make. Ruse's claim, that artificial selection shows only natural selection *can* have big effects, while operating on small variations, is clearly too weak. For artificial selection reveals more than a bare possibility. It demonstrates the actuality of large changes arising through natural selection on small variations. Finally, insofar as we treat artificial selection experiments as a source of data *prima facie* different from the observations of wild populations, they clearly strengthen Darwinian theory's claims for the ubiquity as well as the existence of the mechanism of natural selection.

Reasoning by analogy from artificial selection to natural selection in the initial chapters of *The Origin of Species* (Darwin 1859 , Chapter 1) was of course perfectly in order, and represented a heuristically invaluable means of introducing an initially implausible theory of evolution. Indeed, had Darwin argued from the outset that artificial selection was but a special case of natural selection he would quite plausibly have been accused of begging the question of how the former worked. But we must distinguish the introduction or presentation of the theory from its content; especially when we set out to test it. And it seems a clear consequence of the content of the theory that artificial selection is but a form of natural selection. This strongly augments the bearing of the evidence of its occurrence for the hypothesis of natural selection. (For a detailed account of the introduction of Darwin's theory and the analogical role played therein by artificial selection see Ruse 1979 .)

The unwarranted distinction erected between natural and artificial selection also bedevils Ruse's treatment of the question whether one can devise experiments that will support the theory of natural selection. His answer is a cautious yes. But the caution is more than that invariably invoked against closed laboratory investigations of phenomena difficult to unambiguously monitor in nature. Any laboratory experiment in even so well attested and relatively unproblematical domains like that of Newtonian mechanics is attended by cautions that inferences from its results may seriously underestimate the complexity of the "same" type of phenomenon beyond the laboratory. But Ruse's caution goes further. And again the caution is based on Ruse's supposition that inferences from laboratory selection to natural selection in wild populations are analogies and not confirmational inferences from particular cases to general hypotheses. Thus he writes, "evidence [for the theory] would seem to be forthcoming if one can suggest that one's artificial selection bears analogies (in addition to differential reproduction) to possible cases of natural selection." Thus, if we can establish geographic isolation in the laboratory and thus produce reproductive isolation, or regulate food, space or air temperature, pressure, or wind velocity, and detect changes in the distribution of phenotypes among, say Drosophila, one might, according to Ruse, be able to learn "the direction which natural selection could take, the rate of its effects, and some of the overall consequences." (Ruse 1973, p. 104). It must be admitted that biological experiments are more complex and more removed from natural settings than many physical ones, but surely the differences between selection experiments in the laboratory and, for instance, mechanical experiments employing Atwood's machine are matters of degree and not of kind. If this is so, then either we must say that Atwood's machine experiments merely show that gravitational attraction could be a constant, or we must say that selection experiments in the laboratory show the direction that natural selection does take. For differences of degree in complexity of experiments cannot make for modal differences in the form of conclusions drawn from them. They can only make differences in the strength of our rational beliefs about the conclusions.

The special importance of laboratory selection experiments is that they increase our evidence for the extent of evolution by natural selection. As noted by Ruse, observation of wild populations establishes the occurrence of Darwinian selection, but there are natural obstacles to the human accumulation of evidence as to its extent or ubiquity. By replicating in the laboratory various occurring natural phenomena hypothesized to act as selective forces, we can provide evidence for the extent as well as the existence of natural selection.

While Ruse's account of the bearing of artificial selection by agriculturalists and laboratory selection by evolutionists is too weak, his subsequent account of the differential confirmation of Darwinian theory by comparison to its competitors is far too strong. He considers three alternative theories of evolution, each of which has had responsible biological proponents in the post-Darwinian period, and each of which is on Ruse's view clearly excelled by Darwin's theory in respect of two particularly crucial bodies of evidence. The three theories are Lamarck's hypothesis of the hereditary transmission of acquired characteristics; the saltationist theory according to which evolution proceeds not exclusively or even mainly through selection on small variations, but at least sometimes through the appearance of large heritable changes, so called macromutations; and the orthogenic theory which has it that evolution is the result of changes in heritable variations which invariably take them beyond adaptiveness towards extinction-producing mal-adaption. In our assessment of Ruse's claims about the greater confirmation of Darwin's theory by comparison to each of these, it must be remembered that each of these theories, like Darwin's, asserts the existence of phenotypes, of heritable properties of organisms, and each will require at least the existence of a mechanism to provide for the invariable transmission of hereditary properties in the absence of forces of evolutionary change. Ruse makes disarmingly short work of these three theories: unlike Darwinian theory they all fail to account for the evidence of genetics and cytology, "and this is the major reason why they are disregarded today." (Ruse 1973, p. 113). Assessing this claim involves agreeing on what the "evidence of genetics and cytology" constitutes, and determining the degree to which the theory of natural selection "accounts for" this body of evidence.

It also requires some specification of what the "accounts for" relation to which Ruse appeals comes to. Apparently the notion is supposed to be akin to the sort of confirmation of a hypothesis which is produced by an inference to the best explanation of confirming findings. Thus, if the best explanation for the truth of p involves appeal to q, then the truth of p evidently confirms q. Similarly, if q "accounts for" p, then q is an important or central part of a body of propositions that explain p, and whose truth is therefore confirmed by p. This at any rate is how I shall understand the notion of "accounts for" in what follows.

The cytological evidence to which Ruse refers is evidently the microscopically observed cellular phenomena of mitosis and meiosis. The genetic evidence to which Ruse refers is clearly the transmission and distribution of phenotypes in rough accord with the Mendelian laws of segregation and independent assortment. It is clear that the relation between the existence of mitosis, and meiosis, and the heritability of traits in accordance with Mendelian ratios is not logically incompatible with any of the three competitors to Darwin's theory that Ruse canvasses. Indeed, since all of them require the existence of traits with varying but considerable degrees of heritability, they all require some theory of heredity and some account of the cellular mechanism of hereditary transmission that meiosis provides and of developmental ontogeny that mitosis affords. Moreover, it would take little ingenuity to create bridge principles that establish connections between these non-Darwinian theories and descriptions of cytological and genetic phenomena that are formally identical to those which exist between these latter and Darwinian theory (for an illustration see Rosenberg 1979). In what sense, therefore, does the latter theory account for these cytological phenomena when its competitors do not?

The phenomena of meiosis, and mitosis, the Mendelian laws, and Darwin's theory are interconnected in a vast network of biological findings and theories, as well as auxiliary hypotheses and theories drawn from chemistry, optics, x-ray - crystallography, ecology, etc. The relevant part of the network may be sketched out as follows: The existence of meiosis and mitosis is inferred from indirect observations of theoretically identified organic material, which has usually been stained by mixture with a dye, prepared on a slide, and mounted on a microscope, which is then manipulated until a theoretically calculated degree of resolution is attained. Conclusions about the existence of the cytological phenomena occurring in the nuclei of the cells on the slide turn on independently substantiated assumptions about physical optics, generalizations about the effects of dyes on organic material, and on our observation of this material, and principles of zoological identification and cellular anatomy. Once an agreed description of the topography of these cellular occurrences is in hand, the principles of Mendelian genetics may be appealed to in order to broadly explain the topographical order of events in meiosis and mitosis, and perhaps more importantly to explain the consequences of varying sorts of breakdowns in the processes to be explained. But here again, the explanatory relation between Mendelian genetics and the cytological phenomena is extremely complex and involves appeal to at least the following further assumptions : the findings and methods of biochemical and fine-structure genetics, including the physical regularities governing electron-microscopy, x-ray-crystallography, electropherisis. It is only together with appeal to the vast theoretical edifice of physics and chemistry that Mendelian genetics "accounts for" the cytological phenomena. Of course, there is further independent evidence for the Mendelian laws to be found, for instance, in agricultural experiments of the type that led Mendel to frame these laws, and without such further evidence, of course, the mere fact that together with a vast array of other assumptions Mendel's principles explain the topography of cytological phenomena is little reason to embrace this theory.

Now even if Mendelian genetic theory bears direct evidential or explanatory relation to the theory of natural selection, the latter can at best be no more well-confirmed by cytological evidence than Mendel's laws themselves. That is, if the laws of independent assortment and segregation "can account for" or "are accounted for" by Darwin's theory, then it will share some of the confirmation which the phenomena of meiosis and mitosis indirectly provide these laws. But of course the relations between these laws and Darwinian evolution is very far from being direct and is mediated by further auxiliary assumptions over and above those already invoked to link Mendel's principles and cytological phenomena. The first thing to note is that the undisturbed behavior of genetic material in accordance with Mendel's laws not only does not produce evolutionary change, it precludes it. As Ruse shows in chapter three of Philosophy of Biology, the laws of segregation and independent assortment imply the Hardy-Weinberg law, according to which gene ratios must remain constant across indefinitely many generations, ceteris paribus. Under what additional conditions will Mendel's laws provide support for the theory of natural selection? I have said 'provide support for', not as Ruse seems to have it "be accounted for by" the theory of natural selection, because there are no conditions under which the hypothesis that evolution invariably proceeds by selection over small variations "will account for", will explain, imply, entail or otherwise suggest the laws of independent assortment and segregation of genes. The relation is if anything the reverse. How could a theory which stipulates the conditions for evolutionary change account for two principles which by themselves preclude it? If this cannot be done, then Darwinian theory does not excel its competitors in its ability to account for either the genetic regularities, or the cytological facts which the latter account for. Further consideration of Ruse's discussion requires that we reverse the confirmational relation, and treat his claim as one according to which the Mendelian laws together with additional assumptions account for Darwin's theory, but do not with equal evidential strength account for its competitors.

To arrive at the theory of natural selection, given Mendel's laws, we must add practically the rest of biological, geological, and meteorological theory, known and unknown, as well as the actual and possible findings of experimental and ethological psychology. For natural selection involves the survival of the fittest. But fitness is not to be defined in the theory by appeal to differential reproduction (though it may be so measured). Otherwise the theory would be guilty of the mistaken charge frequently lodged against it that it is circular, tautological, unfalsifiable, etc. Fitness of organisms and plants (or genes for that matter) must be characterized not by appeal to differential reproduction, but by appeal to those biological, geological, meteorological, behavioral and environmental features that determine the fitness of the subjects of selection, be they genes, individual organisms, populations, or whole species. Differential fitness is given by appeal to the consequences of causal variables of all these theories working together for the determination of ability of the subject of selection to reproduce. (Fuller details of this matter are given in Rosenberg 1978 .) It is only at this point that heredity enters, and with it, Mendel's laws. If differences in fitness are hereditary, then in the course of successive generations, small variations in hereditary properties (caused by biochemical mutations in nucleic acids that are random with respect to selection pressures, another auxiliary assumption) will result in Darwinian evolution. Mendel's laws give the mechanism of heredity required to account for evolution, but only when they are added to the rest of this edifice so vast that we will never see its details filled out. That is, while we believe that levels of fitness are, in principle, determinable from considerations about the subjects of selection that have, will, or can be provided by all the rest of the theories describing the causal forces acting on individual organisms, species, or genomes, we have no more reason to suppose that this plainly unattained possibility can be converted to actuality than we have reason to believe in the uniformity and the expressible finitude of nature and its regularities. Because we nevertheless hold the relevant beliefs, we claim that together with this perpetually unfinished body of theories, and the hypothesis that mutations occur, Mendel's theory does account for the theory of natural selection.

The complex relations among Darwin's theory, Mendel's laws and the cytological phenomena may be diagrammed schematically and incompletely as follows, where downward arrows denote Ruse's relation of "accounts for":

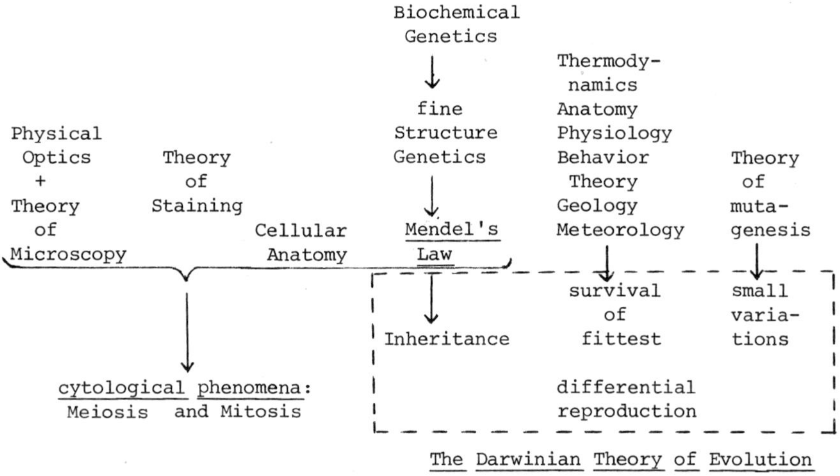

The theoretical and confirmational relations are of course much more complex than here represented. Indeed, a full representation would probably require the resources of Hilbertian vector space, and not two dimensions. I have underlined the three elements of the representation that Ruse concerns himself with. If the diagram is representative, it should be clear that Darwin's theory accounts neither for the cytological evidence nor for Mendel's laws. Furthermore, the latter do not account for Darwin's theory either. In particular, by itself Mendelian genetics accounts for only a portion of the theory of natural selection; the inheritability of characteristics that it requires. Most importantly, since the competitors to Darwin's theory, Lamarckianism, the saltation theory, and orthogenisis, also require the inheritability of characteristics, Mendel's theory accounts for them just as well as it accounts for Darwin's theory. Therefore, with respect to the evidence that Ruse thinks most crucial to Darwinian theory's comparatively greater support, no decision can be taken among the available alternatives.

The conclusion of this argument should not be misunderstood. It is not the claim that the theory of natural selection and its rivals are on an evidential par. As noted in the first sentence of this paper, Darwin's theory seems to me to be established beyond reasonable doubt. Indeed, if my diagram is correct, it is sustained by the joint operation of a larger number of independently confirmed theories, findings and principles than almost any other theory available. This is no surprise considering the fact that it governs the behavior of entities that are also subject to all the laws of physics, chemistry and the rest of biology, while the converse is not true. Of course, had one or more of the findings and theories of the rest of natural science been different from what they in fact are, some other evolutionary theory might have been as well as or more strongly confirmed than Darwin's. Exploring what sorts of differences in the findings and theories of other areas of natural science would make for such conclusions is probably the best means of manageably detailing the evidence which confirms Darwin's theory more strongly than any other. My claim is that explorations that focus on the cytological evidence or on Mendel's laws are less likely to reveal this evidence than either Ruse thinks or than examination of other areas of science provide.

My argument against Ruse has traded on a claim that he and biologists as well are likely to reject. For in this discussion I have treated Darwin's theory and Mendel's as two distinct and separate theoretical edifices. One reason for doing this is that to treat them as a single joint theory is to trivialize the question of whether Darwinian theory is more well-confirmed by or can account better for Mendel's laws than can its competitors. For taken together, their conjunction truth functionally accounts for one of their conjuncts more directly than other theories, and is more directly confirmed by the confirmation of one of its conjuncts.

Despite this potential trivialization of the matter, Ruse assimilates the two theories under the name, the "synthetic theory" of evolution. In this he follows established biological practice to the extent that contemporary accounts of evolution appeal not just to the existence of heritable variations, but to the mechanism of this heritability as described in Mendel's theory and its successors. Contemporary biologists signal their commitment to this theory as the account of the heredity that Darwin's theory requires but does not itself provide, by giving the two distinct theories a single name: 'the synthetic theory'. And Ruse is of course perfectly justified in referring to the conjunction of these two theories under this single name by invariable biological practice. On the strength of this notational practice he might respond to my argument that it turns on an unwarranted separation of two inextricably linked sets of principles, and produces conclusions at variance with Ruse's own by effecting this artificial separation. Of course, if my description of the rationale for bracketing these two theories together under a single name is correct, this objection holds no force. Ruse and the biologists who concur need an argument to show that Mendel's laws are an essential part of the theory of evolution by selection over small heritable variations, no matter apparently how these variations are inherited. In terms of the diagram above, biologists who hold the inseparability of these two theories in the contemporary synthetic theory of evolution are simply drawing the dotted line that surrounds Darwin's theory differently from the way suggested. They are drawing it in such a way that it includes Mendel's laws, and perhaps also the theory of mutagenesis. But if we are to expand the boundaries of the theory of natural selection, why stop at Mendel's laws, why not include all the other theories in biology, chemistry and physics which account for one or another of the features of the theory? Unless good reason can be given to stop where the exponents of the synthetic theory do, it will probably be best to stop at some point before the one they seem to recommend. For if we expand the boundaries of Darwin's theory much beyond the bounds Darwin would himself have recognized,we will end up including findings on which Darwinian theory's competitors may rest as much as it does, or worse co-opting the evidential basis of Darwinian theory into that theory in such a way as to obscure the evidentially secure though frightfully complex grounds on which it actually rests.

References

Darwin, Charles (1859). The Origin of Species. London: John Murray. (As reprinted in The Origin of Species. Philadelphia: University of Pennsylvania Press, 1959.)

Rosenberg, Alexander. (1978). "Supervenience of Biological Concepts" Philosophy of Science 45:364-386.

--------------------- (1979). "Genetics and the Theory of Natural Selection, Synthesis or Sustinence?" Nature and System 1:1-16.

Ruse, Michael. (1973). The Philosophy of Biology. London: Hutchinson University Library.

-------------. (1979). The Darwinian Revolution: Science Red in Tooth and Claw. Chicago: University of Chicago Press.

Part IV

Statistics, Probability and Likelihood

The Philosophical Relevance of Statistics

Deborah G. Mayo

Virginia Polytechnic Institute
and State University

1. Introduction

The foundations of a number of scientific theories are more or less relevant to philosophy. The better a theory is at elucidating the structure of science and scientific method the more a study of its foundations is relevant for philosophers, particularly philosophers of science. But while philosophers have studied probability and induction, statistics has not received the kind of philosophical attention mathematics and physics have. Although the terms "philosophy of induction" and "confirmation theory" are common, the term "philosophy of statistics" is rarely used; and despite the fact that modern statistical methods have been used increasingly in a number of sciences, specific developments of statistics have been little noted in the philosophy of science literature.

Statistics has been fundamental for determining such things as whether a substance causes cancer, which method of psychotherapy gives the most recoveries, whether censorship causes the instability of nations, whether a fossil belongs to a man or a chimpanzee, whether there is regularity in the oscillations of the velocity of gas in the sun's atmosphere, whether an individual authored a piece of writing, whether the economy will improve, and numerous other issues involving uncertainties. More and more sciences have incorporated statistical techniques leading to a number of new fields, such as biometrics, econometrics and psychometrics. With such widespread use of statistics and so many public policy decisions based upon it, getting clear on the nature of statistical reasoning is imperative. But what is the logic involved in making such statistical inferences, and when are statistical inferences valid? Practitioners have been far more concerned with applying statistics and absorbing statistical techniques into their respective sciences than with answering questions concerning the foundations of these tools. This is not surprising, since it is the business of the philosopher and not the scientist to scrutinize

---

PSA 1980, Volume 1, pp. 97-109

the foundations of inference methods. However, philosophers have so far been unsuccessful in adequately answering these foundational questions, and controversy and confusion about which methods to use and how to interpret them are greater than ever, both among philosophers and those who apply statistics.

In this paper I want to show that it is important for philosophers to work on foundational questions in statistics because of their relevance both to theoretical and to what might be termed applied problems of philosophy. I begin by considering the philosophical problem of induction and then go on to discuss the reasoning that leads to causal generalizations in science and how statistics elucidates the structure of science as it is actually practiced. I hope to show that in addition to being relevant for building an adequate theory of scientific inference, statistics provides a link between philosophy, science, and public policy.[1] As philosophy of science moves out of the positivistic framework, and becomes more interested in a framework that is relevant to actual problems in science, a tool for bridging the gap between philosophy and science is required. I claim that statistics can provide such a tool, and if I am right, philosophy of statistics has a significant role to play in developing a new framework for philosophy of science.

2. Problems of Induction

What is sometimes referred to as the "old" problem of induction arose from the empiricist principle that the acceptance and rejection of theories is to be solely the result of observation and experiment, and the recognition by Hume that no finite amount of experimentation can justify general theories. Unable to show that induction could be relied on to produce conclusions that were even highly probable, scientific method appeared unjustifiable. This lead some philosophers to attempt a less rigorous means of justifying induction; namely to show that the basic rules for inductive inference in science correspond to what common sense and science deem intuitively reasonable. But what are the basic rules for inductive inference in science? Attempts to answer this question gave rise to the "new" problem of induction. (See Skyrms 1975.)

Many attempts to set out a theory of inductive inference have proceeded by setting up a quantitative relation M which is to measure the evidential strength between data and a hypothesis, and thus the "new" problem of induction involves the problem of finding relation M. While the hypothesis could not be logically certain given the data, the possibility of it having some degree of probability suggested itself, and hence, the use of probability functions in setting up relation M. Typically, M is defined as measuring degrees of belief, confirmation, support, credibility, plausibility and the like. Setting up a purely formal theory of inductive inference by specifying an evidential measure M was very attractive to Positivist philosophers, and the most complete inference theory of this sort is that of Carnap and his successors. Here, M is to hold between two propositions:

one expressing a hypothesis, the other data; and it measures the degree of confirmation the data afford the hypothesis. A number of difficulties were found to stand in the way of their attempt to reconstruct scientific method by means of a formal system (e.g., the "grue" paradox as set out by Goodman (1965)). The simple languages for which Carnap's theories are designed are incapable of expressing many kinds of hypotheses and data needed in actual scientific inferences. And there is also the problem of choosing which of infinitely many functions to use for M.

In general, the inductive logics proposed by philosophers have been too oversimplified to do the actual work of scientific inference.[2] A more promising approach would be to build a theory of induction upon the already well-worked out techniques of statistical theories, methods of hypothesis testing, and estimation. Such a theory of inference will be far more complicated than one composed simply of an evidential relation M. As such, statistics will not provide a "logic" of inference in the ordinary sense, but I think it can provide systematic methods which can cope with the complexities of actual scientific inferences involving probabilities. While theories based on statistics are not going to resolve the "old" problem of induction better than other theories of inductive inference, they can do a better job with the "new" problem, which involves setting up an adequate theory of inductive inference. Given theories which are adequate for the tasks of scientific inference, as I believe those based on statistics are, it is possible to assess how well they accomplish these tasks and so justify preferring some theories of statistics to others. This would provide a much needed basis for resolving controversies between rival theories of statistics.

3. Controversies in the Foundations of Statistics

The major controversies in the foundations of statistics are between two rival inference philosophies: the <u>orthodox</u> or <u>standard</u> philosophy (e.g., Neyman-Pearson and Fisherian) and the non-standard philosophies (e.g., Bayesian and Likelihood). Despite the fact that most of the statistical methods used in the sciences come from the standard theory of statistics, the standard methods have increasingly come under attack both by philosophers and practitioners. The great majority of philosophers who have taken an interest in this controversy have come to support non-standard methods of inference. Among the most philosophically relevant of the criticisms raised against the standard methods are that they fail to adequately represent scientific inference, that they are unable to perform the tasks to which they are put in science, and that they are not objective. Such criticisms are of relevance to those philosophers interested in the logic of scientific inference. If the criticisms against the standard method are correct, then much of what passes as science is misconceived and one would not want to build a logic of inference upon it. In general, the problems concerning the foundations of statistics are relevant for philosophy of science because any adequate account of inductive inference and scientific method will have to cope with precisely these problems.

Hence philosophy of statistics provides an important source of problems for theories of scientific inference, and any theory of inference which is constructed without an awareness of such problems is likely to be irrelevant to actual scientific practice.

Most of the philosophical criticisms of statistical methods are based upon their formulation in statistical texts, and because these formulations have little relation to how statistics is actually practiced, the resulting criticisms have little bearing on much of the use of statistics in science. This point was illustrated at a recent conference on statistical foundations, where J. Neyman (1977, p. 97) told of the following exchange between an "authority" in non-standard statistics and a practicing statistician. "The Authority: 'You must not use [standard inference methods]; they are discredited!' Practicing Statistician: 'I use [them] because they correspond exactly to certain needs of applied work.'" A major task for philosophers of statistics, as I see it, is to uncover and make explicit that which enables certain statistical tools to "correspond exactly" to the needs of the practicing statistician. Understanding what enables these methods to perform valid and informative inferences when they do, is an important step towards forming a theory of inference adequate to science as it is actually practiced. This would be a significant contribution even for those practitioners who do correctly use statistics, for even they may not correctly comprehend why the methods work. At most they follow "rules of thumb" which, while not presently included in the formal expositions of statistics, might be made precise enough to be incorporated in an eventual philosophical theory of inference. To accomplish this task, a philosophical examination of the aims of science and the role that statistics actually plays in science is needed. Constructing an adequate theory of inference would shed light not only on certain controversies in statistics but on certain problems in philosophy as well.

Many of the problems of statistics are not very different from those with which philosophers of science wrangle. For example, statistics is concerned with the following questions: What should be observed and what may be inferred from the resulting observations? How well does data confirm or fit a model? What effects do the processes of observation and measurement have upon the results observed? How can spurious relationships be distinguished from genuine lawlike ones? How can a causal hypothesis be tested? Although the language in which statistical questions are expressed differs from the terminology of philosophers of science, the concerns of one often parallel the other. For example, where philosophers want to know what evidence is needed to reject a theory, a statistician wants to know what is the best <u>rejection region</u> for a test hypothesis. Similarly, the philosopher's concern about objectivity of observation is parallel to the statistician's concern to obtain unbiased observations by such means as <u>randomization</u>. Statistical problems are relevant to philosophy in two ways: the solutions to the problems statistics has solved may suggest solutions to parallel problems in philosophy (e.g., obtaining objectivity of observation via randomization), and those statistical problems which remain unsolved may suggest analogous philosophical problems not recognized by philosophers. Many of the statistician's concerns are, in a sense, applied versions of some of the concerns of philosophers of science, and as Kempthorne (1976) suggests, statistics may be seen as "applied philosophy of science."

4. Statistics and Causality: Testing a Causal Claim

Few problems have generated as much philosophical interest as the problem of causality and the reasoning behind causal inferences in science. Statistical tests play an important role in establishing causal (and other) generalizations in science, but precisely how they do so is not something revealed in statistical theory itself. Consideration of the relationship between statistical inferences and substantive scientific inferences, such as those involving causality, is typically considered by statisticians to fall outside their proper domain. A philosophical theory of scientific inference which did include such considerations could provide a link between statistical and scientific inferences, and in doing so elucidate the logic of causal inferences. An examination of the role of statistical tests in evaluating causal claims is necessary for such a theory of inference, as well as for assessing the criticism that the (standard) tests are irrelevant for making causal inferences.

Consider for example the claim that saccharin causes cancer in rats. If this claim is true it does not mean that all rats fed saccharin will get cancer. Hence, in testing this claim, an observable prediction will be an assertion about the _average_ or _mean_ number of saccharin-fed rats which get cancer. (The average is an example of a statistic.) However, the average number of saccharin-fed rats who get cancer may be high even if saccharin is causally irrelevant to cancer, since it may be the result of some other cause. What one wants to know is whether on the average the number of cancers in saccharin-fed rats (i.e., _treated rats_) is significantly greater than the number of cancers in rats not fed saccharin (i.e., _control rats_). When possible,[3] this information may be obtained by carrying out a comparative random experiment.

Without going into the details of the experimental design,[4] the idea is roughly to take a random sample of rats who do not have cancer, treat half of them with saccharin (for a suitable period of time) while leaving the other half untreated. At the end of the experiment the average number of rats having cancer in each group is recorded. Let M_t and M_c be the mean number of rats observed to have cancer in the saccharin-treated and control groups respectively. In this case, one wants to know whether M_t is sufficiently greater than M_c to consider that a genuine causal relationship between saccharin and cancer exists. (In other cases one may not be interested in the specific direction of a difference.) To answer this, a statistical test of the significance of differences may be carried out.

In a nutshell, what the test consists of is a rule which designates, before the experiment, which observations are going to be taken to reject the statistical hypothesis under test; that is, the test or null hypothesis h_0. These observations make up the rejection region, and it is specified so that there is a given probability, called the size or the significance level of the test, of obtaining such observations given that the null hypothesis is true. Conventionally, significance levels are chosen to be .05 or .01. One obtains the observation and either rejects or fails to reject h_0 according to whether or not it falls in the rejection region. (In Neyman-Pearson tests, failure to reject h_0 leads to accepting the alternative hypothesis.)

The logic of testing basically follows the pattern of modus tollens, and in the saccharin example it takes the following form. Since, if saccharin is causally irrelevant to cancer (in rats), then there will be no difference in the mean cancer rates in the populations of saccharin-treated and non-saccharin-treated rats, by rejecting the consequent (that the means are equal), it can be concluded that saccharin is not causally irrelevant. Letting μ_t and μ_c be the mean cancer rate(s) in the populations of saccharin-treated and control rats respectively, the null hypothesis can be formally expressed by h_0: $\mu_t - \mu_c = 0$. The alternative may be that $\mu_t - \mu_c$ is either not equal to zero (two-sided test), or is less than or greater than zero (one-sided tests); and in this case one is likely to be interested in the last of these. As the entire population of treated and control rats are unavailable for inspection, a (random) sample of each is taken and the difference between the mean number of cancers is observed. Thus, the data for this test may be modeled by the statistic S: $M_t - M_c$. The following four levels summarize the way the test statistically models the actual observations and the scientific claim:

Scientific Claim: Saccharin is causally irrelevant to cancer in rats.
Statistical Hypothesis: h_0: $\mu_t - \mu_c = 0$

Statistical Data: S: $M_t - M_c$ (test statistic)

Scientific Data: Observations on a random experiment of saccharin-treated and control rats.

The test consists of the rule: reject h_0 just in case the probability of S given h_0 is less than .01 or .05 depending on the signficance level (i.e., just in case S falls in the rejection region).

However, when tests are carried out by mechanically setting up a .05 or .01 significance level and then rejecting or accepting the null hypothesis, the resulting inferences are often misinterpreted, and many of the criticisms of tests are the result of such misinterpretations. To avoid misuses and misunderstandings, testing logic must be better understood. Tests, as I see them, function primarily to detect certain types of discrepancies between observations and hypotheses and between hypotheses. By a suitable choice of test statistic, larger discrepancies correspond to smaller probabilities. One wants to reject h_0 in our example not directly because the observed difference S is very improbable under the assumption that h_0 is true, but rather because S is sufficiently large. While it is often not realized, with enough observations an improbable difference may actually be of trivial magnitude and then the resulting rejection of h_0 is misconstrued. Such misuses may largely be avoided by formulating the test so that it rejects h_0 just in case an observed discrepancy is large enough to indicate that h_0 is false. Precisely how I think such a reformulation of tests can be made to do this is a matter to be taken up in a separate paper. (See Mayo1980).

5. Lawlikeness and Causation by Chance

A related problem of philosophical significance (particularly for resolving the "new" problem of induction) is distinguishing between genuine or lawlike (or projectible) regularities and spurious ones. Although the standard theory of statistical testing plays a major role in rooting out lawlike relationships in science, philosophers have rarely made direct use of these tests in dealing with this problem. This is unfortunate, as I think an examination of the logic of statistical tests and their role in science is quite relevant to the philosophical problem of lawlikeness.

Statistical tests provide a systematic way of determining whether an observed difference is large enough to indicate a lawlike relationship or small enough to be attributed to non-lawlike or chance factors. The null hypothesis typically takes one of the standard forms for asserting that the difference is "due to chance factors". In our example, the null hypothesis models the claim that the relationship between saccharin and cancer is not lawlike by asserting that the difference in means is due to chance. Accepting (or failing to reject) the null is tantamount to concluding that the observed difference is not indicative of a systematic or lawlike relationship. In testing a non-lawlike relationship (e.g., between saccharin and eye-color in rats) one would not want to reject the null hypothesis since it would be true. However if the null is true, there will still be discrepancies between the hypothesized difference in means (i.e., 0) and the observed difference in means. Hence, what is needed is a standard for determining whether or not a discrepancy can be plausibly attributed to chance, and probability models provide such a standard. The basic probability models are derived from observations generated by probabilistic mechanisms such as tossing a fair coin (i.e., probability of "heads" is .5). The type of discrepancy between .5 and the proportion of "heads" observed in tossing a fair coin is a standard measure for a discrepancy that is due to chance.

The notion "due to chance" is in need of careful philosophical scrutiny, but here a few remarks must suffice. It is often wrongly thought that attributing a discrepancy to chance means that the observations stem from a probabilistic mechanism. As I see it, it may simply mean that the type of discrepancy is similar to the type known to arise from a probabilistic mechanism, and hence may be represented by a probability model. It may merely serve as a way of saying that the discrepancy is not large enough to be of interest at a certain stage of research. As the practice of attributing certain discrepancies to chance is frequent in science, any theory of scientific inference must take account of it, and examining statistical tests is relevant for doing so.

If the null hypothesis in our example is rejected, it can be concluded that saccharin is not causally irrelevant to cancer in rats. However, this does not imply the conclusion that saccharin causes cancer in rats. The reason is that even with the randomization techniques of experimental design, not all of the relevant factors can be controlled (or even known), and the difference may be due not to saccharin but to these uncontrolled factors. Still, the statistical hypothesis test may be seen as a step in the reasoning processes that lead to inferring causality. Rejecting the null hypothesis (in an appropriately designed test) at least suggests that further experimentation is warranted; accepting it permits the causal claim to be denied.

Getting clear on the reasoning behind causal inferences is not only relevant to the theoretical problems of causality and lawlikeness, but to applications of philosophy to problems concerning public policy involving statistics. Questions about the justification of such policy (e.g., banning substances claimed to "have been determined to cause cancer in rats") are very much dependent upon being able to analyze the validity of the statistical reasoning on which they rest.

6. The Structure of Scientific Activity

One very general concern shared by both statisticians and philosophers of science is to provide a structure for scientific activity, and our saccharin example suggests how statistics may be seen to provide such a structure. Although the statistician is interested in providing a structure at the "working level", the structure it provides is relevant for the philosopher's interest in the structure of science. That statistics functions to provide a structure for science at all is typically not recognized. I attribute this to the overly narrow conception of the role of a theory of inference both in philosophical theories of inference and in presentations in statistical texts. Philosophical theories of inference typically begin with the assumption that the data and hypotheses are given, and the only job that remains is to construct a measure of relationship between them. In presenting statistical inference, statistical texts also typically start out with the data and hypotheses given; and in addition it is assumed that the data follows a known distribution. In fact, inference does not begin only after data and hypotheses are given, nor is the data known to follow a given distribution. Statistical reasoning enters at a number of different stages of scientific research, and several statistical inferences may be involved in a single research effort.

Statistical reasoning, construed broadly, is involved in the initial planning of experiments, the gathering and modeling of data, and in the construction and evaluation of hypotheses and theories. In other words, statistical reasoning is involved in constructing and linking the four levels sketched in our saccharin example. On the level of the data, statistics (or more exactly, the statistical theory of experimental design) is involved in specifying what should be observed, and how to observe it, and then in modeling the observation by means of a statistic. Statistical inference links this data to the statistical hypothesis, and lastly, the statistical hypothesis is linked to the substantive scientific claim which it models. Statistics may first be used to detect potentially interesting relationships in the data, to indicate which quantities are likely to be relevant in an eventual theory, and hence should be studied further. For example, one would first check whether there was any relationship between saccharin and cancer before going on to test a specific relationship.

As different inference principles may be relevant at different stages of inquiry, an adequate theory of inference is going to have to be more complex than generally thought. While the standard theory of statistics provides a conglomeration of inference principles, there has not been an attempt to unify them, and doing so is an important task for a philosophical theory of inference built upon (the standard theory of) statistics. (It may also be possible to unify some of the standard principles of inference with non-standard principles.) Each stage of inquiry involves a different kind of model, and it is by linking these models that statistics provides a structure for scientific research. Hence a philosophical analysis of the statistical inference principles involved at different stages of research will offer insight into the structure of sciences which employ statistics. This is particularly true for the biological and social sciences which are faced with much less controlled experiments than the physical sciences, and hence have a greater need for statistics.

Experiments in the physical sciences typically enjoy relatively little error in measurement; items to be measured are nearly identical with respect to the property of interest and they are relatively stable over time. As such, in measuring the effect of a change in things like temperature or pressure, one is not faced with the same variability faced by the scientist of sociology, psychology and biology in measuring effects on animals and humans. Understanding the statistical methods for accumulating knowledge in the so-called inexact sciences is relevant for answering a number of philosophical questions concerning these sciences. The following are examples of such questions: Are the social sciences methodologically or logically distinct from natural science? Are the social sciences inferior? Can they be objective? Building a theory of scientific activity on the model of the physical sciences, and failing to recognize the roles of statistics in science, has often lead to narrow-minded answers to these questions.

Statistical considerations are not absent even in those sciences considered exact. The reason is that only a finite sample of data is available, and the accuracy, precision, and reliability of the data are limited by distortions introduced by the processes of measurement and observation. Scientific theories contain theoretical concepts which do not exactly match up with things that can be observed; data are finite and discrete while theories may refer to an infinite number of cases, and continuous variables such as weight and temperature. As such, the data can rarely be expected to agree exactly with theoretical predictions, and because of this, the observable prediction is often a statistical claim. Hence, whether theories are considered to be probabilistic or deterministic, studying them often gives rise to very similar statistical situations. By examining the role of statistics in both social and natural sciences, it might be possible to see the differences between them as simply being a matter of degree. In any case a clearer picture of their relationship may emerge.

7. Conclusion

To summarize, a study of statistics, its problems and its roles in science, is far more relevant to philosophy than is typically recognized. Though here I have only been able to sketch a few of the major reasons for thinking this, hopefully enough points have been raised to suggest a number of possible ways of putting the foundations of statistics to philosophical work. I have indicated the relevance of statistics for the (new) problem of induction and for elucidating the structure of actual scientific practice and the reasoning behind causal inferences. As such, statistics may be seen to be relevant to general problems of philosophy of knowledge. The following remark of Oscar Kempthorne is of interest in this regard: "To resolve the obscurities about probability and inductive inference is equivalent in my opinion to laying out a philosophy of knowledge. Workers in statistics and particularly those working on theories of statistical inference are on the boundaries of philosophy often, and usually, I fear, without being aware of the fact." (1971, pp. 482-483).

In addition to being useful for theoretical problems of philosophy, statistics provides a bridge linking philosophy to applications of science in government and society, where statistics is extensively used. If the positivistic framework of science is to be replaced by a framework that is relevant to actual scientific problems, philosophy of statistics has an important role to play in its construction. Ronald Giere has put it this way: "One may hope that in time philosophers will appreciate the significance of the [statistical] significance test controversy for inductive logic and it will begin to appear in the philosophical literature. Eventually we may have something relevant to say to sociologists and psychologists who, it seems, very much want and need to hear something relevant from someone who understands the full dimensions of the problem." (1972, p. 180).

Notes

[1]I will not here discuss the philosophical relevance of decision theory, which is a branch of statistics, and which has found applications to moral and political philosophy and is relevant to public policy. For a discussion of this see Suppes (1961).

[2]For a good discussion of the inductive theories of a number of philosophers see Giere (1979a).

[3]When such a randomized experiment is not available, as in historical surveys, more complicated statistical arguments are required.

[4]For a detailed discussion of some of the experiments used in determining the relationship between saccharin and cancer in rats, as well as an excellent explanation of statistical methods in science generally, see Giere (1979b).

References

Carnap, R. (1962). Logical Foundations of Probability. 2nd Edition. Chicago: University of Chicago Press.

---------- (1971). "A Basic System of Inductive Logic, Part I." In Studies in Inductive Logic and Probability, Vol. I. Edited by R. Carnap and R. Jeffrey. Berkeley: University of California Press. Pages 33-165.

Finch, P. D. (1976). "The Poverty of Statisticism." In Harper and Hooker (1976). Pages 1-44.

Fisher, R. A. (1959). Statistical Methods and Scientific Inference. 2nd Edition. Edinburgh: Oliver and Boyd.

Giere, R. N. (1972). "The Significance Test Controversy." British Journal for the Philosophy of Science 23: 170-181.

------------ (1976). "Empirical Probability, Objective Statistical Methods, and Scientific Inquiry." In Harper and Hooker (1976). Pages 63-93.

------------ (1979a). "Foundations of Probability and Statistical Inference." Current Research in Philosophy of Science. Edited by Peter D. Asquith and Henry Kyburg. E. Lansing: Philosophy of Science Association. Pages 503-533.

------------ (1979b). Understanding Scientific Reasoning. New York: Holt Rinehart and Winston.

------------ (1980). "Causal Systems and Statistical Hypotheses." In Applications of Inductive Logic. Edited by L. J. Cohen. New York: Oxford University Press. In Press.

Godambe, V.P. and Sprott, D. A. (eds.). (1971). Foundations of Statistical Inference. Toronto: Holt, Rinehart and Winston of Canada.

Good, I. J. (1971). "The Probabilistic Explication of Information, Evidence, Surprise, Causality, Explanation, and Utility." In Godambe and Sprott (1971). Pages 108-122.

Goodman, N. (1965). Fact, Fiction and Forecast. New York: Bobbs Merrill Company, Inc.

Harper, W. L. and Hooker, C. A. (eds.). (1976). Foundations of Probability Theory, Statistical Inference, and Statistical Theories of Science, Vol. II. (University of Western Ontario Series in Philosophy of Science, Vol. 6.) Dordrecht: D. Reidel

Kempthorne, O. (1971). "Probability, Statistics, and the Knowledge Business." In Godambe and Sprott (1971). Pages 470-499.

------------- (1976). "Statistics and the Philosophers." In Harper and Hooker (1976). Pages 273-309.

Lieberman, B. (ed.). (1971). Contemporary Problems in Statistics. New York: Oxford University Press.

Mayo, D. (1980). "Testing Statistical Testing." In Philosophy and Economics. (Western Ontario Series in Philosophy of Science.) Edited by Joseph C. Pitt. Dordrecht: D. Reidel. In Press.

Morrison, D. E. and Henkel, R. E. (eds.). (1970). The Significance Test Controversy-A Reader. Chicago: Aldine Publishing Co.

Neyman, J. (1977). "Frequentist Probability and Frequentist Statistics." Synthese 36: 97-131.

Rosenkrantz, R. (1973). "The Significance Test Controversy." Synthese 26: 304-321.

Skyrms, B. (1975). Choice and Chance. 2nd Edition. California: Dickenson Publishing Co., Inc.

Spielman, S. (1978). "Statistical Dogma and the Logic of Significance Testing." Philosophy of Science 45: 120-135.

Suppes, P. (1961). "The Philosophical Relevance of Decision Theory." The Journal of Philosophy 58: 605-614.

Grounding Probabilities from Below[1]

Ian Hacking

Stanford University

What is at issue between the propensity and the long run frequency schools of thought about objective probability? Recent debate, although vigorous, has contributed little to our understanding of natural law or statistical inference. Yet there is at least one real question, none other than the much maligned topic of emergentism. Although this invokes a number of metaphysical views, such as the unity of science, emergentism primarily poses a question of matter of fact. Is it the case that every stable frequency, correctly represented by a mathematical probability, is 'grounded from below' by probabilities that apply to individuals? That is, does the frequency distribution in the population derive from probabalistic facts about the individuals that compose it? Or are there some stable frequencies that pertain to populations, but do not derive from probabalistic facts about members of the population?

At least two approaches to these questions have been developed. First, one may construct mathematical models of grounding from below, a tradition inaugurated by Poisson's (1835) treatment of what he called the law of large numbers. Secondly, in the manner of Durkheim (1895) one may examine real life stable frequencies and ask, in the difficult cases, whether we do not witness emergent probabalistic phenomena, not reducible to probabilities of individuals. In Section (1) I try to make my questions a little more precise. Then in (2) I describe the classic mathematical results in the field. Then in (3) I report the classic emergentist position on probability. In (4) I introduce another difficult case, chiefly to illustrate how, in real population data, the matters are intricate, messy, important and still receiving serious analysis. Finally in (5) I conclude by suggesting, although not of course proving that some natural phenomena may be accurately described in propensity terms, while others are accurately described only in frequency terms.

1. Binomial Trials

We agree that some facts about what von Mises called 'mass phenomena'

PSA 1980, Volume 1, pp. 110-116

are well represented by mathematical probabilities. A 'collective' is a population which is representable in this way. The best understood case is a sequence of binomial trials, already well analysed by Jacques Bernoulli (1713). In this case the collective consists of a sequence of trials, each of which may result in either the event 'success' or 'failure'. The probability of 'success' is a constant p for each trial. The expectation of the number of successes in n trials is np and as the size of the collective grows without bound, the probability of proportions significantly different from p becomes vanishingly small.

Binomial trials in nature are a good deal more difficult to contrive than one might have thought, but some real phenomena seem well represented in this way. It is entirely natural to ascribe a probability to each individual trial, and to speak for example of the propensity or disposition of 'the coin' to turn up 'heads'. It is also entirely natural to speak of the stable relative frequency on trials of this sort. Philosophical taste has prompted some writers to favour one description over another. Peirce, for example, gave a frequency account in his younger years, and later acquired a preference for propensity. He described this as a special case of maturing from nominalism to scholastic realism. But although I regret that it is offensive to say so, it does not seem that the preference for frequency over propensity, or vice versa, has ever been connected with any question about matters of fact. In the case of binomial trials one can always translate from one idiom to the other.

There are also real phenomena that are represented by urn models, in which when one e.g.,draws a red ball ('catches cholera'), new red balls are added to the urn ('the chance of a new infection has increased.') Here one may say there is no actual stable long run frequency, but rather a curve, or some fluctuations, hopefully declining to almost zero ('the epidemic is defeated.') It may be said that there are only changing propensities of infection. But a nominalist can, without much imagination, readily construct a description in which it is the law-like long run features of the population that are paramount. The talk of propensities is, he will say, merely derivative from the description of the population statistics. I would not want to adjudicate on this issue, for it seems to me to be of no importance.

Indeed I am not concerned with whether all talk of propensities can be restructured in terms of populations, but rather with whether all talk of populations can be restructured in terms of propensities of individuals. In binomial trials and urn models alike it is natural for anyone indifferent to philosophical niceties to speak of probabilities of particular individuals. But are there always such probabilities in the offing to which mass phenomena can be reduced?

My question is: are the probability facts about every collective, which in fact exists in the real world, reducible to probability facts about members of the collective? I put aside one frivolous answer. 150 French men and women per million commit suicide annually. Hence we can construct an entirely derivative kind of trial, in which French people are picked at random; the propensity on trials of this sort, of getting a 1980 suicide, is 0.00015, but obviously this propensity is entirely derived from the population plus a sampling device. I am asking whether there are some probabalistic facts about French individuals, which together are equivalent to the probabalistic fact about the population?

I must also put aside a second topic. For some writers, propensities are probabilities that apply to individuals, or single cases. For other writers, a propensity is whatever-it-is about the individuals that leads to some statistical mass phenomenon. In this latter usage, the propensity of a coin might be described by some physical geometry, some physics and so forth, in short, an occult power which, if only we know more, would enable us to predict probabilities. I am concerned only with probabalistic propensities, and not with the second sense which has been given to the term. A full emergentism would hold that perhaps nothing about individuals fully fixes the statistical facts of the population (i.e.,nothing except the trivial after-the-end-of-the-year-fact, that just these 150 French people killed themselves). I shall not ask whether such all out emergentism makes sense. Here I inquire only whether nothing correctly represented by a probability, applied to individuals, suffices to engender the statistic about the whole population.

2. The Law of Large Numbers

Poisson begins by observing that binomial trials must be distinguished from the case in which a large number of coins is tossed, each coin being tossed once only. It is not realistic to suppose the probability of heads is constant, the same for all coins. However, he insisted that we shall in this case too obtain a statistical stability, and he called this fact the law of large numbers, in deliberate contrast with what he called Bernoulli's theorem.

Poisson's readers have not always found it easy to see just which cases he had in mind, but some of the classical results are tidily set out by Heyde and Seneta (1977). In the simplest case, which is probably what Poisson intended, the expected number of successes in n trials in $n\bar{p}$, where $\bar{p}$ is the average probability per trial. The variance, in such a scheme, is less than the variance in binomial trials, so in a certain sense we expect _more_ stability in heterogenous trials than in homogenous binomial trials.

Naturally such an analysis extends to the case of the population of coins, sampled at random, and with a known average of the probabilities of the individual coins. But nineteenth century workers were more struck by the fact that many empirical frequencies, although stable, tend to show long term trends, e.g.,in the rate of fertility. Or else one is interested in short sharp changes in statistical phenomena, as are suggested by vaccination. Hence instead of Poisson's scheme one may consider that due to Lexis (1877), in which there are e.g.,m blocks of n binomial trials. (Unlike Poisson's case, the variance here is greater than with ordinary binomial trials.)

Lexis called one of his characteristic papers, 'On the theory of the stability of statistical series', (1879) and it will be readily seen that he was addressing just my question: to what extent are stabilities of mass phenomena -- he had already used the term Massenerscheinungen -- grounded from below, in probabalistic facts about individuals. Even if binomial trials never occur in nature, we see that there are all sorts of 'Bernoulli-like' situations which, on analysis, are connected with stable statistics. Hence we are led to expect that many probabilities associated with collectives are indeed grounded from below. Talk of any exact grounding is irrelevant. We are concerned with the way in which one adequate mathematical representation at the level of populations may be connected with another adequate representation, at the level of individuals.

3. Suicide

The classic nineteenth century statistics are furnished by crime and suicide. It was observed quite early that e.g.,the rate of conviction in cases of crimes against property, can be correctly estimated from a far smaller sample, than the rate of male births (the latter statistic becoming stable only in rather large populations). There is an almost sinusoidal law, with suicides peaking in June and at a minimum in December. The methods employed are likewise stable. Parisians drown themselves and inhale carbon monoxide, but Londoners never use CO but prefer hanging and shooting. Hacking (1980) describes the morbid fascination with crime and suicide throughout the century. Dozens of statistical books on suicide occur from the 1820's on. Durkheim's great (1895) is only the culmination of this work. He has since been criticized for an uncritical use of data but from his point of view there was overwhelming evidence on which to draw inferences about suicide.

Now Durkheim was certainly a philosophical emergentist, having learned well from a preceding generation of French teachers such as Boutroux and Renouvier. He was well disposed to notice cases in which the whole might be larger than the parts. But his claim about suicide statistics -- on which he based many of his doctrines about sociological laws -- is a claim about matters of fact. There are some social facts about suicide rates within different classes, and at different levels of urbanization. If there are any 'well-confirmed' statistical laws, they are here. They may form the basis for a remarkable evaluation of the quality of life within a society -- an evaluation technique recently revitalized by Emmanuel Todd (1979). But although we can certainly talk about the probabilities of suicide within a large number of social groups, the frequencies are simply minute within any group large enough to have statistical stability at all.

Incidentally Durkheim was able to discuss the matter quite explicitly in terms of 'propensities'. The phrenologists Spurzheim and Gall introduced a doctrine of individual penchantes associated with bumps on the skull; the English translations of the day render this 'propensity'. (e.g.,Spurzheim 1815). It is precisely the word 'penchante' that is taken over and regularly used by French work on social indicators such as crime and suicide. Durkheim is denying that there is any natural representation of individuals in terms of their 'penchante' for suicide. Perhaps there is some vague qualitative notion -- this man, a petit bourgeois, has a greater penchante for suicide than this woman, a peasant. But there is nothing quantitative which realistically attaches to the individuals, in the way that the probabilities attach to the populations.

Durkheim was certainly prepared to defend a stronger emergentism, in which he described statistical regularities of segments of society 'acting on' their individuals, helping determine their individual actions, in the way that the laws of gravity 'from outside' also act on individuals. But he has a forceful argument only for a weaker sort of emergentism, what one might call the emergence of probability -- a probability that emerges at the levels of society, not being grounded on probabilities that can realistically be ascribed as properties of individuals.

4. Fecundity in Marriage

Turning to more blessed events, we now know as much about fertility as Durkheim knew about suicide. Pity the poor graduate student who would like to study the decline of fecundity with the onset of industrialization. What state, nation, village or county has not now been the object of a PhD dissertation in some department of demography? But the more we know the harder things become. Fecundity declines in every industrializing segment of the Western world, but there uniformity ends. Moreover one begins to notice a phenomenon that might have already been suspected by Durkheim.

Take Knodel's (1972) study, The Decline of Fertility in Germany, 1871-1939, together with some more recent unpublished work. Knodel found that the national decline in birth rate is nicely mirrored in each administrative district or Kreis. The numbers are as regular as any which could be hoped for in demography. But when we pass to smaller units, such as the village, the uniformity collapses. Although every Kreis is doing the same thing as every other, villages within Kreise are all going their own ways, without much in the way of underlying laws. Now suppose that the probability facts about the Kreis are to be grounded from below, in probabalistic facts about individual married couples. What facts? In the reference class consisting of the Kreis we find homogeneity, but the probability of having say 2 children, relative to that class, is certainly not what one would infer from the villages within the class, nor any other plausible social segmentation of the Kreis. It may turn out that there is simply no quantitative propensity to have 2 children that can realistically be ascribed to individual couples, although there is a probability fact about the population at large.

But things are more delicate even than this. Distinguish fecundity in marriage (actual reproduction) from fecundability, the biological ability of couples to have children if they want to. Fecundability is affected by diet, age at marriage, sanitation and so forth. While fecundity is decreasing in the period under study, fecundability is rising. It is possible to make a rough and ready factoring out of fecundability in marriage. Unlike fecundity we can produce quite meaningful reference classes which enable us to speak, in a fairly sensible way, of the fecundability of individual couples. This fits well enough with my own metaphysical inclinations; there really is something biological going on, tempered by social constraints of a widely diffused sort. But the propensity to limit family size may simply not be a number which represents a property of each of the individual couples within the Kreis. There is only a very striking property that applies to the Kreis as a whole, that the birth rate is declining in a systematic and law-like way.

5. Conclusion

I think, then, that there may be collectives whose probabilities cannot be reduced to, or derived from, realistic assumptions assigned to individual members of the collective. There are of course clear-cut cases, analysed by Bernoulli, Poisson and Lexis, and by more sophisticated successors, in which a representation of the properties of the collective may be realistically based on a representation of properties of the individuals. In short: some phenomena may be best understood in terms of a frequency theory approach, while others are well understood in terms of a propensity approach. The propensity account may fit some cases, while frequency fits another, and which fits which is less a philosophical question than a matter of fact.

I do not think my distinction makes the slightest difference to any practical problem of statistical inference. But it does suggest that there may be two genuinely distinct kinds of chance process, one grounded from below, and the other, to imitate Durkheim, imposed from above.

But I acknowledge one defect in this conclusion. It is too much based on numerical probabilities. A useful continuation of Poisson's programme for a law of large numbers would examine the extent to which quantitative frequencies within a collective may derive from purely qualitative and unstructured propensities assigned to individual members of the collective.

Notes

[1] Research supported by National Science Foundation Grant SOC 7906928

References

Bernoulli, Jacques (1713). Ars Conjectandi. Basel: Thurnisiorum.

Durkheim, Emile (1895). Le Suicide Paris: Seuil.

Hacking, Ian (1980). "La Statistique du Suicide au XIXe Siècle." L'Approache Probabaliste en Médecine. Edited by A. Fagot. Paris: Flammarion. In Press.

Heyde, C.C. and Seneta, E. (1977). I.J. Bienaymé, Statistical Theory Anticipated Berlin: Springer-Verlag.

Knodel, John E. (1972). The Decline of Fertility in Germany, 1871-1939. Princeton: Princeton University Press.

Lexis, W. (1877). Zur Theorie der Massenerscheinungen in der Menschlichen Gesellschaft. Freiburg: Wagner.

Poisson, S.D. (1835). Recherches Sur la Probabilite des Jugements Principalement en Matiere Criminelle. Comptes Rendu Hebdomaires des Seances de l'Academie des Sciences, I. Pages 473-494. Also Note Sur la Loi des Grands Nombres, Ibid, II, 1836. Pages 377-382.

Spurzheim, J.G. (1815). The Physiognomical System of Drs. Gall and Spurzheim London: Baldwin, Cradock, and Joy.

Todd, Emmanuel (1979). Le Fou et le Prolétaire Paris: Laffont.

On the Use of Likelihood as a Guide to Truth

Steven Orla Kimbrough[1]

University of Wisconsin-Madison

Much work in confirmation theory is directed towards finding or interpreting a function, c, taking on hypotheses and evidence as its arguments and returning a value by which the degree of support of the evidence for the hypothesis can be assessed. Confirmation functions are generally thought of as probability functions. That is, $c(H,E)$ is thought to obey the axioms of probability theory, where H is an hypothesis and E the evidence being considered. A difficulty with this approach is that the various problems of induction, including Hume's and Goodman's, militate against construing any confirmation function as a probability function. To date, no satisfactory probability function has been found which addresses the general problem of ampliative inference and which is immune to the standard objections of the Humean or Goodmanian type.[2]

These sorts of considerations suggest it would be promising to examine, as candidates for an acceptable confirmation function, functions which are not probability functions. The purpose of this paper is to present an argument tending to favor one such function, the likelihood function. Likelihood is not only an intuitively appealing basis for confirmation, but it is actually used extensively in statistical inference. Hacking (1965) and Edwards (1972) have gone so far as to argue that all legitimate statistical tests are, overtly or covertly, likelihood tests. This alone requires that likelihood be given serious examination as a basis for confirmation theory.

The likelihood of an hypothesis, H, on a body of evidence, E, is defined to be the probability of the evidence conditioned on the hypothesis, or

$$L(H|E) =_{df} P\{E|H\},$$

where the standard definition of conditional probability is

$$P\{E|H\} =_{df} P\{E \cap H\}/P\{H\}.$$ [3]

PSA 1980, Volume 1, pp. 117-128

It is important to understand why likelihood is not a probability function. In the use of the function, various hypotheses, *H1*, *H2*, etc., are compared on the basis of an unchanged body of evidence, *E*. Thus, *E* is a constant while *H* varies. In $L(*|**)$, * holds the place of a variable and ** holds the place of a constant.

To see that $P\{E|*\}$ is not a probability function, allow that we have a coin flipping situation and we are considering three hypotheses:

*H*1: The probability of heads is 0.5.
*H*2: The probability of heads is 0.51.
*H*3: The probability of heads is 0.49.

In our experiment, we toss the coin once and observe that heads results. Now,

$$L(H1,E) = P\{E|H1\} = 0.5$$
$$L(H2,E) = P\{E|H2\} = 0.51$$
$$L(H3,E) = P\{E|H3\} = 0.49$$

Since *H1*, *H2*, and *H3* are mutually exclusive, by an axiom of probability theory their joint probability (if $P\{E|*\}$ is a probability) is just the sum of their individual probabilities. That is,

$$P\{E|H1 \cup H2 \cup H3\}$$

should equal 0.5 + 0.51 + 0.49 = 1.5. This, however, violates another axiom of probability theory. No probability can be greater than 1. The contradiction shows that likelihood is not a probability function.

Interpreting confirmation as a probability function is attractive because the question, "Why should we want to confirm anything?", is readily answered. Confirmation provides high probability and so assures us that by choosing the hypothesis with the highest degree of confirmation we have the best chance of choosing the true hypothesis.[4] With likelihood, it is not always the case that choosing the hypothesis with the highest likelihood gives us the best chance of choosing the true hypothesis.[5] Although other criticisms have been leveled against use of likelihood as a confirmation function, this fact is a most fundamental obstacle to the use of likelihood in confirmation. Giere (1979, p. 508) rightly observes that "The trouble with the likelihood view is that it is difficult to understand why a favorable likelihood ratio constitutes evidence for the favored hypothesis."

I shall now sketch a defense of likelihood against this objection, which may be put as follows:

> Probability is a sound basis for confirmation because it is a reliable guide to the truth. Likelihood is not a sound basis for confirmation theory because it cannot reliably be used to determine if an hypothesis is true.

My argument is in two steps. Step 1 is completely unoriginal. It is a trivial proof and is a commonplace in statistics. I include it because it is short, it is required for Step 2, and it is instructive.

Step 1

I shall demonstrate that, on the assumption of a binomial distribution, the rule of choosing the hypothesis with the maximum (highest) likelihood will, in the limit, provide us with the true value of the binomial parameter. Before proving this, however, let us see what this claim amounts to.

The binomial distribution is a good model for a coin tossing experiment, so I shall illustrate the concepts of the binomial distribution by referring to a coin flipping situation. In order to have a binomial distribution, three conditions must be met:

1. The experiment can have only two possible results, called success and failure. In the case of the coin toss, these are heads (success) and not-heads (failure).

2. The probability of success, q, is constant over time. q is the parameter of the binomial distribution and it is what we are attempting to estimate.

3. The probability of success on any particular experiment is independent of the result of any other experiment. In the case of the coin, this means that the probability of heads on any particular toss (or trial) is independent of the result of any other toss of the coin.

Define a random variable, X, as follows:

X = the number of successes in n trials.

Then, $P\{X = x\}$, read as "The probability that the number of successes in n trials is equal to x, on the [6] assumption of a binomial distribution with parameter q" can be shown to be

$$P\{X = x\} = \begin{cases} nCx(q^{x})(1-q)^{(n-x)}, & n=1,2,\ldots;\ x=0,1,\ldots n \\ 0, & otherwise \end{cases}$$

In performing the experiment, we observe the number of trials, n, and the actual number of successes, x. These constitute our evidence. Thus, $P\{X = x\}$ is the likelihood of the hypothesis that q has a particular value, on the evidence of the observed values of x and n. Allow that we perform an experiment, observe the results, and estimate q in some way. I will now show that if we estimate q by choosing the hypothesis with the highest likelihood on the evidence, then--in the limit--our estimate of q will be the true value of q.

We want to find the maximum value of $P\{X = x\}$ for any given n and x. Note that q is the only variable in the expression (above) for $P\{X = x\}$

and that q is continuous. One way to find the maximum value of $P\{X = x\}$ is to set its derivative (with respect to q) to zero and to solve for q. Because the logarithm function is monotonically increasing, $ln[P\{X = x\}]$ has its maximum at the same value of q as does $P\{X = x\}$. Taking the logarithm of the expression, we have: $ln(nCx) + x ln(q) + (n-x) ln(1-q)$. Taking the derivative and setting it to zero we have:

$$\frac{d\ ln(nCx)}{dq} + \frac{d\ x ln(q)}{dq} + \frac{d(n-x) ln(1-q)}{dq} = 0,$$

or: $0 + x/q - (n-x)/(1-q) = 0$. Rearranging, we have: $x(1-q) = q(n-x)$, leading to: $x-qx = qn - qx$, and finally: $x/n = q$. Thus, the maximum likelihood estimate of q is x/n, exactly what intuition tells us to expect. If we tossed a coin 100 (=n) times and got 54 heads (= x), our intuitively best estimate of the probability of heads for that coin would be 0.54.

In the limit the maximum likelihood estimate of q is $lim\ x/n$, where the limit is taken as n goes to infinity. Under the freqeuncy interpretation of probability, this limit is the true value of q. We may conclude, then, that for a binomial distribution the maximum likelihood estimate of the parameter gives the true value of the parameter in the limit.

This completes the first part of my argument. A second step is needed for two reasons: (1) there are conditions under which following the rule of maximum likelihood will not lead us to the truth, even in the limit; and (2) we cannot assume, without justification, that we have a binomial distribution underlying the phenomena under investigation, and such a justification is blocked by the familiar problems of induction.

Step 2

I now wish to argue that, given any two hypotheses one of which is true, likelihood can in principle be used to determine which hypothesis is true. It is, in addition, required that the two hypotheses be in principle observationally distinct. I will clarify this notion presently. Finally, there are two assumptions being made for the argument that follows:

1. Only one random variable is in question. In the example of the binomial distribution, above, the random variable was X, the number of heads observed on n tosses. Once my argument is understood, the reader should have no difficulty relaxing this restriction and extending the proof to any number of random variables.

2. The random variable in question is of the continuous variety. Random variables may be discrete (as was X in the above example), continuous (i.e., can take on an uncountable number of values in any interval on which it is defined), or mixed (sometimes continuous, sometimes discrete). Extending my argument to cover discrete variables is trivial. Handling the case of mixed random variables should be obvious, once the method of dealing with

discrete and with continuous random variables is understood.

With these points in mind, the second step of my argument can now be presented. Allow that we have defined a continuous random variable, Y, and we have a probability density function (p.d.f.) for $Y, P\{Y = y\}$, analogous to $P\{X = x\}$, discussed above. For notational perspicuity, let $P\{Y = y\} = f(y)$, consider the graph of f, where the abscissa (horizontal axis) is y and the ordinate is $f(y)$. Let $g(y)$ be a second p.d.f. for the random variable Y, and consider its graph. In Figure 1, the graphs of two such arbitrary functions are superimposed upon one another. $f(y)$ and $g(y)$ are our two competing hypotheses. Assume that one of $f(y)$ and $g(y)$ is true,[7] but that we have no information about which is true.

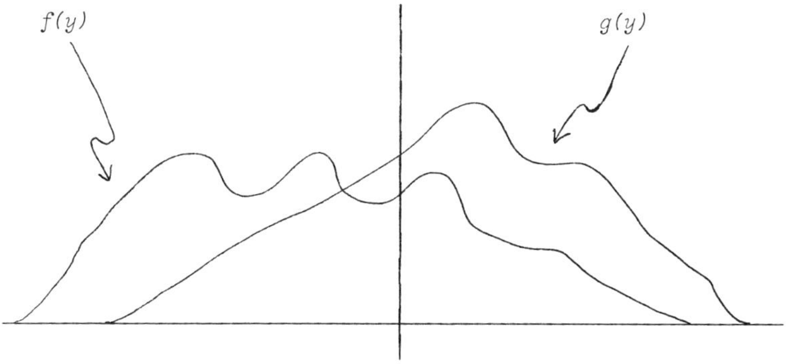

Figure 1

horizontal: y-axis
vertical: f,g (superimposed)-axis

No matter what f and g are, either they have identical graphs or they do not. If they have identical graphs, then f and g are identical hypotheses. Since, by hypothesis, at least one of f and g is true, we arrive at the truth by choosing either f or g.

Consider the non-degenerate situation in which the graphs of f and g are not identical. Because f and g are probability functions, we know that $\int f = \int g = 1$, where the limits of integration are positive and negative infinity. Let a, b be values of Y. For all a, b and for all f, g either $\int f = \int g$ (where the limits of integration are a and b) or not. First consider the case in which they are not identical for some particular values of a and b.

If the hypothesis f is true, then $P\{a \leq y \leq b\}$, i.e., the probability that the observed value of the random variable lies between a and b, is equal to $\int f$, integrated between a and b. A similar statement is true of $\int g$. We may now define a new random variable Z, in the following way:

$$Z = \begin{cases} 1, & \text{if } a \leq y \leq b \\ 0, & \text{otherwise} \end{cases}$$

This in turn leads to two new p.d.f.'s, f' and g' [8]:

$$f'(z) = P\{Z=z\} = \begin{cases} \int f, & \text{(limits of integration are } a,b\text{), if } a \leq y \leq b \\ 1- \int f, & \text{(limits of integration are } a,b\text{), if not } a \leq y \leq b \end{cases}$$

$$g'(z) = P\{Z=z\} = \begin{cases} \int g, & \text{(limits of integration are } a,b\text{), if } a \leq y \leq b \\ 1- \int g, & \text{(limits of integration are } a,b\text{), if not } a \leq y \leq b \end{cases}$$

Integrating between a and b, let $\int f = r$ and $\int g = s$. Notice that f' and g' are really assertions about a binomial distribution with $n = 1$ trial and parameters r and s. That is, we have converted the hypotheses, f and g, into hypotheses f' and g', and these are guaranteed to be binomial distributions. By repeating the experiment, we may increase n to an arbitrary value. Thus, I have shown how to reduce the present case to the binomial case, discussed in Step 1. My argument is:

1. Either f is the true hypothesis or g is the true hypothesis.
2. If $P\{a \leq x \leq b\} = r$, then f is true; if $P\{a \leq x \leq b\} = s$, then g is true.
3. It is possible (in the limit) to determine whether $P\{a \leq x \leq b\}$ equals r or equals s by employing a maximum likelihood test for an appropriate binomial distribution.

---

4. Therefore, it is possible (in the limit) to determine which of f and g is true.

We are, finally, left with the possibility that for all a, b, and integrating between a and b, $\int f = \int g$. This implies that (for all a,b and integrating between a and b) $\int |f - g| = 0$. From analysis, this implies that $f=g$ (a possibility we have already excluded) or that f and g differ by a set of zero measure. This means that the area underneath the points by which f and g differ is equal to zero. But this area is just the probability of the events corresponding to these points occurring. Thus, the probability is zero for observations to occur which distinguish f and g.

A simple example illustrates these remarks. Consider:

$$h(w) = \begin{cases} 1/2 & 1 \leq w \leq 3 \\ 0 & \text{otherwise,} \end{cases}$$

$$i(w) = \begin{cases} 1/2 & 1 \leq w \leq 3 \\ 3/4 & w = 4 \\ 0 & \text{otherwise} \end{cases}$$

Because $h, i \geq 0$ and $\int h = \int i = 1$ (integrating between plus and minus infinity), both h and i are probability functions. h and i are identical except at the point $w=4$, where $h=0$ and $i=3/4$. It is impossible to construct an interval on the w-axis, in the manner described above: for no values of $a, b (a \neq b)$ is it true that $\int h \neq \int i$. Since h and i are identical except at $w=4$, h and i yield exactly the same predictions, except possibly at $w=4$. But, integrating over the set by which h and i differ, $\int h$ (integrated from 4 to 4) equals $\int i$ (integrated from 4 to 4) equals zero. Thus, h and i necessarily yield exactly identical predictions, even though they are distinct hypotheses.

This completes my argument in favor of the claim that (in the limit, by gathering an infinite amount of evidence) likelihood can be used to determine which of two observationally distinct hypotheses is true. There are, of course, a large number of questions to be answered before we have an adequate theory of confirmation based upon likelihood. My goal has been more limited, viz., to show that likelihood can in principle guide us to the truth.[9] In closing, however, I should like to sketch an answer to what must now be the most obvious worry concerning the argument presented here: Even if likelihood guides us to the truth in the limit, why should we use it to estimate the truth when we do not have an infinite amount of information?

Since every pair of hypotheses that we can hope to distinguish observationally can be reduced, as has been shown, to a pair of hypotheses claiming different values for the parameter of a binomial distribution, we shall examine two binomial hypotheses with respect to the question at hand. Let $k(v)$ be the hypothesis that the random variable, V, is distributed binomially with parameter $q=0.5$. Similarly, let $l(v)$ be the hypothesis that the random variable, V, is distributed binomially with parameter $q=0.3$. Let $n=4$. Thus,

$$k(v) = P\{V=v\} = \begin{cases} 4Cv(0.5^4), & v = 0,1,2,3,4 \\ 0, & \text{otherwise} \end{cases}$$

$$= \begin{cases} 0.0625, & v=0 \\ 0.25, & v=1 \\ 0.375, & v=2 \\ 0.25, & v=3 \\ 0.0625, & v=4 \\ 0, & v \neq 0,1,2,3,4 \end{cases}$$

$$l(v) = P\{V=v\} = \begin{cases} 4Cv(0.3^v)(o.7^{n-v}), & v=0,1,2,3,4 \\ 0, & \text{otherwise} \end{cases}$$

$$= \begin{cases} 0.2401, & v=0 \\ 0.4116, & v=1 \\ 0.2646, & v=2 \\ 0.0756, & v=3 \\ 0.0081, & v=4 \\ 0, & v \neq 0,1,2,3,4 \end{cases}$$

We are assuming that k is true or that l is true.

Is there a decision rule such that if we perform an experiment, observe a particular value of V, and choose either k or l on the basis of the decision rule and the observation, then we shall have a reasonably high probability of choosing the true hypothesis? Happily, in this case there is such a rule, namely: Choose k if $V>1.5$; choose l otherwise.

To see how this works, allow first that k is the true hypothesis. If V is greater than 1.5 we will choose k. That is, if V equals 2, 3, or 4 we will choose k. The probability that V equals 2, 3, or 4, given that k is true, is 0.375 + 0.25 + 0.0625 = 0.6875. Conversely, if l is the true hypothesis, the probability of choosing it, given that we follow the decision rule, is just the probability that V equals 0 or 1. From the equations above, this is 0.2401 + 0.4116 = 0.6517. Since k is true or l is true, by using the above decision rule we can be assured that the probability we will choose the true hypothesis is at least 0.6517. Notice that $L(k|V>1.5)$ = 0.6875 and that $L(l|V\leq 1.5)$ = 0.6517.

I trust that it will be granted that in the situation just described we have evidence in favor of whichever hypothesis is chosen by using the decision rule. The situation can be generalized. I shall now prove the following theorem.

Let $H1$ and $H2$ be any two hypotheses such that:

(i) $H1$ and $H2$ differ by a set of nonzero area.

(ii) $H1$ and $H2$ are assertions about the same set of random variables.

Then, for all C, $0<C<1$, there exists a decision rule requiring only a finite amount of experimental information, such that:

(iii) The probability of choosing $H1$, given that $H1$ is true, is greater than or equal to C.

(iv) The probability of choosing $H2$, given that $H2$ is true, is greater than or equal to C.

(v) The decision rule always determines choice of either $H1$ or $H2$, but never both.

Here, C can be thought of as a level of confidence in the decision between $H1$ and $H2$. The theorem implies that we can drive C as high as we wish (so long as $C<1$) and still there will be a test that distinguishes $H1$ and $H2$ when only a finite amount of information is at hand. For example, in the case of binomial hypotheses, so long as C is fixed at less than 1, n (the number of trials) can be finite.

Because I proved above that any two hypotheses satisfying conditions (i) and (ii) can be converted to competing binomial hypotheses, I need prove this theorem only for competing binomial hypotheses. I shall do so by showing how to construct a decision rule which satisfies the theorem.

Let $H1$ and $H2$ be two binomial hypotheses.

$H1$: The random variable X is binomially distributed with probability of success = p.

H2: The random variable X is binomially distributed with probability of success = q.

Without loss of generality, assume that $p<q$.

Using slightly different notation, Feller (1968, pp. 150-1) derives the following equations for a binomial random variable, X.

(3.5) $P\{X\geq r\} \leq r(1-p)/(r-np)^2$, if $r>np$ and p = probability of success

(3.6) $P\{X\leq s\} \leq (n-s)q/(nq-s)^2$, if $r<nq$ and q = probability of success

Let $d = (q-p)/2$, $r = n(p+d)$, and $s = n(q-d)$. This implies that $d>0$, and that $r=s$.

Substituting into (3.5) we get $P\{X\geq r\}=$ $P\{X\geq n(p+d)\} \leq n(p+d)(1-p)/(n(p+d)-np)^2$.

The rightmost quantity equals $n(p+d)(1-p)/n^2d^2$, which in turn is less than or equal to $1/nd^2$, since $0\leq(p+d)\leq 1$ and $0\leq(1-p)\leq 1$.

Substituting into (3.6), we get $P\{X\leq s\}$ = $P\{X\leq n(q-d)\}\leq (n-n(q-d))q/(nq-n(q-d))^2$.

The rightmost quantity equals $n(1-(q-d))q/n^2d^2$, which in turn is less than or equal to $1/nd^2$, since $0\leq(1-q-d))\leq 1$ and $0\leq q\leq 1$.

Summarizing, we have: $P\{X\geq r\}\leq 1/nd^2$, $P\{X\leq s\}\leq 1/nd^2$, and $r=s$. Having fixed C, $0\leq C\leq 1$, we may find a value, M, of n such that $1/nd^2<1-C$, i.e., $M\geq 1/d^2(1-C)$. This is possible because $1-C$, $d^2>0$, and $\lim(1/nd^2)=0$, where the limit is taken as $n\to\infty$ All of this implies that given a finite value of C there exists a finite M such that $M=n$ and $1/nd^2<1-C$.

These facts allow us to construct the decision rule which completes the proof. Given as above *H1*, *H2*, d, r, s, M and C, the decision rule is as follows:

Run the experiment with n trials, where $n\geq M$, and observe the results. If $X\leq r$, accept *H1*. If $X>s$, accept *H2*.

It is easy to see how this rule works. If *H1* is true, then p= the probability of success. We have seen that $P\{X\geq r|H1\}\leq 1 - C$. But, $P\{X\geq r|H1\}= 1 - P\{X\leq r|H1\}$, so $1-P\{X\leq r|H1\}\leq 1 - C$, leading to $P\{X\leq r|H1\}\geq C$. Thus, condition (iii) is satisfied.

Similarly, if *H2* is true, q = the probability of success and $P\{X <s|H2\}\leq 1-C$, leading to $P\{X > s|H2\}\geq C$. Thus, condition (iv) is satisfied. Condition (v) is met because $r=s$, so that the rule always leads to a unique choice. This completes the proof. Notice that $L(H1|X\leq r) \geq C$, and that $L(H2|X>s)\geq C$.

The foregoing is, to repeat, barely the beginning of a theory of confirmation based on likelihood. Many problems remain that can be handled only in other papers. At least, I hope to have made it plausible that likelihood properly used can guide us reliably towards the truth.

Notes

[1]I would like to thank Fred Dretske and Stephen Vincent for their comments on an earlier draft of this paper.

[2]See Suppe (1977, pp. 624-632) and Giere (1979) for reviews of the current state of the art in confirmation theory.

[3]Here, H is discrete (can take on at most a countable number of values). The definition can be extended if H is not discrete. See Feller (1971, p. 157).

[4]Throughout this article I shall be assuming a frequentist interpretation of probability.

[5]See Cox and Hinckley (1974, p. 291). Simpler examples are not difficult to construct, e.g., binomial hypotheses (cf. $H1$, $H2$, below) with $n|p-q| < 0.5$.

[6]$P\{X = x\} = nCx(q^x)(1 - q)^{(n - x)}$ is a claim that a certain thing, here the number of successes, behaves in accordance with a particular probability distribution. $nCx = n!/x!(n-x)!$ is the binomial coefficient. Also, the binomial random variable is often interpreted as having two parameters, q and n. To simplify things, I am treating n as observed and hence the binomial random variable as having only one parameter, q. In this article all the hypotheses I discuss are "statistical" in that they assert something about probability distributions. I do not believe this is any real restriction on the generality of my argument. Even such "universal" hypotheses as "All emeralds are green" can be interpreted as assertions about probability distributions. For example, let G = emerald color, where G is a discrete random variable. Then,

$$f(g) = P\{G = g\} = \begin{cases} 1, & \text{if } g \text{ is green} \\ 0, & \text{otherwise} \end{cases}$$

Consider also $\vec{F} = m\vec{a}$. Let $\vec{F}$ = force acting upon a body, where $\vec{F}$ is a continuous random vector. Then,

$$f(g) = P\{\vec{F} = \vec{g}\} = \begin{cases} 1, & \text{if } \vec{g} = m\vec{a}. \\ 0, & \text{otherwise} \end{cases}$$

[7]I make this assumption for the sake of the exposition. The techniques described in this paper allow one to test a single hypothesis, using likelihood, in order to determine whether it is true. How that

can be done is a more complicated story and one that should be reserved for another paper. Suffice it to say that I believe the assumption, that one of our two hypotheses is true, will not turn out to be an impediment to basing confirmation theory on likelihood.

I am also making a second assumption for the sake of the exposition. The arguments of this paper apply strictly only to simple statistical hypotheses, i.e., statistical hypotheses in which the distribution of the random variable is completely specified. Simple statistical hypotheses have the form $P\{X = x\}$. Statistical hypotheses that are not simple have such forms as $P\{X < x\}$. Again, I believe that this restriction to simple hypotheses will not be an obstacle to a likelihood based theory of confirmation.

[8] f' and g' are not the derivatives of f and g.

[9] Strictly speaking, of course, I have shown that likelihood cannot guide us to the truth because it cannot distinguish between two hypotheses which differ by a set of zero measure. It is in the following sense that likelihood guides us to the truth: by using likelihood we can choose an hypothesis which is in principle observationally identical with the true hypothesis. Notice that this problem does not arise for discrete random variables.

References

Cox, D.R. and Hinkley, D.V. (1974). Theoretical Statistics. London: Chapman and Hall.

Edwards, A. W.F. (1972). Likelihood. Cambridge: Cambridge University Press.

Feller, William. (1968) An Introduction to Probability Theory and Its Applications, Volume I. 3rd ed. New York: John Wiley & Sons, Inc.

---------------. (1971). An Introduction to Probability Theory and Its Applications, Volume II. 2nd ed. New York: John Wiley & Sons, Inc.

Giere, R.N. (1979). "Foundations of Probability and Statistical Inference." In Current Research in Philosophy of Science. Edited by Peter D. Asquith and Henry E. Kyburg, Jr. East Lansing, Michigan: Philosophy of Science Association. Pages 503-533.

Hacking, Ian. (1965). Logic of Statistical Inference. Cambridge: Cambridge University Press.

Suppe, Frederick. (1977). The Structure of Scientific Theories. 2nd ed. Urbana: University of Illinois Press.

Part V

Literature and the Philosophy of Science

Philosophy of Science and Theory of Literary Criticism: Some Common Problems

Walter Creed

University of Hawaii at Manoa

Almost since the theories of Einstein and the quantum physicists came into being, a small but significant number of novelists, poets, and playwrights have turned to them for everything from casual metaphors to justification for the philosophies underlying their works. Until fairly recently, however, literary critics have tended to avoid the new physics, finding it uncongenial if not irrelevant. But over the last twenty-five years an increasing number of them, especially critical theorists, have turned to physics and even philosophy of science in the belief that these disciplines offer epistemological and methodological models applicable to the problems of literary criticism.

The primary reason for this increased interest in philosophy of science is the rise of critical formalisms, particularly structuralism and its variants. Structuralism usually regards Russian formalism as its most significant precursor, but Northrop Frye's *Anatomy of Criticism* (1957) must also be seen as an important forerunner, since it provided formal criticism with a rationale. In his "Polemical Introduction" Frye argues that criticism should be accorded the status of a separate discipline, "a structure of thought and knowledge existing in its own right, with some measure of independence from the art it deals with." (1957, p. 5). Calling for "a coherent and comprehensive theory of literature, logically and scientifically organized" (1957, p. 11), Frye repeatedly holds up science and mathematics as disciplines to imitate. He notes that as soon as astronomers ceased to regard "the movements of heavenly bodies as the structure of astronomy" and instead sought "a mathematical theory of movement [as their] conceptual framework" (1957, p. 15), astronomy as a science, in the modern sense of the term, became possible. Frye then describes criticism as "badly in need of a coordinating principle, a central hypothesis which, like the theory of evolution in biology, will see the phenomena it deals with as parts of a whole" and by means of which the "containing forms" of literature will be found (1957,

PSA 1980, Volume 1, pp. 131-140

p. 16).

In fact Frye posits several hypotheses. The most famous derives from seasonal myths and provides a rationale for organizing traditional genres (comedy, romance, tragedy, and satire or irony) into an ingenious cyclical schema. No one, Frye included, would claim that he did for literature what Copernicus did for astronomy or Newton for physics—or even what Darwin did for biology; but he has shown that through the systematic application of plausible hypotheses, literature can be seen not merely as an agglomeration of works written at different times by different authors in different languages, but as a complex *network* displaying subtle interconnections among the various works and implying that each work can be fully understood only in relation to a number of other works having characteristics in common with it.

Frye's dazzling erudition, his genuine concern for literature *as* literature, and his ingenious schemata have earned *Anatomy of Criticism* a lasting place among the major critical texts of our era. Yet until recently one of Frye's basic ideas, that literary criticism should become more like a science, made most critics uneasy if not hostile, and only in the past fifteen years have some of them taken the idea seriously.

Structuralists have done just this, agreeing with Frye that the systematic investigation of a self-contained, coherent literary universe should be carried out with the help of techniques and assumptions derived from science. Nevertheless, structuralists reject the specific techniques Frye uses, particularly the historical, ethical, archetypal, and rhetorical schemata he uses to organize and analyze works of literature. These schemata are, structuralists claim, either superficial, resulting in mere taxonomies, or brought in from other disciplines having no special relevance to literature. (See especially Tzvetan Todorov 1970, chapter 1.) As Robert Scholes says in his introductory survey of structuralism, "Frye's system had the considerable virtue of persuading readers of the possibility of a systematic study of literature without convincing them that it had been achieved. It was a system that rang deeply true in its essence while being plainly wrong in much of its substance." (Scholes 1974, p. 118).

Structuralism, too, initially derived many of its principles from other disciplines, most of all from linguistics (primarily the pioneering work of Ferdinand de Saussure) and structural anthropology (most notably the work of Claude Lévi-Strauss). But linguistics, structuralists maintain, is integrally related to the study of literature; and structural anthropology both is related to it (through its analyses of myths) and provides a highly suggestive *model* for literary study. Moreover, structuralism has rapidly developed its own principles, modifying or assimilating those it has borrowed.

Attempting to define structuralism is a perilous task, since it developed rapidly and at the hands of many critics, some of them uncomfortable with its basic premises and ready at any moment to subvert them. Nevertheless, I will try to suggest one or two aspects that may help characterize structuralism. It rejects the fundamental tenet of the "new criticism" (a quasi-formalist movement that dominated American criticism from the late 1940s until fairly recently) and phenomenology, that one can study a poem (or other work) in isolation. Structuralism sees each aspect of a text as deriving its significance from various intra- and intertextual relationships. The phenomenological assumption that a work of literature can be "bracketed", or temporarily isolated from all presuppositions and categories that might define or restrict one's perception of it, goes against a fundamental tenet of structuralism, that nothing has meaning in or by itself.

Structuralism is not primarily concerned with the interpretation of individual texts but with the "codes" or conventions of literature that enable one to make sense of and enjoy individual works. To take a simple example of a code: most of Dickens' novels involve mysteries of identity, in which the resolution of the hero's problems depends in part on learning who his rightful parents are. As the plot unfolds, clues are offered which enable the astute reader to guess the parents' identity before it is explicitly revealed. The structuralist is not concerned with the mystery or with its resolution, but with those elements of the narrative—the codes—that enable him to read the clues *as* clues and not simply as miscellaneous narrative details. Such codes are not consciously written into the work, they are aspects of narrative that have evolved as the novel has evolved as a genre, the particular codes of a given novelist evolving as he writes. Learning to recognize and interpret these codes requires reading a number of novels *and* becoming aware (if only tacitly) of their existence and function. Learning them, like learning other codes, can presumably be taught (though they are often difficult to analyze); and this makes structuralism a potentially effective tool in teaching literature.

Structuralism is an interdisciplinary critical method, since it not only derived some of its principles from other disciplines but is part of a (loosely coordinated) movement aimed at working out epistemology and methods applicable to all of the "human sciences", epistemology and methods similar to those of the natural sciences. Structuralism is therefore a critical movement of some interest to philosophers of science. Of particular interest, however, are some of the problems that structuralism has engendered, and that I see as implicit in some philosophies of science as well.

The first grows out of the assumption that literature comprises a self-contained universe having no necessary connection with the world of everyday experience—an idea that is hardly new but that structuralists tend to push to its logical extreme. Frye does this, too. In fact, in the "Tentative Conclusion" to *Anatomy of Criticism* he advances one of the more sophisticated arguments for it that I know of, using an analogy between literature and mathematics. "[J]ust as in mathematics we have to go from three apples to three, and from a square field to a square," he suggests, "so in reading a novel we have to go from literature as reflection of life to literature as an autonomous language. Literature also proceeds by hypothetical possibilities, and though literature, like mathematics, is constantly useful—a word which means having a continuing relationship to the common field of experience—*pure* literature, like pure mathematics, contains its own meaning." (Frye 1957, p. 351; italics added). Two paragraphs later Frye takes a neo-Cartesian view of pure mathematics and (by implication) pure literature: "pure mathematics exists in a mathematical universe which is no longer a commentary on an outside world, but *contains that world within itself*," and it regards the content of the "objective world...as being itself mathematical in form." (1957, p. 352; italics added).

Structuralists take the position that a text's meaning does not and cannot lie in its relationship to the external world but only in the internal relationships and relationships to other texts I mentioned earlier. "A book of a certain kind will always appear as a window through which [the outside] world is clearly visible," Terrence Hawkes admits. But the "transparency" of the window is just "an illusion" (Hawkes 1977, p. 143). According to Fredric Jameson, a critic skeptical of certain tenets of structuralism, what takes place when a work seems to provide a view of the external world is actually "a kind of peculiar structural deflection of that impulse which, on its way towards the real and towards some genuine referent, strikes a mirror instead without knowing it." (Jameson 1972, p. 205). Jameson also describes a form of radical self-reference in which, according to Todorov, "Every work, every novel, tells through its fabric of events the story of its own creation, its own history," and in which "the meaning of a work lies in its telling itself, its speaking of its own existence." (Todorov 1967, p. 49; quoted in Jameson 1972, p. 200). Jameson comments: "We may thus say that the essential content of a Simenon novel is the act of writing novels, but that this content is itself concealed by the detective story form... ." (1972, p. 205). In the ultimate variety of self-reference implicit in the work of Jacques Derrida, the external world is not simply lost sight of or "contained in" a work of literature, it ceases to exist altogether. Or so Michael Wood says of Derrida (and I believe Wood is right): "What Derrida is suggesting is not that language (or literature) doesn't refer to reality, but that reality is so unstable a category that nothing can simply 'refer' to it; that reality is a fabric of references, a web of signs which point to *each other* and *not* to a God or to a 'real' reality behind the appearances." (Wood 1977, p. 29; italics added). What we—perhaps carelessly—call 'reality' becomes, by this reasoning, merely the illusion created by networks of signs.

My objection to such extremes (which I can only sketch in here and which is based on a belief in realism) is that all literature grows out of experience, real and imagined, direct and vicarious (that is, through other literature), and evokes comparison with the experience of the reader, again direct and vicarious. At the same time, works of literature constitute a literary universe, separate to some extent from the real world and answerable to its own (formal) laws. I will try to make this clearer and perhaps more convincing by invoking Henry Margenau's distinction between "formal" and "epistemic" connections among the concepts (or "constructs") of sub-atomic physics: electron, energy, mass, and so on. In *The Nature of Physical Reality* Margenau describes both the *formal* relationships among the constructs themselves, and the *epistemic* relationships between some (not all) of the constructs and the world of experience. Both relationships are necessary to physics; neither is imaginary. (See Margenau 1950, especially chapter 5.) Just so with the works comprising the literary universe, I would argue.

Mathematics, too, is sometimes regarded as a purely self-referring universe, as Frye indicates in his analogy between pure literature and pure mathematics, and as mathematicians, physicists, and philosophers of science are well aware.[1] There would seem to be little reason to regard physics as comprising such a universe, so obviously does it refer to the external world. Nevertheless, the conventionalism of Duhem and Poincaré, as well as later variations on their philosophies, implies something like hermeticism in insisting that there is no necessary connection between the laws of physics and the natural world. (See Duhem 1914 and Poincaré 1901.) I can make this point clearer by turning to another problem that structuralism has brought about.

The interpretation and judgment of a work of literature have never been empirically testable (and hence verifiable or corroborable) in the same way that a physical theory is—or until recently was regarded as being. The problematic nature of interpretation and judgment has been especially apparent since the poets of the romantic movement substituted fidelity to the poet's state of mind or emotions for fidelity to classical models and canons of judgment, and even the exclusive attention that the new critics paid to the text itself (at the expense of all "extrinsic" considerations) could not resolve such issues. Nor is it likely or even desirable that any interpretation or judgment, based on any method, will be free from controversy. Structuralists are well aware of this, and some have attempted to incorporate indeterminacy into their theory of interpretation. A useful way to go about interpreting a text, they argue, is to construct a *model* comprising the characteristics one experiences when reading it, then to test the resulting interpretation against the judgment of other readers. "Though there is no automatic procedure for determining what is acceptable," Jonathan Culler says in describing this process, "that does not matter, for one's proposals will be sufficiently tested by [other] readers' acceptance or rejection of them... . The meaning of a poem within the institution of literature is not...the immediate and spontaneous reaction of individual readers but the meanings which they are willing to accept as both plausible and justifiable when they are explained." (Culler 1975, p. 124[2]). This seems to me good empiricism, especially since Culler avoids the dangers of strict empiricism, warning (later in the book) that one must not "take too seriously the actual and doubtless idiosyncratic performance of individual readers" (Culler 1975, p. 258); yet there are dangers in it. One danger, which comes from the way many critical theorists understand the use of models in scientific explanation, is that too little attention may be paid, in theory and in practice, to feedback. The model becomes an end in itself and is taken as unalterable, on the assumption that the whole process is arbitrary.

This stress on arbitrariness has a definite parallel in physics *and has been influenced by* philosophy of science, specifically conventionalism and Kuhn's model of scientific progress. This is a complex issue, and I cannot do justice to it here, but the idea behind it is that nature can be said to be "lawful", not because the laws we think we "discover" are inherent in it, but simply because we can formulate laws that appear for a while to describe it—or that, more precisely, the relevant scientific community temporarily accepts as adequately describing it. In this way physics can be thought of as a self-contained discipline. Similarly, a text has meaning, not because that meaning is inherent in it, but because a critic can formulate an apparent meaning, which the relevant community (competent readers, other critics) for a time will accept as valid. "Man is not just *homo sapiens*," Culler remarks, "but *homo significans:* a creature who gives sense to things." (Culler 1975, p. 264). In the absence of enduring criteria of validity, both philosophy of science and critical theory are susceptible to the threat of anarchism, which some theorists—Feyerabend in philosophy of science, Derrida in critical theory, for example—have turned into more than just a threat. (See Feyerabend 1975, Derrida 1967.)

The third problem that structuralism has engendered is the opposite of the last one, yet closely related to it, as Charybdis is to Scylla: the diminution of individuality. The loss of individuality is a theme in much modern fiction, but only recently has it emerged as a critical principle.[3] It is founded on the (valid) premise that reading literature in such a way as to make sense of it *as* literature is a process governed by implicit rules, which one learns to follow by becoming familiar with the codes and conventions I mentioned previously. This "tacit knowledge" (to use Michael Polanyi's appropriate term [see Polanyi 1958]) is acquired by reading and discussing a good many works of literature. Thus the greater one's competence in reading literature, the more one reads, on one level, in the same way that other competent readers read. The reading "subject," Culler says, "is an abstract and interpersonal construct," "constituted by a series of conventions," so that "The empirical 'I' is dispersed among these conventions which take over from him in the act of reading." (1975, p. 258). This level of impersonal activity can be seen as simply one aspect of the development of the human sciences, which (again according to Culler) "begin by making man an object of knowledge, [but] find, as their work advances, that 'man' disappears under structural analysis." Culler then quotes Lévi-Strauss's remark that "The goal of the human sciences is not to constitute man but to dissolve him" (Lévi-Strauss 1962, p. 326; quoted in Culler 1975, p. 28), and then follows with Michel Foucault's belief that the concept "man" is only two centuries old and "will disappear as soon as [our] knowledge has found a new form." (Foucault 1966, p. 15; quoted in Culler 1975, p. 28).

I think it will be some time before 'man' disappears as a concept (provided he survives as a species); I also think that even though reading is governed by rules and should be cultivated as a process that is to some extent interpersonal, it can never become completely interpersonal, or even as interpersonal as some structuralists imply. In showing us how skilled readers read a text, structuralists have contributed immensely to our understanding of both the reading process and the sophisticated ways in which texts are structured, or seem to structure themselves. But in doing this they have concentrated unduly on one aspect of reading at the expense of others, and at the same time invited excessive opposite reactions from critics such as Derrida, who sees interpretation as an infinite progression from one arbitrary reading to another, more arbitrary reading.

And philosophy of science? If Kuhn's account of the way science is done is correct (and I believe that in many respects it is), then scientists are constrained to some degree by their education, by the paradigm they operate within, and by peer-group pressures, particularly from referees and editors of professional journals. Moreover, these constraints are augmented, perhaps even brought into existence (some of them at least), when a scientific community accepts Kuhn's description of its behavior as accurate. The effect of these constraints is (in Kuhn's words) that "the members of a scientific community see themselves and are seen by others as the men uniquely responsible for the pursuit of a set of shared goals, including the training of their successors. Within such groups communication is relatively full and professional judgment relatively unanimous." (Kuhn 1970, p. 177). Such closed communities may be necessary in the sciences, given the rate at which established sciences advance—or seem to advance. Nevertheless, when Kuhn applies similar criteria to disciplines outside science, he seems to me to circumscribe the activities of these disciplines too narrowly, to demand too severe allegiance to their reigning paradigms. "What does the group collectively see as its goals," he asks; "what *deviations*, individual or collective, will it *tolerate*; and how does it *control* the *impermissible aberration*?" (1970, p. 209; italics added). It may be simply that I am reading too much into Kuhn's words, but I think not.

I have presented several problems which philosophy of science and critical theory share, though sometimes in different forms. I have said little about how to resolve these problems, offering instead brief comments on their consequences. I believe they *can* eventually be resolved, and have given some thought to ways of resolving them. But the point I want to make now is that an interchange of ideas, an opening up of channels of communication between the two disciplines, will quicken the search for solutions and enhance critical discussion of proposed solutions. Such discussions are not likely to promote harmony between philosophers of science and critical theorists, but tension instead—*creative* tension that helps keep each discipline from thinking of itself as answerable only to its own laws.

Notes

[1]Since at least the time of Descartes, mathematicians, physicists, and philosophers of science have tried to explain the relationship between mathematics and the physical world. Several worthwhile commentaries on this question are Poincaré (1901), Einstein (1921), Hardy (1940, especially sections 16 and 20 through 28), Wigner (1960), Bochner (1965), Kline (1972, especially chapter 36), Putnam (1975), Ulam (1976, chapter 15), and Dyson (1978).

[2]Fredric Jameson has argued that a model or paradigm borrowed from another discipline often governs an entire critical movement; that, for example, organicism provided the basic model for the romantic movement and linguistics the model for structuralism (see Jameson 1972, pp. vi-ix; also see my comment on this idea [in Creed 1980]).

[3]In 1925 José Ortega y Gasset published an essay titled *The Dehumanization of Art* (see Ortega 1925), which might be said to have started this movement in criticism. However, the principle I discuss here is quite different, in that it involves the depersonalization of the reader rather than of the writer (though structuralism holds that the writer, too, is partially constrained by rules).

References

Bochner, Salomon. (1965). "Why Mathematics Grows." Journal of the History of Ideas 26: 3-24.

Creed, Walter. (1980). "Is Einstein's Work Relevant to the Study of Literature?" In After Einstein: Proceedings of the Einstein Centenary Conference held at Memphis State University, March 14-16, 1979. Edited by P. Barker and C.G. Shugart. Memphis: Memphis State University Press. In Press.

Culler, Jonathan. (1975). Structuralist Poetics: Structuralism, Linguistics and the Study of Literature. London: Routledge & Kegan Paul.

Derrida, Jacques. (1967). L'Ecriture et la difference. Paris: Editions du Seuil. (Translated by Alan Bass as Writing and Difference. Chicago: University of Chicago Press, 1978.)

Duhem, Pierre. (1914). La theorie physique: son objet et sa structure. Paris: H. Riviere. (Translated by P.P. Wiener as The Aim and Structure of Physical Theory. Princeton: Princeton University Press, 1954.)

Dyson, Freeman. (1978). (Review of Benoit Mandelbrot, Fractals: Form, Chance, and Dimension.) Science 200: 678.

Einstein, Albert (1921). "Geometrie und Erfahrung." Preussische Akademie der Wissenschaften, Sitzungsberichte, 1921, pt. I. Pages 123-130. (As reprinted in Ideas and Opinions by Albert Einstein. New York: Crown Publishers, 1954. Pages 232-246.)

Feyerabend, Paul. (1975). Against Method: Outline of an Anarchistic Theory of Knowledge. Atlantic Highlands, New Jersey: Humanities Press.

Foucault, Michel. (1966). Les Mots et les choses. Paris: Gallimard. (Translated as The Order of Things. New York: Pantheon Books, 1970.)

Frye, Northrop. (1957). Anatomy of Criticism: Four Essays. Princeton: Princeton University Press.

Hardy, G.H. (1940). A Mathematician's Apology. Cambridge: Cambridge University Press.

Hawkes, Terrence. (1977). Structuralism and Semiotics. Berkeley: University of California Press.

Jameson, Fredric. (1972). The Prison-House of Language: A Critical Account of Structuralism and Russian Formalism. Princeton: Princeton University Press.

Kline, Morris. (1972). Mathematical Thought from Ancient to Modern Times. New York: Oxford University Press.

Kuhn, Thomas. (1970). The Structure of Scientific Revolutions. (International Encyclopedia of Unified Sciences, Vol. II, number 2.) Second Edition. Chicago: University of Chicago Press.

Levi-Strauss, Claude. (1962). La Pensee sauvage. Paris: Plon. (Translated as The Savage Mind. Chicago: University of Chicago Press, 1966.)

Margenau, Henry. (1950). The Nature of Physical Reality. New York: McGraw-Hill.

Ortega y Gasset, Jose. (1925). La Deshumanization del arte e Ideas sobre la Novela. Madrid: Revista de Occidente. Translated by Helene Wayl as The Dehumanization of Art and Notes on the Novel. Princeton: Princeton University Press, 1948.

Poincare, Henri. (1901). La Science et l'hypothese. Paris: Flammarion. (Translated by G.B. Halsted as Science and Hypothesis. In The Foundations of Science. New York: The Science Press,1913.)

Polanyi, Michael. (1958). Personal Knowledge: Towards a Post-Critical Philosophy. Chicago: University of Chicago Press.

Putnam, Hilary. (1975). "What is Mathematical Truth?" In Mathematics, Matter and Method. (Philosophical Papers, Vol. I.) Cambridge: Cambridge University Press. Pages 60-78.

Scholes, Robert. (1974). Structuralism in Literature: An Introduction. New Haven: Yale University Press.

Todorov, Tzvetan. (1967). Litterature et signification. Paris: Larousse.

-----------------. (1970). Introduction a la litterature fantastique. Paris: Editions du Seuil. (Translated by Richard Howard as The Fantastic: A Structural Approach to a Literary Genre. Ithaca: Cornell University Press, 1975.)

Ulam, S.M. (1976). Adventures of a Mathematician. New York: Charles Scribner's Sons.

Wigner, Eugene. (1960). "The Unreasonable Effectiveness of Mathematics in the Natural Sciences." In Communications on Pure and Applied Mathematics 13: 1-14. (As reprinted in Symmetries and Reflections: Scientific Essays of Eugene P. Wigner. Bloomington: Indiana University Press, 1967. Pages 222-237.)

Wood, Michael. (1977). "Deconstructing Derrida." New York Review of Books 24: 27-30.

Progress in Literary Study

Edward Davenport

John Jay College, The City University of New York

Literary study, along with other social sciences, has been thought by many to be incapable of progress because it, like the other social sciences, is culturally defined. The aims and standards of literary study always depend on the value assumptions and preconceptions of the culture which fosters literary study--for literary study is a social institution. Because literary study and other social sciences are social institutions, obvious difficulties arise with transplanting the aims and standards of social science in one society into another society. The idea of progress, however, appears to demand that we have aims and standards which apply independently of particular social contexts, and independently of the boundaries of a culture or an age. Thus the cultural definition of literary study has sometimes been thought to prevent progress in this field.

Max Weber proposed a strategy for working toward objective knowledge and progress in the social sciences generally, which can be applied to literary study. The Weberian strategy attempts to transcend particular cultural definitions of aims and standards of the social sciences by making explicit the value assumptions and preconceptions behind each different cultural definition of social science.

Applying the Weberian strategy to literary study enables one to see different histories, theories and criticisms of literature as representing different and competing value assumptions, and different standards for progress. When these competing standards and assumptions behind literary schools of thought are made explicit, one has already a more objective knowledge of literary study in the sense that one sees a possible order where before all seemed chaos. It gives one a more objective view of literary study to be able to see, instead of a chaos of inconsistent and apparently pointless work, a more orderly possibility of competing schools of thought, each of which may be viewed as an experimental research program into literature.

PSA 1980, Volume 1, pp. 141-148

Each of these different research programs has its own preconceptions, its own value assumptions, and its own aims and standards for literary study. The competition between the various research programs, the various schools of thought, can only be judged by a more universal set of standards for literary study, and this more universal set of standards is in turn a reflection of value assumptions and preconceptions which can be made explicit.

There may be a problem, however, with calling such increased articulation of our values and ideas "progress". The problem is that the Weberian strategy has often been criticized as involving us in an infinite regress if we seek knowledge which is true regardless of context, and as being unnecessary if we are satisfied with truth which is relative to a context, since our choice of a context must be arbitary.

The idea that conventionally chosen contexts for knowledge must be arbitrary, so that there can be no progress in culturally defined sciences, was the idea of cultural relativism. Taken to an extreme the idea of cultural relativity means that all cultures and all cultural activities are equal--or at least that any statement to the contrary is merely an expression of cultural bias--an arbitrary choice of an interpretive context.

Weber did not accept this extreme interpretation, even though he agreed that all cultural activities could be understood only with respect to some context. Weber did not agree with relativists that the absence of a single universally true context left only the option of some arbitrary context for interpreting culture. He thought scholars could discuss, compare, and perhaps improve upon the conventionally chosen contexts for knowledge in each field. Completely objective knowledge, at least in the social sciences, remained elusive, but Weber's view was that the attempt to make progress in social science--the attempt to create a growing body of increasingly objective knowledge--was not a vain attempt, although a difficult one.

Weber's strategy has been attacked by cultural relativists for holding to the concept of objective knowledge, even as an ideal, and attacked by positivists for being too skeptical of objective knowledge. His position remains the best one for working toward objective knowledge and progress in the social sciences, because he does not slight either the problem of cultural bias or the possibility of objective evidence. I turn now to the area of literary study in particular, and I come shortly to the application of the Weberian strategy to this field.

Can there be progress in literary study? Literary study is a combination of history of literature, theory of literature, and criticism of literature, so my question can be divided into three parts:

(1) Can literary history be seen to progress? Is there growth of historical knowledge about literature apart from mere discovery of documents and isolated facts?

(2) Can theory of literature progress? Can some literary theories, beginning with those of Plato and Aristotle, be refuted in whole or part, and replaced or improved by better theories?
(3) Can literary criticism progress? Are the interpretations and evaluations and analyses of literature capable of being compared and evaluated in their turn, in such a way that we can say we have a growing body of knowledge rather than an accumulation of mutually irrelevant opinion?

My answer to all three questions is "yes." We can make progress in literary study with the help of the Weberian strategy. I will explain how, but first, since the Weberian strategy is a method of solving problems raised by relativism, I will give the relativist view of literary study, and the problems to which this draws attention.

(1r) For the relativist there can be no progress in literary history because we cannot reject one history in favor of another. At least we cannot do so except in cases of flagrant misuse of documentary evidence. In most cases we are not able to choose among a large body of competing histories of literature, which differ not as to documentary evidence but as to the interpretation of that evidence. Admittedly such choices can be and are made by literary historians--but the relativist questions the basis on which such choices are made. Such choices must always be made with respect to some conceptual framework, some context of historical knowledge, which is conventional, and therefore, says the relativist, arbitrary.

(2r) Similarly there can be no progress in literary theory, because we have only a chaos of mutually exclusive or mutually irrelevant theories of literature, which are based on different values and preconceptions, and therefore incommensurable with one another. Relativists readily admit that some debate between theories does take place, and sometimes traditional theories, such as those of Plato and Aristotle, are considered refuted in some point by most scholars. But such consensus does not prove truth, and such a change of official views may not, the relativist insists, be progress.

(3r) Finally, in the relativist view there is no progress in literary criticism. It is impossible for unjustly neglected authors to receive a more just and higher regard from later generations. From the extreme position of cultural relativism, all authors from Shakespeare to the freshman composition student write equally well, or at least any statement to the contrary must be based on a conventionally chosen context, and must therefore be arbitrary. There can be no progress in literary criticism because there can be no objective evaluation of an author's merits: there can be no unjustly neglected authors in any case.

Relativism raises interesting questions about how much we can know about literature, but relativism has no positive program for literary study. It reduces literary study to an arbitrary cult. For this reason those who want to improve their knowledge of literature seek answers to the relativist questions. The Weberian strategy supplies one such answer.

By the Weberian strategy the context in which we must make all the choices which would make progress possible is a conventionally chosen context, and it is not natural or universal. Nevertheless this strategy dissents from the conclusion that all such choices based on convention must be arbitrary. Rather a conventionally chosen context of knowledge may be more or less objective than other alternative contexts, for reasons which we can perhaps articulate if we try. Applying the Weberian strategy to literary study allows the social scientist to conjecture that there may sometimes be real differences in literary merit between authors, despite fads and fashions in literary taste, and even though the particular standards used to measure those differences must be culturally influenced, and must be viewed as relative to some particular research program into literature.

Applying the Weberian strategy to each of the branches of literary study may enable the social scientist to make progress in each. That the members of competing research programs can explain their reasons for thinking they are making progress, and can debate this point with members of other research programs, is certainly no guarantee of progress. Indeed the phrase "academic debate" may well bring to mind futile scholastic controversies of the past which have seemed especially unprogressive. The Weberian claim for progress in literary study is therefore a claim that progress may occur--not that it must occur. This is a skeptical dissent from the dogmatic know-nothingism of the relativist, but it is not an independent promise of a great leap forward in literary criticism. That result is yet to be tested.

According to the Weberian strategy, new historical approaches can be seen as either being in certain traditions of historical writing, or as changing certain traditions: progress in various styles of history may go on simultaneously, according to the traditions of those doing the writing. Moreover, the students of historical method can attempt to compare various methods of doing literary history, according to some goals which all or many of these methods have in common. In this way we can test and try various methods within larger and smaller contexts, in an experimental fashion, to learn what we can about the possibilities of each.

We can also divide literary theories into different traditions, and say that progress in literary theory is being made within the context of a given tradition or group of traditions with respect to common goals. Such progress in literary theory depends on the possibility of refuting aspects of "classical" literary theory, and this may be a sticking point for some in literary study. Scholars who would laugh at the idea of Aristotle's ideas of physics being taken seriously today are reluctant to admit that Aristotle's ideas about literature might be equally wrong in one or two points.

I am not concerned to refute such classicism, which is but another symptom of the widespread feeling that no progress can be made in literary study. More important for my interest here is the problem of how to evaluate the many competitors for succession to the classicists' Plato and Aristotle as solvers of the problems of literary theory. There are many who claim to have made progress in literary theory, including Marxists, structuralists, phenomenologists, new critics, neo-classic critics, and Freudian critics, to name but a handful. What methods can we use to choose between the claims of one school and the claims of another?

Empirical evidence, contrary to what one might expect, is often available and relevant to literary disputes. Claims about textual meaning, authorial intent, or ideological context can often be refuted by clear empirical evidence. But such evidence, though available and relevant, is usually overshadowed in significance by other kinds of evidence in debates over literary theory.

In evaluating Plato's theory of literature, for example, one might go to the text of Homer's Iliad to try to refute Plato's claim that Homer taught men to be cowards about death--but textual evidence would not be the most significant matter here. Rather it is more important to examine Plato's interpretation of the evidence in book III of The Republic. We must ask: A) What is Plato's reason for claiming that Homer teaches anything at all? B) What is Plato's reason for assuming that what Homer teaches--if he teaches--is wrong? and C) What is Plato's reason for assuming that the state censor should forbid wrong works, and deport wrongheaded poets, like Homer? These are the crucial questions about Plato's theory of literature, and they cannot be answered by empirical evidence alone. Therefore, it does not help much to point out that Plato also distorts Homer's text, and that there is more documentary evidence that Homer teaches courage, if he teaches anything, than that he teaches cowardice. This point may suggest that Plato is unscrupulous, or at least careless, but it is a minor point when considering the claims of Plato's theory of literature. How then, if not empirically, can Plato's theory be criticized?

Here again we can follow the Weberian strategy. We can make the value assumptions and preconceptions underlying Plato's theory of literature more explicit, and in this way discover that Plato's interpretation of Homer is based not only on a theory of literature which we may disagree with in part or whole, but is based on a set of standards for what makes a good theory of literature, which we may share only in part, if at all.

In the field of literary theory we find that there is competition not just between different theories, but between standards for theories. Indeed individual literary schools are usually characterized not by a particular theory but by a set of standards for theories: this allows the school the flexibility to develop or even revolutionize their theories in the light of what their experimental researches reveal. There were once a number of different romantic theories of literature, just as today there are a number of different Marxist theories. The different schools or research programs into literature are competing, each with its own standards for a good or improved theory of literature, and consequently each with its own theory of how to make progress in literary study. We are still left with the problem of choosing among them, and indeed the problem of deciding whether any of them be correct, and we are forced to consult our own cultural bias, or conventional context, in order to make these decisions. Nevertheless the Weberian strategy enables us to examine competing claims about literary theory with a critical eye and to understand with reference to what values each claim is made.

As with history and theory of literature, so with literary criticism, we have competing answers to problms, and competing standards for progress in solving problems. Applying the Weberian strategy, and making the value assumptions of different schools of criticism explicit, we can see with reference to what basic values and what metaphysical outlook a particular piece of literary criticism is formulated.

Frederick Crews' The Pooh Perplex (1963), a well-known spoof on doctrinaire literary criticism, paradoxically also shows the strength of such criticism, which openly acknowledges its metaphysical preconceptions and its problem-orientation. Crews spoofs doctrinaire critics by showing the absurd lengths to which their metaphysical outlooks may lead them in interpreting a text, but in doing so he provided an early and exciting example of critical comparison of literary schools. He makes this possible by caricaturing and thus making even more explicit the metaphysical outlook or value assumptions of each school. Crews provides a merciless spoof, which ought to delight the heart of any classicist, since it makes such telling points against modern criticism, yet at the same time he shatters forever the idea that literary criticism can count on all critics reading the same thing when they read the same book.

This shattered assumption of the reliability of the text was a basic assumption of the classicists, and its destruction was initially the work of cultural relativism. This discovery, that what one reads depends on the context (Freudian, Marxist, classicist) in which one interprets what one reads, is also a fundamental assumption of the Weberian strategy, with the added qualification that the choice of context need not be arbitrary, that readers can learn new contexts, learn to switch between contexts, and simultaneously contemplate alternative readings based on two or more contexts. Thus, context may be seen as a necessary structure, but not a prison house. Crews also seems persuaded that intelligent readers ought to be able to appreciate more than one context.

A second major aim of Crews' spoof, evident throughout the book, is to draw attention to the distinction between important and trivial literary problems. Because Crews writes satirically, his distinction is not a sharp or clear one, but nevertheless the need for the distinction is made clear. In this way too The Pooh Perplex was a harbinger of a new interest in progress in literary study.

The Weberian strategy makes progress possible in the social sciences by means of focusing attention on one problem: how can we improve our social scientific institutions, such as literary study, conventional though they be? The Weberian strategy makes possible answers to this problem, though always conjectural answers--never justified answers. Thus we may view progress in social science as a matter of doing without the certainty of dogmatic truth and also doing without the certainty of dogmatic relativism: instead we seek to improve and thus go beyond the conventional or conjectural framework of our current knowledge.

References

Agassi, Joseph. (1975). Science in Flux. (Boston Studies in the Philosophy of Science, Vol. XXVIII.) Dordrecht: D. Reidel.

--------------. (1977). Towards a Rational Philosophical Anthropology. The Hague: Martinus Nijhoff.

Davenport, Edward. (1978). "Why Theorize About Literature?" In What is Literature? Edited by Paul Hernadi. Bloomington: Indiana University Press. Pages 35-46.

Crews, Frederick C. (1963). The Pooh Perplex. New York: E.P. Dutton and Company.

Plato. The Republic. (As printed in Great Dialogues of Plato. Translated by W.H.D. Rouse. Edited by Eric H. Warmington and Philip G. Rouse. New York: New American Library, 1956.)

Popper, Sir Karl. (1962). Conjectures and Refutations. New York: Basic Books.

----------------. (1972). Objective Knowledge. Oxford: Oxford University Press.

Weber, Max. (1949). The Methodology of the Social Sciences. Translated by E.A. Shils and H.A. Finch. New York: The Free Press. (Translation and collection of three articles: "Der Sinn der 'Wertfreiheit' der soziologischen und okonomischen Wissenschaft." Logos 7(1917): 40-88; "Die 'Objectivitat' sozialwissenschaftlicher und sozialpolitischer Erkenntnis." Archiv fur Socialwissenschaft und Sozialpolitik 19 (1904): 22-87; "Die protestantische Ethik und der 'Geist' des Kapitalismus." Archiv fur Socialwissenschaft und Sozialpolitik 20 (1905): 1-54.)

Part VI

Reduction in Biology and Psychology

The 'Reduction by Synthesis' of Biology to Physical Chemistry [1]

Monique H. Levy

1. The 'Classical' Conception of Reduction and Its Problems

The problem of the reduction of biology to physical chemistry has raised numerous controversies for several years. The reader is referred to Ayala & Dobzhansky (1974), Hull (1972, 1973, 1974, 1976), Polanyi (1967, 1968), Roll-Hansen (1969, 1974), Ruse (1971, 1976), and Schaffner (1967, 1969, 1974, 1976). Supporters and opponents of reduction are continually exchanging arguments and it is almost impossible, even for an objective reader, to discern who is correct.

The partisans of reduction insist that physical chemistry contributes a very useful explanatory schema for the understanding of a certain number of biological processes. For instance, in genetics, great progress has been realized because of the contribution made by chemistry in the determination of the genetic code and in the understanding of the molecular mechanisms governing the laws of heredity. Likewise, the processes of digestion and respiration, the study of hormones, the synthesis of proteins, the enzymatic catalysis, the importance of some ions in the cellular metabolism, are all domains where application of chemistry is made to living matter.

The opponents of reduction insist on the deficiency of the proposed explanations. It is claimed that many areas of biology are still not touched by physical chemistry and reductive explanations are relatively inadequate when they do exist. This relative inadequacy is the consequence of limits inherent in the use of models. The reduction of biology is most frequently realized by the isolation of a biological object and the construction of a model connecting this problematical biological object to the physiochemical formalism most likely to explain it.

These models constitute the crux of the controversy between reductionists and antireductionists. True, these models are

---

PSA 1980, Volume 1, pp. 151-159

explanatory, but they are also, as is every model, only approximations of the perceived reality; so their usefulness is only within certain limits. Consequently, depending on whether emphasis is put on the possibilities of the models or on the limitations of the models, one falls heir to the position of being a supporter or an opponent of reduction.

As the future relations between biology and physical chemistry will probably consist of the construction of a larger and larger number of models, some authors will have even more motives to insist on the value of the reduction, whereas others, upholding the approximative character of these models and their limited power with respect to the biological phenomena, will have good reason to protest against any affirmation of reduction. So stated, the problem of the reduction of biology to physical chemistry seems to be incapable of resolution.

Why is this? Simply because the problem rests on a conception of reduction (which could be called 'classical'), which is not adequate. This conception repeats in its broadest outline Nagel's approach (1961). This conception defines reduction as an asymmetric relation, joining two historical theories or sciences with a bond of a deductive nature (in a strong or weak sense). This relation, which is both explanatory and predictive due to its deductive nature, systematizes a domain by integrating partially or completely the reduced theory or science into the frame of the reducing theory or science. Though reductions of this sort principally show the continuity of two theories or sciences, they simultaneously introduce a new point of view.

This 'classical' conception of reduction is the one which at present underlies the discussions about the relations between biology and physical chemistry; for instance in the attempt made to deduce, according to Nagel's criteria, Mendel's genetics from molecular genetics. But this approach, which undeniably describes some aspects of reduction, is not satisfactory because it does not reflect scientific practice.

2. Reduction by Synthesis

The inadequacy of the 'classical' concept with regard to the actual evolution of science can easily be shown by studying a case of reduction considered as ideal and non-problematical: the relations between chemistry and physics. This study, which has been carried out in detail elsewhere (Levy 1979), leads to the introduction of a new concept of reduction, which repeats some insights of the 'classical' concept, but which also gives a new interpretation to the concept of reduction. This new approach, called 'reduction by synthesis', conforms to scientific practice since it describes the effective relations between the two sciences, one of which is universally recognized as reduced to the other.

'Reduction by synthesis' is an asymmetrical relation: this means there is an indisputable, theoretical priority of the reducing discipline which offers the reduced discipline its formalism, its conceptual frame, its principles, its explanatory and predictive power, and its power of systematization. But this asymmetrical relation is not of a deductive nature. On the contrary, it assigns specific theoretical roles to each discipline, and this is the reason why we speak of synthesis.

The reduced discipline functions as follows:

(1) It tests the value of the explanations and predictions offered by the reducing discipline (this is a classical aspect).

(2) It interprets and adapts to the formalism of the reducing discipline so that this formalism is really efficacious in describing the phenomena which the reduced theory treats. This adaptation is realized by specially constructed intermediary domains having the purpose of allowing a more elaborate communication and as complete an exchange of information as possible between the two sciences. For instance, in the case of the reduction of chemistry to physics, three theories (quantum chemistry, chemical thermodynamics and statistical mechanics applied to chemistry) assure the transmission of the theoretical and experimental indications from one domain to the other. These theories apply the physical formalism to the chemical problems: study of the molecule and of the chemical bond in quantum chemistry; introduction of thermodynamical functions adapted to the chemical situations (Helmholtz free energy, Gibbs free enthalpy, chemical potential...); introduction of the statistical collision theory in kinetics, etc... . In this way, these three theories of physical chemistry concretize the first theoretical function of the reduced theory.

(3) The reduced theory plays a second theoretical role. It guides the reducing theory when this theory is unable to predict or even to explain a phenomenon of the reduced domain. This guidance is carried out by supplementary ad hoc hypotheses which transmit information arising from the reduced domain and enable the reducing theory to have explanatory power. For instance, thermodynamics predicts the occurrence of many reactions which in fact do not occur spontaneously. This contradiction between physical prediction and experimental fact is explained by the following supplementary hypothesis, originating in experimental chemistry: in order to occur, these problematical reactions need a certain amount of energy, called 'activation energy', which is supplied, for instance, by heat. Expressed in the vocabulary of the reducing theory, these hypotheses do not belong to this theory, but supply indications and constraints which complete the theory and which originate from the reduced domain. These hypotheses formalize the second theoretical part of the reduced domain.

As a consequence of the specific role played by each of the disciplines partaking in the reduction, every 'reduction by synthesis' proceeds not by annexation of a domain in the frame of another, but by interaction, and by reciprocal enrichment. This kind of reduction involves construction of theoretical domains composite and intermediate between the reduced and the reducing theories. These domains are charged on the one hand to transmit the formalism of the reducing discipline and on the other hand to convey the experimental and theoretical information from the reduced discipline. In the case of the reduction of sciences, these domains realize their purpose with the help of models, which can establish communication by simplifying the relevant data of each science. This passage through models is necessary because two sciences having very different approaches and subjects cannot communicate directly.

Finally, 'reduction by synthesis' is based upon a very important presupposition; namely, such a reduction can effectively be enriching and not mutilating for the reduced theory if both domains are, previous to any attempt of reduction, in a certain state of development. The reducing theory must be in a position to give an original theoretical analysis of the main concepts of the reduced theory, and this theory in turn must be suited to receive and to adapt to a complex and foreign formalism. This maturity condition of both disciplines is necessary to ensure a fruitful communication and particularly to preserve the specificity of the domain to be reduced.

3. The Relationships of Biology and Physical Chemistry

'Reduction by synthesis' provides an interesting basis for analyzing the case of the reduction of biology to physical chemistry. From its perspective, the question is not that of forcing biology to enter into the explanatory and conceptual frame of physical chemistry, but to realize a synthesis between the explanatory propositions of the reducing science and the requirements, practical as well as theoretical, of biology.

At the present time, biology already shows one of the basic elements of 'reduction by synthesis': explanatory models which, contained in the intermediary theories which define reduction (biochemistry, molecular biology, molecular genetics...), concretely realize the communication of the two sciences. Having in both a part of formalism of the reducing disciplines and one or some supplementary hypothesis from the reduced discipline aimed at guiding this formalism toward a particular problem, these models demonstrate synthesis between the two domains.

In spite of the existence of these models, the 'reduction by synthesis' of biology to physical chemistry is far from being achieved: the reducing science is not currently in a position to form the basis for an original theoretical analysis of the basic concepts of biology (and this requirement constitutes a fundamental condition without which a reduction by synthesis cannot be achieved.)

As a matter of fact, biology treats essentially not of physico-chemical mechanisms, but of well-defined and organized entities: living systems, which are the loci of these physico-chemical processes. The reducing science, although it gives valuable information about these processes, is not suited to consider either the basic entity of biology: the organism, or the basic concept of biology, organization. Consequently,'reduction by synthesis', has not been achieved because of the insufficient development of the reducing theory, which cannot integrate and formalize adequately the fundamental intuitions of the reduced theory.

This conclusion would seem to admit that, in the above controversy, the antireductionists are right. This is not so, as many arguments lead one to believe that biology will reduce, some day, to physical chemistry (in the sense that is here given to the term 'reduction'). In view of what has just been said, the first requirement imposed by this kind of reduction is the widening of the reducing science; and some facts show that this process is already being effected, and opens the way to a more complete 'reduction by synthesis'. These facts are:

(1) 'Reduction by synthesis', which corresponds to the establishment of close bonds permitting the transmission of the abstract formalism of one discipline to the concrete domain of another discipline, takes place in the actual evolution of the different sciences which develop in the direction of a more and more profound interaction. In particular, chemists are fully conscious of the necessity to develop their science in the direction of biology in order to adapt the basic results of chemistry to this more complex domain.

(2) A second argument, much stronger than the first, consists in establishing an analogy between the present relations of biology to physical chemistry and a past situation typical of the relations between chemistry and physics. As a matter of fact, at the beginning of the 20th century, chemistry was, with respect to physics, in a position similar to the one that biology occupies today with regard to physical chemistry. In the former, the two sciences had numerous points of contact, chemistry applying to its own domain the results provided by physics (for instance by thermodynamics), though the reduction was not yet realized because the most basic concept of chemistry, the concept of atom, had not received a signification in physics. It was only when physics, following the fortuitous event of the discovery and study of radiation, had integrated into its own frame the concept of atom that one was able to speak of reduction. As a matter of fact, this integration of the implications of 'atom' in physics introduced a veritable 'Copernician revolution' and led not only to the new signification of the concept of atom, but also to the construction of a whole theory (quantum mechanics) to account for the structure of matter. And this theory, because of its origin and its domain of study, had explanatory and predictive power with respect to chemistry, which made many authors incorrectly conclude that chemistry reduced to quantum mechanics.

This example of the reduction of chemistry to physics shows how a reducing discipline was obliged to develop considerably before being able to account adequately for the basic intuitions of the domain to be reduced. As the history of science shows that such a mechanism already had taken place, nothing prevents one from assuming that physical chemistry, in turn, could not be altered in order to account for the basic concepts of biology. The consequences, nature, and importance of such an alteration cannot be predicted.

(3) Furthermore, the preceding hypothesis has already received a certain realization. The basic concept of biology has received an initial signification in chemistry as well as in physics.

In chemistry, and particularly in organic chemistry, the notion of organization is indispensable to describe the structure of complex molecules. Thus, numerous compounds can have the same molecular formula (they have the same atoms in the same proportions) and nevertheless they have structural formulae and properties which are very different. These compounds, called isomers, can be divided into two categories: the stereo-isomers, which are diffentiated only by the spatial orientation of the atoms in the molecule, and the structural isomers, which are differentiated by the order in which the atoms are joined together in the molecule. It is evident that the spatial ordering of the atoms plays a considerable part in the kind of reactions and in the reactive characteristics of a molecule, and this is the reason why the concept of structure of an 'organized entity' has a great importance in organic chemistry. This is recognized also by philosophy for, as Whitehead says, the basic chemical entities: the atom, the molecule, the crystal...are organizations.

In physics, studies have recently been made to attempt to formalize the biological notion of organization. These studies are based upon the 'information theory' of Shannon. Very early, biology and cybernetics understood the usefulness of 'information theory' in the study of living beings. According to one supporter of this approach (Atlan 1972) the organization of a living or non-living system is a function of the rate of change of the quantity of information. This quantity of information of a system, defined by the H function of Shannon, is the uncertainty removed by the realization of this particular system among a certain number of possibilities. It is the sum of two functions, the first one being the maximum quantity of information, and the second being the redundancy (or the diminution of the quantity of information by symbol) whose part is to oppose the errors during the transmission of the information message. In the specific case of living systems, two parameters acquire importance: the initial structural redundancy R_o, which determines the self-organizing character of the system and which is responsible for its capacity of adaptation, and the reliability, or probability of occurrence of an error at the output of the system. This reliability characterizes the existence in the self-organized systems of memories with high-fidelity able to make invariable reproductions with very low rates of error.

With the help of this explanatory (and indeed simplified!) schema and of the notion of organization, Atlan is in a position to account not only for the existence and functioning of the self-organized systems, but also for the logic of evolution.

As with other contemporary studies aimed at expressing the organization of the systems on the basis of 'information theory', this approach shows how one of the most fundamental and qualitative concepts of biology is on the verge of receiving an original signification thanks to a physical theory. Although all the problems are far from being resolved, the mechanism of the 'rapprochement' of biology and physics, as it has just been sketched, clearly goes in the direction indicated by 'reduction by synthesis'. As a matter of fact, biology adopts and adapts a physical formalism (the adaptation is realized through the introduction of data specific for living systems); and in turn, physics will have to develop more and more to account fully for the phenomena of the science to be reduced.

The three arguments which have just been discussed all suggest profound interaction between biology and physical chemistry, and show how each science is developing in the direction of the other. Of course, it would be an error to conclude that an accomplished reduction between the two domains currently exists. Much work remains to be done. Some biologists could, for instance, allege that the concept of organization is not the only specific concept of biology, and that it is necessary to account also for the concept of 'life', etc..., and this must be recognized. But the fact remains that although present day biology does not reduce to present day physical chemistry, one must not forget that the two sciences are in continual evolution, and that this evolution, far from being anarchical, points in the direction of a thorough communication (this has just been shown). Consequently, according to the preceding definition of 'reduction by synthesis', it seems that firm and well-grounded reasons indicate that the 'reduction by synthesis' of biology to physical chemistry has already begun, and that it actually goes on in the attempt made by chemistry, physical chemistry and physics to extend and adequately contain the specific concepts of biology.

This new conception of reduction will enable the reductionists and antireductionists to re-examine their positions in the light of a criterion more consistent with scientific practice than the 'classical' criterion. The aim of reduction is no longer to deduce biology from physical chemistry, or to enclose the reduced science in a foreign frame, but to give rise to a development and an enrichment of each science by the other. As development of the sciences is continual,'reduction by synthesis', which rests upon a mutual questioning, will always acquire more importance and will never cease. This type of reduction results in the construction of intermediary disciplines and in that way will play a great part in the progress of science. Biology can now follow the example of chemistry in constructing those intermediary domains characteristic of reduction, without fearing mutilation; for a discipline is enriched when approached by the method of 'reduction by synthesis'.

Notes

[1]I wish to thank Professor Jean Ladriere of the University of Louvain for helpful comments on materials presented in this paper. Thanks also to Sister Fay Trombley for the correction of the English version of this text.

References

Atlan, H. (1972). L'organisation biologique et la theorie de l'information. Paris: Hermann.

Ayala, F.J. and Dobzhansky, T.G. (eds.). (1974). Studies in the Philosophy of Biology: Reduction and Related Problems. Berkeley: University of California Press.

Hull, D. (1972). "Reduction in Genetics: Biology or Philosophy?" Philosophy of Science 39: 491-499.

-------. (1973). "Reduction in Genetics: Doing the Impossible." In Methodology and Philosophy of Biological Sciences. (Logic, Methodology and Philosophy of Science, Vol. 4.). Edited by P. Suppes et al. Amsterdam: North-Holland. Pages 619-635.

-------. (1974). Philosophy of Biological Sciences. Englewood Cliffs: Prenctice Hall.

-------. (1976). "Informal Aspects of Theory Reduction." In PSA 1974. (Boston Studies in the Philosophy of Science, Vol. 32.). Edited by R.S. Cohen et al. Dordrecht: D. Reidel. Pages 653-670.

Levy, M.H. (1979). "Les relations entre chimie et physique et le probleme de la reduction." Epistemologia II, 2: 337-370.

Nagel, Ernest (1961). The Structure of Science. New York: Harcourt, Brace & World, Inc.

Polanyi, M. (1967). "Life Transcending Physics and Chemistry." Chemical and Engineering News 45: 54-66.

----------. (1968). "Life's Irreductible Structure." Science 160: 1308-1312.

Roll-Hansen, N. (1969). "On the Reduction of Biology to Physical Science." Synthese 20: 277-289.

--------------. (1974). "Toward a More Historical Conception of Biology?" Inquiry 17: 131-142.

Ruse, M.E. (1971). "Reduction, Replacement, and Molecular Biology." Dialectica 25: 39-72.

---------. (1976). "Reduction in Genetics." In PSA 1974. (Boston Studies in the Philosophy of Science, Vol. 32.). Edited by R.S. Cohen et al. Dordrecht: D. Reidel. Pages 633-651.

Schaffner, K.F. (1967). "Antireductionism and Molecular Biology." Science 157: 644-647.

--------------. (1969). "The Watson-Crick Model and Reductionism."

British Journal for the Philosophy of Science 20: 325-348.

---------------. (1974). "Unity of Science and Theory Construction in Molecular Biology." In _Philosophical Foundations of Science. (Boston Studies in the Philosophy of Science,_ Vol. 11.). Edited R.J. Seeger and R.S. Cohen. Dordrecht: D. Reidel. Pages 497-533.

---------------. (1976). "Reductionism in Biology: Prospects and Problems." In _PSA 1974. (Boston Studies in the Philosophy of of Science,_ Vol. 32.). Edited by R.S. Cohen _et al._ Dordrecht: D. Reidel. Pages 613-632.

The Formal Structure of Genetics and the Reduction Problem

A. Lindenmayer and N. Simon

Theoretical Biology Group, University of Utrecht
Padualaan 8, Utrecht 3508 TB, The Netherlands

1. Introduction

An often heard remark is that "discovery of the molecular basis of heredity means that all life processes are now or will shortly be explainable on a molecular basis." This extreme view is defended with surprising vehemence by one faction of biologists. The opposing view, just as extremely expressed is that "no interesting biological phenomena have yet been explained by molecular mechanisms, not even genetic phenomena." These two positions have been called the "reductionist" and "anti-reductionist" points of view. Since "reductionism" is given in the philosophy of science many different meanings, we choose the one which in our opinion could shed some light on the problems underlying this controversy. In our view all the emotion-laden epithets aside, "reductionism" has still primarily to do with "theory reduction" in the logical empiricist (or positivist) sense. Clearly this concept is inextricably tied up with that of "explanation", and fundamentally it is explanations that this controversy is about.

Thus we must first of all state our position in regard to theory reduction, before we can say anything about our views on the biology - chemistry - physics relationship. In our opinion the only firm and precisely defined structure in the philosophy of science is that of an "axiomatized theory". This is the only structure which provides us with an exact specification of a (potentially) infinite set of statements by a finite description (namely the axioms of the theory and of logic, and the rules of inference). Thus within the framework of such a structure an inexhaustible variety of theoretical, observational and mixed theoretical - observational statements can be formulated, proven and tested; and at the same time sharply delimited against other sets of statements. No other proposal for the description of a scientific discipline has this power and elegance.

Concepts like paradigms, research programs and fields all suffer

PSA 1980, Volume 1, pp. 160-170

from a vagueness as to their components and as to their boundaries between what they include and exclude. In addition, because of a lack of internal (derivational) structure, the statements (or texts) of these structures cannot be distinctively organized according to their position within themselves (what belongs to the "core" for instance), thus one does not know which omission or addition would change which part of the structure.

Having stated this (and we could motivate this much more extensively), we can now ask what the relationship could be between two axiomatized theories. Clearly, they can be totally different from each other in all their terms and statements; or they can partly overlap each other in the sense that there are some terms which are present in both theories, and perhaps there are also some statements which are derivable in both theories; or one of the theories (T_1) can be shown to be a subtheory of the other (T_2), satisfying the connectability and derivability criteria of Nagel (1961). The last case is that which is usually referred to by saying that T_1 is reduced to T_2.

As long as we are dealing with scientific theories (such as theories of physics) which have been axiomatized, no serious problems arise concerning their relationships. The problems that have exercised philosophers of biology have arisen because no part of biology or of chemistry has been adequately axiomatized. It is thus entirely justified to ask whether the reducibility of biology to chemistry, or of some part of biology to chemistry, is possible within the above sketched framework. We are purposely asking this question with reference to chemistry, rather than to physics, because this is the more immediate question. Furthermore, it is not at all certain whether chemistry is any more reducible to physics than biology is.

This brings us to the following remarks, preliminary to our discussion of the situation in genetics.

There is a misconception rampant among scientists and philosophers about the explanatory power of physical theories. It is often assumed that if a certain phenomenon can in principle be explained on the basis of some physical laws, then it can also be effectively explained by them. This assumption is put forward also in cases in which it can be quite easily shown that the actual computation would take immensely long time to complete. Thus while theoretically the situation may be well understood, there is little hope to eventually obtain a useful prediction in such cases. This remark applies equally to chemical as to biological explanations and predictions. For instance, we know that at the present time quantum - mechanical calculations concerning molecular structures and interactions can be effectively carried out only for small molecules consisting of a few atoms. For reactions of large molecules we must in fact rely on empirical physico - chemical rules which are not derivable by quantum - theoretical means. It is thus so far unjustified to speak of chemistry as a subtheory of physics.

If we now consider the possibility of a physico - chemical description of a single living cell, then the situation is still more hopeless. A cell is composed of several thousand kinds of different molecules, most of which can react with some others, and they are present in varying concentrations in different cell compartments with diffusion and transport processes going on between them. No effective computation exists for a molecular description of a cell, even based on some approximative formulas of reaction kinetics and thermodynamics. Having said this we must add immediately that chemistry and physical chemistry has provided many extremely important contributions to our understanding of cellular metabolism, heredity, cell movements, and other aspects of cell biology. Thus many particular processes of cells can be explained by organic and physical chemistry in its classical form, and occasionally there have even been explanations based on quantum chemistry.

We might characterize our position with respect to the reducibility of biology to chemistry and physics as that of "non-vitalist organicism". This position (its name comes from Haraway 1976) can be very briefly summed up by stating it as the joint denial of the reductionist claim "all biological phenomena can (eventually) be explained by laws of chemistry and physics," and of the vitalist one "there are some biological phenomena which contradict the laws of chemistry or physics." This formulation of the reductionist and vitalist positions agrees more or less with that of Kemeny (1959).

2. The Structure of Classical Genetics

The original Mendelian theory has been variously axiomatized. In an axiomatization following a version given by Woodger (1959), this theory to begin with contains the primitive terms 'gametic fusion', 'production of gametes' and 'development of zygotes into organisms'. Instead of the original symbols used by Woodger, we introduce the following notations:

(1) gametes x and y fuse and form zygote z ———— U (x,y,z)
(2) zygote z develops into organism y ———— z Dev y
(3) organism y produces gamete x ———— y Pr x

With the use of these primitive terms we can easily define the following sets:

(4) The set of organisms which originate from the fusion of a gamete from class α with a gamete from class β :

$$x \in D(\alpha,\beta) \text{ iff } \exists u\, \exists v\, \exists z\, (u \in \alpha \text{ and } v \in \beta \text{ and } U(u,v,z) \text{ and } z \text{ Dev } x)$$

(5) The set of gametes produced by organisms of the sort $D(\alpha,\beta)$:

$$x \in G(\alpha,\beta) \text{ iff } \exists y(y \in D(\alpha,\beta) \text{ and } y \text{ Pr } x)$$

(6) The set of organisms which originate from the fusion of a gamete produced by a $D(\alpha,\beta)$ organism with a gamete produced by a $D(\gamma,\delta)$ organism (in other words the organisms of the second filial generation) :

$z \in Fil_2 (\alpha,\beta;\gamma,\delta)$ iff $\exists x \exists y \exists u \exists v \exists w$

$\big(x \in D(\alpha,\beta)$ and x Pr u and $y \in D(\gamma,\delta)$ and y Pr v and $U(u,v,w)$ and w Dev $z\big)$

In a Mendelian experiment two mutually exclusive gamete classes α and β must be obtainable from pure bred strains of interbreeding organisms.

(7) We introduce the primitive 'α All β' to indicate that these gamete classes are distinguished from each other by the presence of allelic hereditary factors.

If it is the case that α All β then half of the gametes produced by $D(\alpha,\beta)$ organisms should be in class α and half in class β (the law of segregation), the gametes produced by $D(\alpha,\alpha)$ organisms should be all α's, and the gametes of $D(\beta,\beta)$ should be all β's (in other words α and β must be hereditary classes under the operation G). These assumptions are incorporated in the following axiom:

Axiom 1. If α All β then $\alpha \cap \beta = \Lambda$ and $G(\alpha,\beta) \in (\tfrac{1}{2}\alpha \cap \tfrac{1}{2}\beta)$ and

$$G(\alpha,\alpha) \subseteq \alpha \text{ and } G(\beta,\beta) \subseteq \beta$$

The notation we use here for numerical relationships in sets was introduced by Woodger (1952) and is based on the following definition:

$$X \in pY \text{ iff } \frac{N(X \cap Y)}{N(X)} = p$$

where $0 \leq p \leq 1$ and N(X) is the number of elements in set X, and pY is the set of all sets of which the p-th fraction is in Y. Thus '$X \in \tfrac{1}{2}\alpha$' means that half of the set X is in α.

In order to derive the Mendelian ratios in a monohybrid cross, we need to define the following four-term relation:

(8) Gend(α,β,P,Q) iff

$$\alpha \text{ All } \beta \text{ and } P \cap Q = \Lambda \text{ and } D(\alpha,\alpha) \subseteq P \text{ and } D(\beta,\beta) \subseteq Q$$

Here P and Q are sets of organisms distinguishable by observable characters, such that all $D(\alpha,\alpha)$ organisms are in P and all $D(\beta,\beta)$ organisms are in Q (the abbreviation 'Gend' is meant to indicate that here we are dealing with genetic structures of diploid organisms).

The terms 'All' and 'Gend' come in place of Woodger's original notions of 'phen' and 'genunit'. We need two further definitions concerning the randomness of mating between gametes of the classes α and β and the randomness of development of the resulting zygotes. These definitions :

(9) $\{\alpha,\beta\} \in$ rand (Mt)

(10) $\{\alpha,\beta\} \in$ rand (Dt)

are easily given in probabilistic terms.

In the following theorem concerning the 3 : 1 ratio we assume allelic classes α_1 and α_2. The conclusion depends on four premisses, where the second expresses the dominance of character P over Q.

I. Gend (α_1,α_2,P,Q)

II. D $(\alpha_1,\alpha_2) \subseteq P$

III. $\{\alpha_1,\alpha_2\} \in$ rand (Mt)

IV. $\{\alpha_1,\alpha_2\} \in$ rand (Dt)

$\therefore$ $Fil_2(\alpha_1,\alpha_2;\alpha_1,\alpha_2) \in (\frac{3}{4}P \cap \frac{1}{4}Q)$

Thus the 3 : 1 ratio is derivable from the first law of Mendel under some special conditions. By finding this ratio in a test concerning interbreeding organisms of kinds P and Q we support the hypothesis that the genetic factors giving rise to the characters (or classes) P and Q are allelic with respect to each other. In modern terms we would say that the characters which define classes P and Q are determined by a single pair of allelic genes. Other ratios can be similarly derived, for instance the 1 : 1 testcross ratio.

In Mendelian theory it is clearly expected that the allelic relation is transitive and symmetric. We state this as:

Axiom 2. All is transitive and symmetric.

Thus we can form quasi-equivalence classes of gamete classes in the field of the All relation. (These are not equivalence classes because All is irreflexive).

The second law of Mendel (the law of independent assortment) is a limiting case (when $k = \frac{1}{4}$) of the following rule:

Axiom 3. If α_1 All α_2 and β_1 All β_2 and not α_1 All β_1,

then there is a constant k, such that $0 \leq k \leq \frac{1}{4}$ and $G(\alpha_1 \cap \beta_1, \alpha_2 \cap \beta_2) \in$

$\{(\frac{1}{2}-k)(\alpha_1 \cap \beta_1) \cap (\frac{1}{2}-k)(\alpha_2 \cap \beta_2) \cap k(\alpha_1 \cap \beta_2) \cap k(\alpha_1 \cap \beta_2)\}$.

This statement says that for each dihybrid cross involving $(\alpha_1 \cap \beta_1)$ and $(\alpha_2 \cap \beta_2)$ gametes there exists a number k with values between 0 and 0.25 such that the frequencies of gametes of the parental combinations will be $(\frac{1}{2}-k)$, and the frequencies of the other two combinations will be k.

If the Mendelian theory is to be extended to cover linkage groups and maps (in order to construct a "linkage genetics" or "Bateson-Morgan-

Mendel genetics"), then a position (locus) on a linear map has to be assigned to each quasi-equivalence class under the All relation:

$$\{\alpha_1,\alpha_2,\ldots\},\ \{\beta_1,\beta_2,\ldots\},\ \text{etc.}$$

The linear maps are represented by a number of separate line segments (according to the number of linkage groups), and these segments can be either open ended or circularly closed. We might then state as our axiom:

Axiom 4. The constants k which obtain for pairs of loci must be additive in the sense that they yield linear maps of open or circular line segments.

The resulting theory is what is usually called "classical genetics". It has received empirical support from chromosomal studies in the sense of the identification of recombination frequencies (k values) with the crossing-over frequencies of chromosomes. The theory provides us with linkage groups which can be tested against haploid numbers of chromosomes in given cells and with linkage maps some features of which can be occasionally identified with band structures on chromosomes. The correlation of some mutations with chromosome breakage points has reinforced this support. It cannot be said however that this empirical support from cytological observations represented a reduction of genetics to cytology. Obviously, classical genetics had many observational and theoretical statements, about phenotypic characters, gametes, organisms, which were outside the domain of the cytology of chromosomes.

It can certainly be said that the original Mendel theory is a subtheory of the Bateson-Morgan-Mendel theory (in the sense that the second law of Mendel is a limiting case of Axiom 3). The old Mendelian theory was not falsified, the original observations are still repeatable and they can still be explained by it, but the new theory can explain many more observations. To support our contention that this is a reasonably complete statement of classical genetics we quote Morgan's original definition:

> We are now in a position to formulate the theory of the gene. The theory states that the characters of the individual are referable to paired elements (genes) in the germinal material that are held together in a definite number of linkage groups; it states that the members of each pair of genes separate when the germ cells mature in accordance with Mendel's first law, and in consequence each germ cell comes to contain one set only; it states that the members belonging to different linkage groups assort independently in accordance with Mendel's second law; it states that an orderly interchange - crossing-over - also takes place, at times, between the elements in corresponding linkage groups; and it states that the frequency of crossing-over furnishes evidence of the linear order of the elements in each linkage group and of the relative position of the elements with respect to each other (Morgan 1926, p. 25).

3. The structure of modern genetics

Next in the evolution of genetics came the rise of "fine-structure genetics". This was necessitated by evidence that crossing-over can occur not only between genes but also within them. The evidence for intralocus crossing-over and for subunits of genes, came from microbial genetic studies at the end of the 1940's and the beginning of the 1950's. It so happened that this development was soon overshadowed by the publication of the Watson-Crick model in 1953, which lent sudden and dramatic support to the idea of subgene components in linear arrangements.

The position we would like to defend is that present-day genetics is a variety of fine-structure genetics such that the purine and pyrimidine nucleotides have been identified with subgene components. The evidence for their linear arrangement has been supplied by chemistry. Thus not only genes are linearly arranged but also their four kinds of subunits are linearly arranged. This allows for the extension of the explanatory power of genetic theory beyond that of the classical theory. Furthermore the purine-pyrimidine pairing mechanism explains the duplication of genes.

The structure of this new genetics can be expressed in short in the following way:

(1) There are four subgene components: the nucleotides adenine, guanine, thymine and cytosine.
(2) These nucleotides form linear macromolecular structures called "DNA strings" through deoxyribose-phosphate links between them.
(3) Specific H-bond pairing can take place between adenine and thymine, and between guanine and cytosine.
(4) A new allelic relation is introduced to be able to express the change of a nucleotide in a given DNA string (the All relation can then be defined in terms of this new relation).
(5) Genes are segments of DNA strings with specific markers delineating their beginnings and their end points.
(6) DNA strings are duplicated by producing complementary strings obtained through the specific nucleotide pairings given by (3); in this way complementary double strings arise.
(7) Chromosomes are composed of double DNA strings and other macromolecules, and can undergo synapsis (specific pairing) with their "homologous" counterparts, - where "homologous" means complementarity or allelic relationship between the nucleotides of the DNA strings.
(8) Crossing-over between chromosomes consists of simultaneous breaking and healing of DNA double strings at corresponding points on homologous chromosomes.

We do not need to introduce terms like "messenger RNA", "protein", "coding", and so on, in order to derive a conceptual basis for fine-structure genetics. In fact for this purpose, we do not need to know more about DNA than the facts listed above. According to our view a detailed chemical description of these molecules and their relationship to protein synthesis is not necessary for a framework of modern genetics. For this reason we do not call it "molecular genetics". Our contention is that it would be a mistake to speak of a reduction of classical genetics to chemistry, just as before we refused to speak of a reduction of Mendelian genetics to cytology. Chemistry only provides empirical support to fine-structure genetics, it does not provide a complete framework of the theory of heredity. Genetics still has many terms and statements which cannot be given chemical definition, like gametes, fusion of gametes, development of zygotes, and the various phenotypic characters of the resulting organisms. In fact one might rather say that genetics has been 'reduced' to linguistics by these developments.

In any case, classical genetics has been reduced to fine-structure genetics. Again the previous theory has not been falsified, the observations which could be explained by it are still repeatable and its explanations still hold. The new theory can express much more, and at the same time provides constraints within which the previous theory still holds (primarily these constraints have to do with the distinction between interlocus and intralocus crossing-overs).

4. Discussion

This paper is based on the view that the concept of heredity must be central to genetics, that is, that genetics has primarily to do with the distribution of inherited factors and characters in populations. The operators 'G' and 'D' which Woodger has introduced in his attempt to formalize classical genetics are characteristic of hereditary relationships. Consider for instance the statements : '$G(\alpha,\alpha) \subseteq \alpha$' and '$D(\alpha,\alpha) \subseteq P$'. The first one says that all gametes produced by organisms which result from the fusion of two gametes from class α must also be in class α. This represents the closure of the set α under operation G, or in the terminology of the <u>Principia</u> (Whitehead & Russell 1910, p. 544) : " α is a hereditary class with respect to G." (Incidentally if one wishes to modify the requirement for strict heredity in order to allow for mutations, one can change the notation in the following way: '$G(\alpha,\alpha) \subseteq \alpha$' is equivalent to '$G(\alpha,\alpha) \in 1\alpha$' in the mixed numerical-algebraic notation that we introduced. Instead we can write '$G(\alpha,\alpha) \in (1-m)\alpha$' in which m is a constant standing for the mutation frequency). The second expression states that all organisms of the above kind are in set P. In other words, the operator 'D' maps sets of hereditary factors into phenotypes. The biochemical mechanisms which are involved in this mapping need not be known in order to specify it.

The axiomatization of classical genetics which is outlined in this paper is based on Woodger's (1959) axiom system but differs from it in the sense that there is an explicit theoretical term introduced by us, namely the notion of the allelic relation (All). Woodger, on the other hand, has attempted to axiomatize Mendelian genetics in terms of observables only, corresponding to his stated desire to prefer statements which are "epistemically prior" (Ruse 1975). He succeeded in this attempt except for his primitive term 'phen', which denotes phenotypes. The use of this term is necessary to be able to express the genotype-phenotype distinction with reference to classes or characters of organisms. This distinction is an essentially theoretical concept. Another difference between our system and that of Woodger lies in our having omitted all reference to environmental effects on genetic processes. This was done merely to keep the presentation simple, the system could easily be extended to include environment.

In our view, present-day genetics is nothing more than a version of fine structure genetics and is based on the same concepts as classical genetics. The theoretical framework for mutations is also the same in both theories, only the positions of mutational changes can be identified with greater precision.

Molecular genetics, as considered for instance by Hull (1974, 1979), Schaffner (1974, 1977), Goosens (1978), Kimbrough (1979) and others, can be separated according to our analysis into two parts. One part is contained in the fine-structure genetics which we have discussed above. The other part consists of a non-formalized area and includes generalizations concerning the genetic code, the Jacob-Monod model and large parts of cellular metabolism. As stated before, we do not consider this part of the theory as essential to a theory of heredity. It is possible that in the future there will arise a (formalized) theory which will include both of these parts, in which it will also be understood under what conditions a DNA segment can serve as a gene, and how phenotypic characters arise as a result of the molecular activities of such genes. (This would imply finding necessary and sufficient conditions for a statement like 'α All β' to hold). But such a larger theory may better be referred to by some other name that by the term 'molecular genetics'.

The main objection raised by Hull (1974, 1979) to reduction having taken place from classical to molecular genetics can now be reconsidered in this light. His argument against reduction has primarily been based on the fact that the development of organisms and thereby the expression of their phenotypes cannot be explained by molecular mechanisms. We have attempted to show that this argument cannot be used if we restrict ourselves to the question of the reducability of classical to fine-structure genetics. In this case molecular mechanisms for the expression of phenotypes do not enter into the theories at all. This applies to notions such as "dominance", "epistasis", "regulatory genes", "operons" and "cistrons", as well. Hull's remarks remain valid only if they are applied concerning the reducibility of classical genetics to the larger "molecular genetics", which is not yet available.

The redefinitions of reduction proposed by Kimbrough (1979) are not pertinent to the restricted reduction question we are posing.

Schaffner (1974, 1977) and Goosens (1978) have been arguing for "real" reduction having taken place. They made various qualifying proposals, partly in the structure of the theories involved, and partly in the definition of reduction, in order to achieve this goal.

In our view it is most fruitful to keep the definition of reduction as simple as possible. We are proposing a modification of genetic theories and in this way we are able to demonstrate how reduction can be achieved in a simple way. We are basing this analysis on the concept of heredity which must lie at the basis of any "research program" by which genetics is distinguished from other scientific areas. This concept is a specifically biological one by which we mean that heredity cannot at the present time formally be defined in physico-chemical terms. This analysis represents an elaboration of our previously stated point of view concerning the reducibility of biology to chemistry, namely the view of non-vitalist organicism. As far as research in biology is concerned, this view seems to be a feasible and fruitful one. It is made possible by the logical empiricist method by which theories can be delimited, and their changes and their relationships can be defined.

After having written this paper, we found that the work of Dawe (1979) is closely related to ours and in fact agrees with various aspects of our conclusions. He constructed set-theoretical predicates for three genetic theories (Mendelian, linkage and fine-structure genetics) which we also distinguish, and exhibited reduction functions between them. These, which represent reduction in the sense of Nagel, he calls "weak reduction functions". The relationship between fine-structure genetics and molecular genetics is considered by him to be of another kind of reduction ("strong reduction") defined as an inclusion of the intended application ranges of the two theories.

References

Dawe, C.M. (1979). Intertheoretic Relations in Genetics. Ph.D. Dissertation, University of London, Birkbeck College.

Goosens, W.K. (1978). "Reduction by molecular genetics." Philosophy of Science 45: 73-95.

Haraway, D.J. (1976). Crystals, Fabrics, and Fields. Metaphors of Organicism in Twentieth Century Developmental Biology. New Haven, CT: Yale University Press.

Hull, D.L. (1974). Philosophy of Biological Science. Englewood Cliffs, N.J.: Prentice Hall.

---------- (1979). "Discussion: reduction in genetics." Philosophy of Science 46: 316-320.

Kemeny, J.G. (1959). A Philosopher looks at Science. Princeton: Princeton University Press.

Kimbrough, S.O. (1979). "On the reduction of genetics to molecular biology." Philosophy of Science 46: 389-406.

Morgan, T.H. (1926). The Theory of the Gene. New Haven, CT: Yale University Press.

Nagel, E. (1961). The Structure of Science. London: Kegan Paul.

Ruse, M. (1975). "Woodger on genetics, a critical evaluation." Acta Biotheoretica 24: 1-13.

Schaffner, F.K. (1974). "The Peripherality of Reductionism in the Development of Molecular Biology." Journal of the History of Biology 7: 111-139.

--------------- (1977). "Reduction, Reductionism, Values and Progress in the Biomedical Sciences". In Logic, Life, and Laws. (University of Pittsburgh Series in Philosophy of Science, Vol. 6.) Edited by R. Colodny. Pittsburgh: University of Pittsburgh Press. Pages 143-17

Whitehead, A.N. and Russell, B. (1910). Principia Mathematica Volume I. Cambridge: Cambridge University Press.

Woodger, J.H. (1952). Biology and Language. Cambridge: Cambridge University Press.

------------- (1959). "Studies in the foundations of genetics." In The Axiomatic Method. Edited by L. Henkin and P. Suppes and A. Tarski. Amsterdam: North-Holland Publishing Company. Pages 408-428.

Reductionist Research Programmes in Psychology[1]

Robert C. Richardson

University of Cincinnati

1. Introduction: Function and Structure

Contemporary psychology has been dominated by two distinct paradigms, each of which have given rise to comprehensive and fruitful research programmes. The functionalist programme purports to explain our cognitive capacities -- what Noam Chomsky (1965, pp. 3 ff.) dubs our "competence" -- in terms of the operations and transformations effected on the available "information". Explaining, e.g., an ability such as color perception means isolating the cues on which we rely in making our judgments as to color, and proceeding to explain how, on the basis of information available from such cues, we could come to reach our actual judgments. This will involve explaining, for example, how it is that our perceptions of color are constant and reliable despite substantial variation in the intensity of light impinging on the retina -- particularly since the intensity of such light seems to be the only relevant parameter which could guide our judgments. This puzzle now seems well in hand (see Land 1959 and 1977); one particularly interesting fact is that the answer is one arrived at independently of any substantial concern for the structure of the visual system. The trick lies in finding out how the available information is used.

A reductionist programme, by way of contrast, would seek to isolate the neural structures which subserve specific functions, detail the interactions of the subsystems, and, ideally, illustrate how the structures in question are able to perform their specific functions. One of the most promising strategies for carrying through a reductionist account of cognitive functions has been through the study of behavioral abnormalities resulting from the destruction, either through surgery or accident, of tracts of brain tissue. (For a summary of some of the available experimental techniques, see Rosenfield 1978.) The paradigm is, of course, Paul Broca's discovery during the mid-

PSA 1980, Volume 1, pp. 171-183

nineteenth century that lesions in the frontal lobe of the dominant hemisphere -- in the region that now bears his name -- gave rise to a speech disorder now known as nominal aphasia.

One of the most significant factors which has lent credence to reductionism in psychology is the recognition that there are significant and sophisticated perceptual and cognitive capabilities present at birth -- far more complex than we might have otherwise expected. It is, of course, the more fixed, and less plastic, abilities for which we might expect discrete structural correlates. Steven Rose remarks: "Specificity determines the characteristics of the species and the population; plasticity, the irreplaceability and inimitability of the individual" (1973, p. 172). As Richard Held and Alan Hein (1963) have so elegantly demonstrated, however, even the most basic capacities are not immune from environmental impact: no traits, behavioral or morphological, are wholly fixed by the genes. The most that reductionism can claim or demand is that "canalized" traits have structural correlates; that is, that those traits -- including behavioral characteristics -- whose developmental progress and manifestation is relatively constant despite minor genetic or environmental variations have a physiological realization which is common to "normal" members of the species. The traits at issue are, then, those manifesting a developmental homeostasis: there is an evolved tendency for individuals to correct deviations from phenotypic targets in such a way that phenotypic constancies will occur across the species despite what would otherwise be relevant environmental variation. There is no reason to think such corrective mechanisms would hold only for gross morphological traits; indeed, behavioral regularities and cognitive capacities are, potentially, even more significant from an evolutionary perspective. (The crucial riders -- that the individuals and the environment must be "normal" or within the range of "normalcy" -- can escape triviality only by reference to the conditions under which the trait in question evolved and was selected for.)

Localizationist strategies, thus, are committed to an endorsement of the following hypothesis:

> H1: There are relatively uniform and structurally discrete neural systems which are responsible for canalized traits and capacities present in normally healthy individuals.

An almost immediate corollary which is reflected clearly in the experimental methodology is this:

> H2: Deficiencies or abnormalities in these neural systems, whether genetically or environmentally induced, will serve to explain abnormal behaviors.

Given discrete neurological structures, it should be possible to ascertain their functions by examining the behavioral deficits which result from their damage.

It is, naturally, not this straightforward. As the programme has since been elaborated, it has incorporated a number of insulating mechanisms which serve to protect it from adverse experimental results, e.g., in the 1940s, Karl Lashley systematically studied the effects of cortical ablation on rat's capacities for recalling spatial tasks. He found no region specific to such capacities. In response to these results, theorists have postualted everything from equipotentiality to functional redundancy, and D.O. Hebb (1949) has even suggested that memory traces might be anatomically diffuse (though still fundamentally structural in nature). Reductionist programmes have simultaneously assumed a more specific theoretical content, in the form of more detailed hypotheses, corresponding to H1 and H2.

2. Localization of Function

In his classic analysis of aphasic syndromes -- that is, of speech disorders resulting from brain damage -- Wernicke relied on a number of case studies for which he had independent information concerning behavioral deficits and neurophysiological abnormalities. Broca had suggested over a decade earlier that aphasic syndromes were due to destruction in the lower and posterior portion of the frontal lobe of the dominant hemisphere. By the time young Wernicke published "The Symptom-Complex of Aphasia" in 1874 it was clear that Broca'a views were incomplete. (A clear and brief exposition is given in Geschwind 1967.) According to Broca's views, the relevant aphasic deficiencies were expressive, and not indicative of comprehension disorders. The location of Broca's area is, of course, precisely what we would expect given these symptoms; for (and this was known before Broca published his first work in 1861) it is adjacent to the motor control areas lying anterior to the fissure of Rolando. Wernicke brought out what he thought was a new clinical syndrome: aphasias involving comprehension disturbances, or "receptive aphasias". The differences between Broca's aphasia and Wernicke's aphasia are striking. In case of Broca's aphasia, speech, if present at all, is labored: articulation is severly impaired, but what language there is is sensible and relevant to the context. It is known that the expressive deficiencies are not due to peripheral disorders; for organs of speech can still be used for other functions, e.g., though unable to speak or sing, such aphasics can still accurately reproduce the melody of song without the words. Even more importantly, in Broca's aphasia, comprehension is unimpaired for both written and spoken language. Wernicke's aphasia, by way of contrast, involves no impairment of the speech mechanisms; in fact speech is copious, and actually difficult for the subject to inhibit. Though well-articulated and, grammatically, acceptable, this aphasia is characterized by speech that is semantically deviant; moreover, the subject exhibits substantial comprehension deficiencies.

As his model of the functional organization of the brain has since been developed, in large part by Norman Geschwind (see especially his 1965, and his 1973 and 1979 for less complete but more recent treatments; see also Penfield and Roberts 1959 for a number of elegant results), the relevant aphasic syndromes can all be explained using six basic components. The primary auditory cortex (PAC) and the primary visual cortex (PVC) are the first receptors for input from the peripheral organs. At least in humans and most primates, the primary receptive areas have no connections save with the adjacent regions in the association cortex. This is what Geschwind terms "Flechsig's principle": "the primary receptive areas (the kinocortices) have no direct neocortical connections except with immediately adjacent, 'parasensory' areas, the 'association areas'." (Geschwind 1965, p.112). (In the majority of subprimates, this is not true, since the primary projection areas and the association cortex are less discrete.) The output of the primary auditory cortex is to Wernicke's area (WA); the output of the primary visual cortex is routed through the angular gyrus (AG) to Wernicke's area. Wernicke's area has connections via the arcuate fasciculus to Broca's area (BA), which projects finally to the motor cortex (MC). (Figures One and Two display the relevant functional and anatomical schemata.)

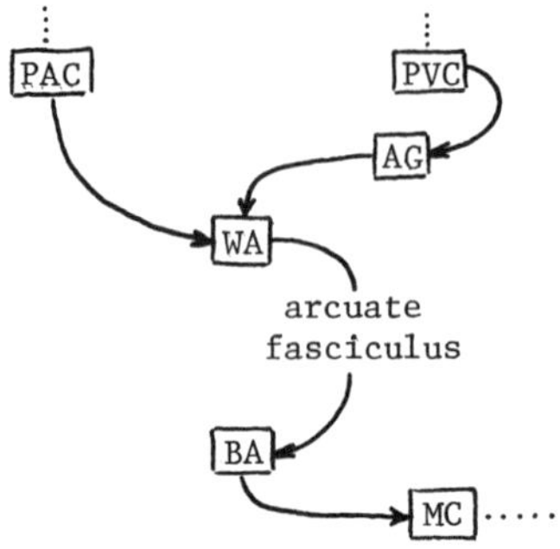

Figure One

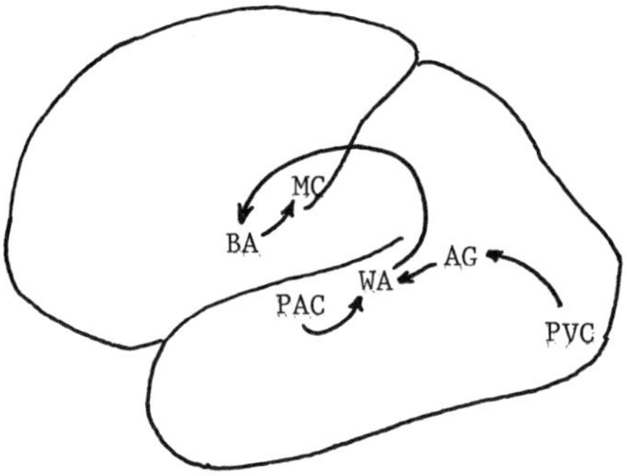

Figure Two

Functionally, the angular gyrus is taken as a visual word-memory store (Geschwind 1965, p. 152), or as audio-visual word association. Wernicke's area is taken as a general memory store for auditory representations of language. It, if anything, is (perhaps together with the angular gyrus) the locus for language comprehension. Broca's area serves as, at least, the locus for phonetic representation. (See Figure Three.)

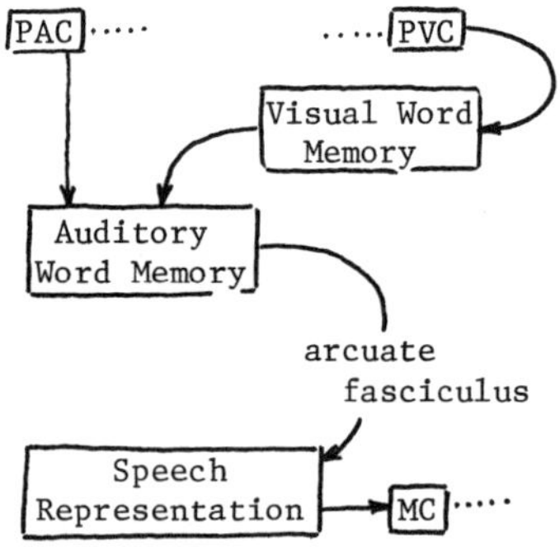

Figure Three

Even this much should make it evident that localizationist approaches to reduction assume the truth of the following hypothesis:

> H3: The structural subsystems in the brain exhibit a substantial functional independence; that is, these structural subsystems should be capable of carrying out the relevant functions independently of the action of other subsystems (save, of course, as input or output to the subsystem).

It is only if this is the case that studies of disconnection syndromes will have any possibility of success; for the presumption in such studies is that structural damage will yield discrete functional and behavioral deficiencies. This does not rule out functional redundancy, though such redundancy would make the use of lesion studies more complicated: we already know there is substantial redundancy, at least insofar as each cerebral hemisphere in an intact brain replicates functions in its counterpart. All that is demanded by H3 is the functional integrity of each structural subsystem.

It may be profitable to note that assumptions tantamount to H3 are also present in much functionalist research. Thus, it is common in studies of speech perception to attempt to obtain a grammatical parsing of input while ignoring semantic information.[2] This amounts to assuming the processing of speech is carried on in (at least) two parallel modes, one semantic and one syntactic. More generally, the assumption is that our cognitive capacities are a patchwork of more specialized systems. Daniel Dennett (1980) remarks: "If we are designed by evolution, then we are almost certainly nothing more than a bag of tricks, patched together by a <u>satisficing</u> Nature."

We cannot be too sanguine about finding structural correlates for just any functions a system might exhibit. Some standard is needed for ascertaining which functions are plausible candidates for localization. Consider a well-known objection raised by R.L. Gregory:

> Although the effects of a particular type of ablation may be specific and repeatable, it does not follow that the causal connection is simple, or even that the region affected would, if we knew more, be regarded as functionally important for the output -- such as memory or speech -- which is observed to be upset. It could be the case that some important part of the mechanism subserving the behavior is upset by the damage although it is at most indirectly related, and it is just this which makes the discovery of a fault in a complex machine so difficult (Gregory 1961, p. 323).

Gregory points out that simplistic uses of functional deficit analysis would lead us, e.g., to conclude that a resistor in a radio is a hum-supressor because the radio hums when it is removed (Gregory 1968, p.99). It is, in brief, not true that finding a deficiency in a given structure in the brain (or any highly integrated system) will directly warrant a conclusion that the function of that element in the intact system is ascertainable by an examination of behavioral abnormalities.

The problem, evidently, is that not all functions need be assigned to a discrete structural subsystem. Even Geschwind comments: "The 'localizationist' approach to genetics has many of the same potential problems as a localizationist approach to the higher [cognitive] functions... . Clearly it is not reasonable to expect that every nameable feature will have a chromosomal or a cortical localization." (Geschwind 1969, p. 432).

The most plausible approach toward obviating Gregory's objection would be one which would seek out an independent criterion capable of isolating the relevant functions. It is clear, in any case, that the attempt to find structural correlates for psychological capacities will involve a complicated interaction of functional and structural hypotheses. Barbara Von Eckhardt Klein suggests that localizationist programmes might be precisely what is needed to show that the functional analyses of cognitive abilities offered by linguists and psychologists are genuine explanations of human cognitive capacities. Her reasoning is quite simple: a functional analysis of a capacity shows only how a system with the proposed functional components could manifest the capacity in question; it does not show that only something with that organization could do so; and, accordingly, what is needed is a guarantee that the functional analysis coincides with structure -- that, in her terms, the analysis into constituent capacities is "structurally adequate" (Klein 1978, p. 41). It is precisely research into the localization of neural structure that provides such a guarantee. If so, then it will, at least, be true that attaining structural correlates for postulated psychological functions would serve to vindicate the functional analysis. It is equally clear that some preliminary analysis into functional components is necessary before a localizationist programme can be instituted. An analysis of a given capacity which treats it as the result of the sequential or parallel operation of functional subsystems -- e.g., following Wernicke, treating language use as explicable in terms of subsystems responsible for word-retreival and for phonetic (grammatical) representation -- could then be used, provisionally, to warrant the search for structural correlates. In short, the operative hypothesis is this:

> H4: A proper analysis of our cognitive capacities will yield functional units which are realized in structurally discrete regions of the brain.

The interaction of functional studies and structural research would, no doubt, be a complicated matter involving substantial feedback between physiological and psychological studies. It is surely unreasonable to think we must choose between "top down" approaches which take a psychological analysis to be unimpeachable and "bottom up" approaches which accord a similar status to neurophysiology (cf., Churchland 1980). A better perspective allows for mutual adjustment. Though this would violate more naive reductionist approaches, there seems to be no substantive objection, and many explanatory advantages, in such methodological interdependency (cf.,Wimsatt 1976 and Richardson 1980).

3. Near Decomposability and Functional Integrity

The third hypothesis isolated above is tantamount to the assumption that the brain is what Herbert Simon terms a "nearly decomposable system". Within what Simon calls "hierarchic systems" -- systems, for the most part, which are systems of subsystems in which there is no subordination of function (cf.,Simon 1969, pp. 87-88) -- a *decomposable system* is one in which, when compared to the interactions *within* subsystems, any interaction *between* or *among* subsystems is negligible. A system will count as *nearly decomposable* if the interactions between subsystems are relatively weak, though not negligible. (Cf.,Simon 1969, pp. 99-100.)

It may well be the case that an assumption of near-decomposability is essential to any reductionist programme. Simon remarks: "If there are important systems in the world that are complex without being hierarchic, they may to a considerable extent escape our observation and our understanding."(1969, p. 108). William Wimsatt (1980) has elegantly argued that the assumption of near-decomposability is integral to the research heuristics of reductionist programmes, and has urged that it is an excessive oversimplification, at least in the context of sociobiology. There are substantial reasons for doubting that the reductionistic assumptions can be maintained in the case we are concerned with as well.

We have already remarked that, according to Flechsig's principle, connections of the primary projection areas are exclusively to the adjacent (primary) association areas. In subhuman primates such as the macaque, cross-modal transfer is exclusively through limbic (subcortical) routes (Geschwind 1965, p. 128). Geschwind remarks that the "monkey brain thus probably contains visual, auditory, and somesthetic regions...operating on the whole independently." (Geschwind 1964, pp. 97-98). Thus, in subhuman primates, the condition of near-decomposability may well be satisfied. This stands in sharp contrast to the account Geschwind offers of the prerequisites of human language. Geschwind notes that, in the evolution of *Homo Sapiens*, the most significant increase is in the association areas, and not the primary projection areas; more specifically, he suggests the major increase is in the direct (i.e., non-limbic) associative systems (Geschwind 1964, p. 98). These are the *secondary association areas*, or, in Geschwind's terms, "the association areas of association areas". It is Geschwind's view that the direct cross-modal transfer afforded by the secondary association areas is requisite even for the learning of general and proper names (Geschwind 1964, p. 97; Geschwind 1965, pp. 126, 148-149). The implication is that, in humans, the brain is not amenable to near-decomposability, and the structures responsible for functional integration are the structures necessary for language learning. No language-using system will be nearly decomposable.

A similar point can be developed from a second, independent perspective. Broca's aphasia, recall, is characterized by halting speech; that is, speech which may be semantically acceptable, but is _syntactically_ primitive and uncomplicated. Wernicke's aphasia, by contrast, is characterized by fluent and _syntactically_ acceptable speech that is semantically aberrant. Geschwind describes speech for Broca's aphasics as "telegraphic". Asked about a dental appointment, a patient might respond: "Yes...Monday...Dad and Dick...doctors." Wernicke's aphasia, on the other hand, is semantically abnormal. Geschwind comments: "A patient who was asked to describe a picture that showed two boys stealing cookies behind a woman's back reported: 'Mother is away here working her work to get her better, but when she's looking the two boys looking in the other part. She's working another time."(1979, p. 186). Geschwind writes: "Wernicke's area...is apparently involved in the recognition of auditory patterns of language ... . This region is also involved in formulating the linguistic message which is then sent forward to Broca's area. ...Broca's area... seems to be involved in recoding the message received from Wernicke's area into its articulated form. It also seems to be involved in the grammatical structure of language."(1973, p. 71). In short, Wernicke's area gets assigned semantic functions, and Broca's area the syntactic functions. (It should be added that there is some reason to allow that the non-dominant [usually right] hemisphere has limited linguistic capacities; in particular, there is a minimal capacity for word recognition (cf.,Sperry 1974). This is further supported by techniques which estimate the level of involvement by measuring blood flow (cf., Lassen, _et al_. 1978; Risberg, _et al_. 1975).) Unfortunately, it is not true that syntactic and semantic processing are carried out independently. Garrett, Bever, and Fodor (1966) have shown that, in fact, imputed grammatical (or syntactic) structure is partially determinative of the segmentation of sentences into words. (See Neisser 1967, Chapter 7; Fodor and Bever 1965; and Fodor 1968, pp. 79-86.) Syntactic and semantic processing are interdependent. If so, the functional integrity of the subsystems that is postulated in accordance with the hypothesis of near-decomposability is simply not to be had.

4. Functional Specificity and Cognitive Abilities

The fourth hypothesis, introduced in Section 2, is also suspect. Let us suppose, contrary to what was indicated in Section 3, that we are concerned with a system with subsystems possessing functional and structural integrity. Let us suppose, furthermore, that we have an appropriate functional analysis. What could explain the evolution of such a system?

It is well known that there are a number of evolutionary strategies available which might be adopted by a species, or a population, in response to environmental demands. (Speculations concerning such strategies, taken in the abstract, are suspect (see Lewontin 1978 and 1979), but, I think, are defensible if care is taken.) The most appropriate strategy will, of course, be partly a function of the environmental demands, e.g., whether a species is confronted with a stable environment or a (spatially or temporally) variable environment will make a great deal of difference. In the former case (an example might be intestinal parasites), monomorphism is an appropriate response: since the demands are relatively constant, the response can be as well. In the latter case (it is here, surely, that Homo Sapiens belongs), a constant response is ill-suited; the appropriate response will be a balanced polymorphism within the breeding population (see Lewontin 1961 and Roughgarden 1979, Chapters 12 and 13 for an elegant discussion) and/or the development of homeostatic mechanisms.

It seems most plausible that, in the case of human cognitive and linguistic capacities, the last option is the most likely. Homeostatic mechanisms, which induce a kind of "artificial stability", may range from "closed" to "open" programs. In response to variable environments, the "closed programs" take the form of specialization of function (a clear example, thinking of the colony as an organism, is the eusocial insects); the "open programs" will favor increased "plasticity". If asked to explain how systems consisting of subsystems characterized by the functional and structural integrity postulated by reductionist programmes in psychology could develop, the answer must, evidently, be that the subsystems evolved independently, each subserving discrete functions. The scenario envisioned would be that sort of coevolution which has traditionally been acknowledged as integral to mutally adapted species, e.g., the advantage of a memory store would be enhanced by an increasingly complex syntactic processing capability (and vice versa). In social interactions, the parallel would be symbiosis (cf.,Wilson 1975, ch. 17). There is little doubt that some subsystems are best explained in this, or a closely analogous, way. It would seem that when we are concerned with functions that are universally demanded in a relatively fixed form, closed programs would take on increasing significance. Clear examples would be sensory capacities and motor control. And as John O'Keefe and Lynn Nadel remark, "the nearer one is to the periphery the better this procedure [of analyzing functional deficits on the basis of neural lesions] works."(1978, p. 236).

When concerned with isolating the function of more central structures (e.g., the Hippocampus, which is the target of O'Keefe's and Nadel's penetrating study), the case is much more problematic. The central question is whether what evolved, and was selected for, was a highly specific and specialized functional capacity or a more general and modifiable capacity. A functional analysis would, in conformance to the fourth hypothesis, be expected to be reflected in structures across the species only if what was selected for were such discrete and specialized capacities.

The only line of approach I am aware of that would appear to lend any support to such a view is again due to Herbert Simon. He demonstrates quite clearly that "complex systems will evolve from simple systems much more rapidly if there are stable intermediate forms than if there are not."(Simon 1969, pp. 98-100). If a system S* is a hierarchically organized system consisting of subsystems $S_1,\ldots,S_n$, and each S_j which is a component of S* is a hierarchically organized system, it can be shown that the construction of such a system will prove to be less difficult as the number of levels in the system increase. (See, especially, Simon's parable, in his 1969, pp. 90-92.) So, it seems a system that segments into discrete subsystems would be more likely to evolve than one which does not. But, as Simon seems aware (cf.,his 1969, p. 110), this scenario will only work given a system whose component subsystems lack a great deal of functional differentiation. This point betrays a limitation on the concept of a hierarchically organized system which Simon's original definition does not display. Simon's hierarchically organized systems cannot allow functional differentiation of the components when the component functions are interdependent. Just as when there is functional subordination, the (near) decomposability of such a system will not facilitate its evolutionary development. The reason is simply that, from an evolutionary perspective, even strict decomposability confers no advantage when the subsystems are functionally interdependent (or subordinated).

My own conjecture would be that the evolutionary scenario would favor the development of plastic, and highly modifiable, systems. If there are language-specific cognitive structures, these would be subject to a short-term developmental plasticity. Such plasticity is generally favored in coarse-grained environments; that is, when environmental variability is low in comparison to the range of individual organisms. Some credence is lent to this possibility by the fact that children suffering brain damage which affects language are more likely to relearn language than are their adult counterparts (see Lenneberg 1967, pp. 142-150). If, as perhaps seems even more likely, there are no language-specific cognitive structures, but only some more basic cognitive ability which itself has a firm developmental sequence (e.g., mode of cognitive organization), we would have more generalized plastic ability. Such plasticity would be favored in fine-grained environments; that is, when environmental variability is high in comparison to the individual's range. In either case, a functional analysis of our linguistic capacities would fix the level of abstraction at too specific a level; to expect neural correlates for the functional units is to expect too much.

Notes

[1]Preliminary research relevant to this work was supported by grants from the University Research Council and the Taft Committee at the University of Cincinnati. I am thankful for their support.

[2]The point was driven home to me in a presentation by Mitch Marcus at the Institute for the Advanced Study of the Behavioral Sciences in March of 1980.

References

Chomsky, Noam. (1965). Aspects of the Theory of Syntax. Cambridge: M.I.T. Press.

Churchland, Patricia. (1980). "A Perspective on Mind-Brain Research." The Journal of Philosophy 77: 185-207.

Dennet, Daniel C. (1980). "Three Kinds of Intentional Psychology." Forthcoming in a Thyssen Philosophy Group volume. Edited by R.A. Healey.

Fodor, Jerry A. (1968). Psychological Explanation. New York: Random House.

-------------- and Bever, T. (1965). "The Psychological Reality of Linguistic Segments." Journal of Verbal Learning and Verbal Behavior IV: 414-420.

Garrett, M., Bever, T., and Fodor, J. (1966). "The Active Use of Grammar in Speech Perception." Perception and Psychophysics 1: 30-32.

Geschwind, N. (1964). "The Development of the Brain and the Evolution of Language." In Monograph Series on Language and Linguistics. Volume 17. Edited by C.I.J.M. Stuart. Washington: Georgetown University Press. Pages 155-169. (As reprinted in Geschwind (1974). Pages 86-104.)

------------. (1965). "Disconnexion Syndromes in Animals and Man." Brain 88: 237-294 and 585-644. (As reprinted in Geschwind (1974). Pages 105-236.)

------------. (1967). "Wernicke's Contribution to the Study of Aphasia." Cortex 3: 449-463. (As reprinted in Geschwind (1974). Pages 284-298.)

------------. (1969). "Problems in the Anatomical Understanding of the Aphasias." In Contributions to Clinical Neuropsychology. Edited by A.L. Benton. Chicago: Aldine. Pages 107-128. (As reprinted in Geschwind (1974). Pages 431-451.)

------------. (1973). "The Brain and Language." In Communication, Language, and Meaning. Edited by George A. Miller. New York: Basic Books. Pages 61-72.

------------. (1974). Selected Papers on Language and the Brain. (Boston Studies in the Philosophy of Science, Vol. 16.) Dordrecht: D. Reidel.

------------. (1979). "Specializations of the Human Brain." Scientific American 241(No. 3): 180-199.

Gregory, R.L. (1961). "The Brain as an Engineering Problem." In Current Problems in Animal Behavior. Edited by W.H. Thorpe and O.L. Zangwill. Cambridge: Cambridge University Press. Pages 307-330.

-------------. (1968). "Models and the Localization of Function in the Central Nervous System." In Key Papers in Cybernetics. Edited by C.R. Evans and A.D.J. Robertson. London: Butterworths. Pages 91-102.

Hebb, D.O. (1949). The Organization of Behavior. New York: John Wiley and Sons.

Held, Richard, and Hein, Alan. (1963). "Movement-Produced Stimulation in the Development of Visually Guided Behavior." Journal of Comparative and Physiological Psychology 56: 872-876.

Klein, Barbara von Eckhardt. (1978). "Inferring Functional Localization from Neurological Evidence." In Walker (1978). Pages 27-66.

Land, Edwin. (1959). "Experiments in Color Vision." Scientific American 200(No. 5): 84-99. (As reprinted in Perception: Mechanisms and Models. Edited by R. Held and W. Richards. San Francisco: W.H. Freeman, 1972. Pages 286-298.)

-----------. (1977). "The Retinex Theory of Color Vision." Scientific American 237(No. 6): 108-128.

Lassen, N.A., Ingvar, D.H., and Skinhoj, E. (1978). "Brain Function and Blood Flow." Scientific American 239(No. 4): 62-71.

Lenneberg, R.C. (1967). Biological Foundations of Language. New York: John Wiley & Sons.

Lewontin, R.C. (1961). "Evolution and the Theory of Games." Journal of Theoretical Biology 1: 382-403. (As reprinted in Topics in the Philosophy of Biology. Edited by M. Grene and E. Mendelsohn. Dordrecht: D. Reidel, 1976. Pages 286-311.)

-------------. (1978). "Adaptation." Scientific American 239(No. 3): 212-230.

-------------. (1979) "Sociobiology as an Adaptationist Program." Behavioral Sciences 24: 5-14.

Neisser, Ulric. (1967). Cognitive Psychology. New York: Appleton-Century-Crofts.

O'Keefe, John, and Nadel, Lynn. (1978). The Hippocampus as a Cognitive Map. Oxford: Clarendon Press.

Penfield, W., and Roberts, L. (1959). Speech and Brain Mechanisms. Princeton: Princeton University Press.

Richardson, Robert C. (1979). "Functionalism and Reductionism." Philosophy of Science 46: 533-558.

--------------------. (1980). "Intentional Realism or Intentional Instrumentalism." Cognition and Brain Theory, III.

Risberg, Jarl et al. (1975). "Hemispheric Specialization in Normal Man Studied by Bilateral Measurements of the Regional Cerebral Blood Flow." Brain 98: 511-524.

Rose, Steven. (1973). The Conscious Brain. New York: Alfred A. Knopf.

Rosenfield, David. (1978). "Some Neurological Techniques for Assessing Localization of Function." In Walker (1978). Pages 219-228.

Roughgarden, Jonathan. (1979). Theory of Population Genetics and Evolutionary Ecology. New York: MacMillan.

Sperry, R.W. (1974). "Lateral Specialization in the Surgically Separated Hemispheres." In The Neurosciences Third Study Program. Edited by F.O. Schmitt and F.G. Worden. Cambridge: M.I.T. Press. Pages 5-19.

Simon, Herbert. (1969). The Sciences of the Artificial. Cambridge: M.I.T. Press.

Walker, E. (ed.). (1978). Explorations in the Biology of Language. Montgomery, VT: Bradford Books.

Wilson, E.O. (1975). Sociobiology: The New Synthesis. Cambridge: Harvard University Press.

Wimsatt, William. (1976). "Reduction, Levels of Organization and the Mind-Body Problem." In Brain and Consciousness. Edited by G. Globus, G. Maxwell, and I. Savodnik. New York: Plenum Press. Pages 199-267.

----------------. (1980). "Reductionistic Research Strategies and their Biases in the Units of Selection Controversy." In Scientific Discovery, Vol. 2: Case Studies. Edited by Tom Nickles. Dordrecht: D. Reidel. In Press.

Part VII

Evolutionary Epistemology and the Sociology of Knowledge

Against Evolutionary Epistemology[1]

Paul Thagard

University of Michigan-Dearborn

By "evolutionary epistemology" I mean Darwinian models of the growth of scientific knowledge. Such models rely on analogies between the development of biological species and the development of scientific theories. Recent proponents of evolutionary epistemology include the psychologist Donald Campbell (1974a), the sociobiologist Richard Dawkins (1976), and philosophers of science Karl Popper (1972), Stephen Toulmin (1972), and Robert Ackerman (1970). I shall argue that the similarities between biological and scientific development are superficial, and that clear examination of the history of science shows the need for a non-Darwinian approach to historical epistemology.[2]

The neo-Darwinian model of species evolution consists of Darwin's theory of natural selection synthesized with twentieth century genetic theory. The central ingredients of the neo-Darwinian model are variation, selection and transmission.[3] Genetic variations occur within a population as the result of mutations and mixed combinations of genetic material. Individuals are engaged in a struggle for survival based on scarcity of food, territory, and mating partners. Hence individuals whom variation endows with traits which provide some sort of ecological advantage will be more likely to survive and reproduce. Their valuable traits will be genetically transmitted to their offspring.

Evolutionary epistemology notices that variation, selection and transmission are also features of the growth of scientific knowledge. Scientists generate theories, hypotheses, and concepts; only a few of these variations are judged to be advances over existing views, and these are selected; the selected theories and concepts are transmitted to other scientists through journals, textbooks, and other pedagogic measures. The analogies between the development of species and the development of knowledge are indeed striking, but only at this superficial level. I shall try to show that variation, selection, and transmission of scientific theories differ significantly from their counterparts in the evolution of species.

PSA 1980, Volume 1, pp. 187-196

First consider variation. The units of variation in species are genes, with variation produced by errors in the process by which genes are replicated. Since the changes in genes are generally independent of the individual's environmental pressures, genetic variation is often said to be random. A better characterization is that of Campbell, who discusses blind variation (Campbell 1974a, p. 422). He outlines three important features of blindness: variations emitted are independent of the environmental conditions of the occasion of their utterance; the occurrence of trials individually is not correlated with what would be a solution to the environmental problem which the individual faces; and variations to incorrect trials are not corrections of previous unsuccessful variations.

It is immediately obvious that the development of new theories, hypotheses and concepts in science is not blind in any of these respects. One does not have to suppose there is some algorithmic logic of discovery to see that when scientists arrive at new ideas they usually do so as the result of concern with specific problems. Hence unlike biological variation, conceptual variation is dependent on environmental conditions. Whereas genetic variation in organisms is not induced by the environmental conditions in which the individual is struggling to survive, scientific innovations are designed by their creators to solve recognized problems; they therefore are correlated with a solution to a problem, in precisely the way in which Campbell says blind variations are not. It is also common for scientists to seek new hypotheses which will correct errors in their previous trials, as in Kepler's famous efforts to discover a formula to describe the orbit of Mars (Hanson 1958, pp. 733ff.). Thus the generation of the units of scientific variation does not have any of the three features of blindness which Campbell describes as characteristic of evolutionary variation.

Let us examine in some detail the process by which new theories are developed.[4] The non-randomness of theory generation has been most interestingly discussed by C.S. Peirce and N.R. Hanson. Peirce describes a form of inference called "abduction" which yields explanatory hypotheses (Peirce 1931-1958, Vol. 2, para. 776). Faced with a puzzling phenomenon, we naturally seek a hypothesis which would explain it. The form of abductive inference can be represented as follows:

(S1) Phenomenon P is puzzling.
Hypothesis H would explain P.
∴ H is plausible, and should be subjected to test.

Arguments for the existence of abduction are of two kinds. First, as a matter of historical fact, it seems that abduction is often used by scientists. Besides the example of Kepler already mentioned, we could cite the developments leading up to Darwin's discovery of the theory of natural selection. He describes being struck by the character of South American fossils and the geographical distribution of species there and on the Galapagos archipelego, and states that these facts are the "origin. . . of all my views." (Darwin 1959, p. 7; cf.,Darwin 1887, p. 42). These phenomena led him to believe that species had become modified, and after fortuitously reading Malthus he conceived how a struggle for survival could lead to natural selection. But the theory of natural selection was on no account a blind variation, since it served to account for phenomena which Darwin had been worrying about for years.

The second argument for the existence of abduction is that without some such sort of reasoning scientific growth would be impossible. For there would be no way of winnowing the unlimited set of possible hypotheses which would have to be considered and tested if hypotheses were generated randomly or blindly (Peirce 1931-58, vol. 5, para. 591; cf. Rescher 1978, ch. 3). If scientific theories and concepts were developed randomly, we would rarely come up with good ones, since the number of possible hypotheses is unmanageably great. Peirce hypothesized the existence in humans of an abductive instinct which innately aids our construction of hypotheses. But regardless of the existence of any special instinct, it is easy to see that a process wherein scientists intentionally strive to come up with hypotheses with certain characteristics will arrive at such hypotheses much more quickly than scientists generating hypotheses blindly. As Rescher notes (1978, p. 56), evolutionary epistemology is unable to account for both the existence and the rate of scientific progress.

N.R. Hanson (1961) discusses a form of reasoning akin to Peirce's abduction, which involves the conclusion that a sought for hypothesis is likely to be of a certain kind. The form of this reasoning is:

(S2) Phenomenon P is puzzling.
Similar phenomena have been explained by hypotheses of kind K.
∴ It is likely that the hypothesis we need to explain P will be of kind K.

Narrowing our search to certain kinds of hypotheses is obviously much more economical than blindly developing a huge variety of hypotheses. That Darwin arrived at a theory in which selection was a crucial concept was not accidental: he had earlier been struck by similarities between modifications in domestic species produced by artificial selection. Kepler's discovery was preceded by his conviction that the orbit of Mars was probably some sort of ellipse. Thus arguments that hypotheses are likely to be of a certain kind are a useful preliminary to the abductive inference that a particular hypothesis is worthy of investigation.

As Toulmin notes (1972, p. 337f.), in the history of science variation and selection are "coupled", in the sense that the factors responsible for selection are related to those responsible for the original generation of variants. Scientists strive to come up with theories which will survive the selection process. The criteria used in looking for a new theory in accord with (S1) and (S2) above are also relevant to arguments that a theory be accepted: at both levels, we want a theory which explains puzzling facts and which has analogies with accepted theories (Thagard 1978a). In contrast, species variation and selection are "uncoupled": the factors which produce genetic modification are unrelated to the environmental struggle for survival, except in special cases where the environmental threat is unusually mutagenic. The coupling of variation and selection for scientific theories makes theory choice a much more efficient procedure. If variation were blind, we would be faced with the necessity of choosing among an unmanageably large number of theories. Instead, the intentional, quasi-logical process by which hypotheses are generated narrows the range of candidates which must be considered for selection. That theoretical variation and selection are coupled is a serious flaw in the Darwinian model of the growth of knowledge.

Another possible objection to evolutionary epistemology concerns the magnitude of the advance which variations achieve over their predecessors. It might be said that variations in theories and concepts can involve substantial leaps, whereas in neo-Darwinian biology the development of species is gradualistic. However, I shall not press this point, because of the difficulty of assessing the relative size of leaps in such disparate spheres. Perhaps relativity theory does represent a "revolutionary" improvement over Newtonian mechanics, of a magnitude unparallelled in current biology which eschews saltations. But critical comparison is prevented by the indeterminacy of criteria for estimating magnitude of change and for distinguishing between revolution and evolution.

A clearer difference between biological and scientific development is that the rate of theoretical variation seems to be partly dependent on the degree of threat to existing theories. In Kuhnian terminology (Kuhn 1970), there is more likely to be a proliferation of new concepts and paradigms when a field is in a state of crisis. The rate of biological variation is not similarly sensitive to degree of environmenal pressure on organisms.

This completes my argument that theoretical variation is substantially different from biological variation. The main differences have concerned blindness, direction and rate of variation, and coupledness of variation and selection. It is ironic that the great merit of Darwin's theory - removing intentional design from the account of natural development - is precisely the great flaw in evolutionary epistemology. The relevant difference between genes and theories is that theories have people trying to make them better. Abstraction from the aim of scientists to arrive at progressively better explanations of phenomena unavoidably distorts our picture of the growth of science. I shall now argue that this is as true of the selection of theories as it is of the origin of theories.

The differences between epistemological and biological selection arise from the fact that theory selection is performed by intentional agents working with a set of criteria, whereas natural selection is the result of differential survival rates of the organisms bearing adaptive genes. Nature selects, but not in accord with any general standards. Nature is thoroughly pragmatic, favoring any mutation that works in a given environment. Since there is such an enormous range of environments to which organisms have adapted, we can have no global notion of what it is for an organism to be _fit_. Fitness is not inherently a property of an organism, but is a function of the extent to which an organism is adapted to a specific environment.

In contrast, theory and concept selection occurs in the context of a community of scientists with definite aims. These aims include finding solutions to problems, explaining facts, achieving simplicity, making accurate predictions, and so on (Kuhn 1977, ch. 13; Laudan 1977; Thagard 1978a). Perhaps at different times different aims are paramount, so that there may be inconstancy and even subjectivity in the application of criteria for theory choice. Certainly the application of such criteria is extremely complex, and there is nothing approaching an algorithm for determining which of competing theories deserves acceptance. Nevertheless, when scientists are advocating the adoption of a new theory, they appeal to some of a basic set of criteria according to which their theory is superior to alternatives. (See Thagard 1978a for illustrations.) Perhaps the criteria themselves have evolved, but since the seventeenth century there seems to me to have been agreement at the general level about what new theories should accomplish in explanation, problem solving and prediction, even if the application of these general aims in particular cases has been very controversial. But the controversy derives from the complexity of the set of criteria, not from any fundamental disagreement about the whole range of desiderata. Defense of this claim would take more space than is available here. If it is true, then selection of theories is strikingly different from the selection of genes. Survival of theories is the result of satisfaction of global criteria, criteria which apply over the whole range of science. But survival of genes is the result of satisfaction of local criteria, generated by a particular environment. Scientific communities are unlike natural environments in their ability to apply general standards.

Progress is the result of application of a relatively stable set of criteria. Progress is only progress _with respect to_ some general set of aims, and results from continuous attempts to satisfy the members of the set in question. Since scientists do strive to develop and adopt theories which satisfy the aims of explanation and problem solving, we can speak of scientific progress. In contrast, there is no progress in biological evolution, since survival value is relative to a particular environment, and we have no general standards for progress among environments. We could perhaps say that evolution of homo sapiens is progressive given our environment and our extraordinary ability to adapt to it, but our species may well someday inhabit an environment to which so-called lower animals are much better adapted. A post-nuclear war environment saturated with radioactivity would render us less fit than many less vulnerable organisms. Biological progress might be identified with increase in complexity, control over the environment, or capacity for acquiring knowledge, but none of these is a universal trend in evolution. As G.G. Simpson summarizes (1967, p. 260): "Evolution is not invariably accompanied by progress as an essential feature." Hence the Darwinian model of development employed in evolutionary epistemology lacks a concept of progress essential in historical epistemology. (For further discussion, see Ayala 1974 and Goudge 1961 on biological progress; and Laudan 1977 on progress in science.)

Thus selection is a stumbling block to evolutionary epistemology with respect to the conscious application of general criteria and the achievement of progress. Let us now consider biological and epistemological transmission.

Modern genetic theory provides us with an account of how genes which increase the fitness of an organism are preserved and transmitted to the organism's offspring. Preservation and transmission of conceptual survivors is quite different. A beneficial gene is replicated in specific members of a population, but a successful theory is immediately distributed to most members of a scientific community. Preservation is by publication and pedagogy, not by any process resembling inheritance. Dissemination of successful theories is much more rapid than dissemination of beneficial genes. This is one of the reasons why conceptual development seems to be so much more rapid than biological development. (The others include the intentional aspect of theoretical variation, and the progressive aspect of theory selection, already discussed.) Thus at the level of transmission of units of variation, as well as at the levels of variation and selection, the growth of knowledge is very different from the evolution of species.

Even the <u>units</u> of variation and transmission have very different properties. Dawkins (1976) postulates "memes" as the conceptual replicating entities analogous to genes. But this postulation is gratuitous since we already have notions which describe the entities which develop in scientific and cultural change. These entities include theories, laws, data, concepts, world views, and so on. Talk of memes does nothing to overcome the immense problems of explicating the nature of theories, concepts and world views. We know very little in detail about the nature of these entities, although they are clearly more complex and interconnected than are genes. A historical epistemology which is faithful to the actual history of science will have to go beyond misleading biological analogies.

What should a model of historical epistemology look like? Two possible alternatives to a Darwinian account of the growth of knowledge can quickly be seen to be inadequate. A Lamarckian model is superficially attractive since theories are passed on like acquired characteristics and there is progress in science, as Lamarck thought there was in natural evolution (Lamarck 1809 Goudge 1961). But a Lamarckian view would neglect competition and selection of theories as well as the way that progress comes about, not through any internal purpose of theories, but through the aims and intentions of scientists. Hegel's dialectic has much to add to historical epistemology, since he was probably the first philosopher to emphasize the historical nature of knowledge, and his notion of <u>Aufheben</u> is useful in conceptualizing how new stages of thought both supersede and preserve their predecessors

(Hegel 1807). However Hegel seems to have made precisely the opposite mistake of evolutionary epistemologists who suppose that the inception of conceptual variants is blind: for Hegel, each stage of knowledge is the logically necessary result of the stage that preceded it. Variation is not blind, but, contra Hegel, it is not wholly determined by context either. There is a subjective, psychological element in discovery along with an aim-oriented, methodological element.

Hence we are not in a position to borrow a model for the growth of knowledge from Lamarck, Hegel, or Darwin. A model needs to be constructed. Our discussion has shown that it should take into account at least the following factors:

1) the intentional, abductive activity of scientists in initially arriving at new theories and concepts;

2) the selection of theories according to criteria which reflect general aims;

3) the achievement of progress by sustained application of criteria; and

4) the rapid transmission of selected theories in highly organized scientific communities.

Evolutionary epistemology fails because it neglects all of these factors.[5]

Notes

[1] I am grateful to Daniel Hausman and B. Holly Smith for suggestions.

[2] My critique of evolutionary epistemology is not concerned with the claim that human biology may be relevant to epistemology in more direct ways, for example in debates concerning innate ideas (cf., Campbell 1974a). Nor do I address the "genetic epistemology" of Piaget (1950). Another important issue omitted here concerns the extent to which the growth of scientific knowledge is not a purely internal matter but is conditioned by social forces.

[3] For summaries of the neo-Darwinian theory of evolution see Lewontin (1974), Simpson (1967), Patterson (1978), and Ruse (1973).

[4] For more extensive discussions of questions related to logical factors in discovery, see Thagard (1977), (1978b), (in press).

[5] Since writing this paper, I have become aware of Skagestad (1978) which covers some of the same ground.

References

Ackerman, R. (1970). The Philosophy of Science. New York: Pegasus.

Ayala, F. (1974). "The Concept of Biological Progress." In Studies in the Philosophy of Biology. Edited by F. Ayala and T. Dobzhansky. Berkeley: University of California Press. Pages 339-355.

Campbell, D. (1974a). "Evolutionary Epistemology." In The Philosophy of Karl Popper. Edited by P. Schilpp. La Salle, Ill.: Open Court. Pages 413-463.

-----------. (1974b). "Unjustified Variation and Selective Retention in Scientific Discovery." In Studies in the Philosophy of Biology. Edited by F. Ayala and T. Dobzhansky. Berkeley: University of California Press. Pages 139-161.

Darwin, C. (1887). The Autobiography of Charles Darwin and Selected Letters. Edited by F. Darwin. London : John Murray. (Edited by Nora Barlow. Reprinted New York: Dover, 1958.)

---------. (1959). "Darwin's Journal." Edited by G. de Beer. Bulletin of the British Museum (Natural History), Historical Series, London, vol. 2, no. 1.

Dawkins, R. (1967). "Lamarck, Chevalier de." In Encyclopedia of Philosophy. Vol. 3. Edited by P. Edwards. New York: Macmillan. Pages 376-377.

----------. (1976). The Selfish Gene. New York: Oxford University Press.

Goudge, T.A. (1961). The Ascent of Life. Toronto: University of Toronto Press.

Hanson, N. (1958). Patterns of Discovery. Cambridge: Cambridge University Press.

---------. (1961). "Is There a Logic of Discovery?" In Current Issues in the Philosophy of Science. Edited by H. Feigl and G. Maxwell. New York: Holt, Rinehart and Winston. Pages 20-35.

Hegel, G. (1807). Die phänomenologie des geistes. Wurzberg: J.A. Goebhardt. (Translated by J. Baille as The Phenomenology of Mind. New York: Harper & Row, 1967.)

Kuhn, T. (1970). The Structure of Scientific Revolutions. 2nd ed. Chicago: University of Chicago Press.

-------. (1977). The Essential Tension. Chicago: University of Chicago Press.

Lamarck, J. (1809). Philosophie zoologique, ou exposition des considerations relatives a l'histoire naturelle de animaux. Paris: Dentu. (Translated by H. Elliot as Zoological Philosophy. New York: Hafner, 1963.)

Laudan, L. (1977). Progress and Its Problems. Berkeley: University of California.

Lewontin, R. (1974). The Genetic Basis of Evolutionary Change. New York: Columbia University Press.

-----------. (1977). "Sociobiology - A Caricature of Darwinism." In PSA 1976, Vol. 2. Edited by F. Suppe and P. Asquith. E. Lansing, MI: Philosophy of Science Association. Pages 21-31.

Patterson, C. (1978). Evolution. London: British Museum (Natural History).

Peirce, C. (1931-1958). Collected Papers. Edited by C. Hartshorne, P. Weiss, and A. Burks. Cambridge: Harvard University Press.

Piaget, J. (1950). Introduction a l'epistemologie genetique. Paris: Presses universitaires de France. (Translated by E. Duckworth as Genetic Epistemology. New York: Columbia University Press, 1970).

Popper, K. (1972). Objective Knowledge. London: Oxford University Press.

Rescher, N. (1978). Peirce's Philosophy of Science. Notre Dame, IN: Notre Dame University Press.

Ruse, M. (1973). The Philosophy of Biology. London: Hutchinson.

-------. (1977). "Karl Popper's Philosophy of Biology." Philosophy of Science 44: 638-661.

Simpson, G.G. (1967). The Meaning of Evolution. Revised Edition. New Haven: Yale University Press.

Skagestad, P. (1978). "Taking Evolution Seriously: Critical Comments on D.T. Campbell's Evolutionary Epistemology." Monist 61: 611-621.

Thagard, P. (1977). "The Unity of Peirce's Theory of Hypothesis." Transactions of the Charles S. Peirce Society 13: 112-121.

----------. (1978a). "The Best Explanation: Criteria for Theory Choice." Journal of Philosophy 75: 76-92.

----------. (1978b). "Semiotics and Hypothetic Inference in C.C. Peirce." VS: Quaderni di studi semiotici 19/20: 163-172.

----------. (1980). "The Autonomy of a Logic of Discovery." In Pragmatism and Purpose. Edited by J. Slater, W. Sumner, and

F. Wilson. University of Toronto Press. In Press.

Toulmin, S. (1972). Human Understanding. Princeton: Princeton University Press.

Levels of Reflexivity: Unnoted Differences within the "Strong Programme" in the Sociology of Knowledge[1]

Edward Manier

University of Notre Dame

Although Barry Barnes (1974, 1977) and David Bloor (1976, 1978, 1981) are co-workers in the Science Studies Unit at the University of Edinburgh, there are hitherto unnoted but fundamental differences in their programs in the sociology of science.

Bloor defines his "strong programme" in the sociology of science by three basic tenets ("causality," "symmetry," and "reflexivity") and also claims to establish laws and to test a general theory of the causal links connecting cognitive and social factors in the history of science (Bloor 1976, pp. 4-5). In other words, the "strong programme" aims to be: (1) Causal, i.e., concerned with the conditions which bring about beliefs or states of knowledge. (2) Symmetrical, i.e., appealing to the same sorts of causes and to the same patterns of causal explanation for all beliefs and cognitive claims, whether regarded as true or false, rational or irrational, successful or unsuccessful. (3) Reflexive, i.e., using the explanatory resources of the sociology of science in the critical evaluation and explanation of the sociology of science itself.

In marked contrast, Barnes (1977, pp. 85-6) denies the possibility of laws and general theories in the sociology of science and restricts his program of investigation to case-by-case analyses. He seeks to distinguish "legitimate" and "ideological" interests contributing to the growth of knowledge, while Bloor does not. For Barnes, "legitimate" interests lead to systems of belief primarily explicable in terms of the esoteric requirements of relevant technical communities (e.g., requirements for adequacy and accuracy of evidence, skill in the use of relevant instrumentation, and so on). Such communities seek explicitly and exclusively to justify cognitive claims by reference to criteria calling for successful prediction and control of the explained event. Such aims are limited, of course, by the scientific and other cultural materials which are historically available. "Ideological" interests, on the other

PSA 1980, Volume 1, pp. 197-207

hand, conceal or fail to acknowledge the criteria by which they assign positive values to cognitive structures. In addition, interests may be characterized as ideological if they are illegitimate. While Barnes provides no general explication of a relevant sense of illegitimacy, he does provide some illumination for the distinction of "legitimate" and "ideological" interests through the analysis of particular case studies (Barnes, 1974, pp. 130-39; Shapin and Barnes, 1979; MacKenzie and Barnes, 1979). Bloor makes no comparable effort and generally proceeds as if a distinction of legitimate and ideological interests would violate his symmetry requirement.

A basic question confronting any program in the sociology of science is: "Can it be reflexive without being self-destructive?" Can the cognitive claims put forward by the sociology of science withstand the uniquely skeptical atmosphere generated by that discipline? Does the thesis that all cognitive claims are socially determined, e.g., as more or less well-disguised attempts at social control, dissolve the grounds of its own credibility? This paper deals with these questions by means of a critical comparison of Barnes' and Bloor's divergent programs in the sociology of science.

1. The "Strong Programme" and the Debate over Implicit Meanings

Bloor (1978) seeks to use Mary Douglas' (1970, 1975) views on "classification and control" in offering sociological explanations for the development of the dialectical method of proofs and refutations in 19th century mathematics (Lakatos 1976). He argues as follows: (a) Eighteenth century German universities were closed and collegial, retaining ancient guild privileges insuring the dominance of group loyalties. These groups, however, were divided by internal conflict and charges of corruption and immorality. The social pressure within such groups explains group members' anxious attitudes toward anomalies and counter-examples. (b) After 1806, Prussian universities were reorganized according to impersonal bureaucratic principles inadvertently insuring that the social conditions of work in the sciences and the humanities would be individualistic, pluralistic, and competitive. Social forms of this sort exert pressure in favor of innovation and novelty, encouraging transactions across the boundaries of existing classificatory schemes, and eventually dissolving and replacing such boundaries. The social pressure within such groups explains a new attitude toward mathematical anomalies, i.e., a tendency to embrace counter-examples for the sake of deepening rather than refuting theorems. This analysis implies that social conditions determine the behavior of individual scientists, but Bloor neither explains the mechanism of such determination nor indicates the means for dealing with the plethora of counter-examples associated with all such claims (Boon 1979; Freudenthal 1979; Worrall 1979).

Bloor's (1981) most recent work uses the findings of J.R. Jacob (1972) and M.C. Jacob (1976) to break out of what he calls the "trivial circle of ideas" dominating the work of historians and philosophers of science concerning seventeenth century beliefs in the complete passivity of matter. Putatively, the factors explaining these beliefs are the social uses of cosmology in the struggle for power dividing religious latitudinarians, dissenters, Catholics and atheists in seventeenth century England. While Bloor criticizes historians and philosophers of science (McMullin 1978) who overlook the Jacobs' accounts of the social context within which Boyle and Newton worked, he makes the inverse error of failing to come to terms with the task of determining the scientific meaning of the thesis that matter is completely inert. P.M. Heimann (1978) offers a relevant set of distinctions. He concurs with M.C. Jacob's recognition that (1) Newton's natural philosophy cannot itself be held to be determined by its social relations. He distinguishes that claim from alternatives (2) that the social struggle for religious power *explains* the triumph of Newtonianism in England, and (3) that these social and political issues played a *vital role* in the acceptance of the Newtonian natural philosophy.

Unfortunately, Bloor's discussion of concrete case materials confuses all three sorts of claims. We never find out whether he is offering sufficient causal determinants or indispensable contributing conditions for his explananda. Nor is it clear whether the explananda themselves are individual beliefs, major archival structures, or the characteristic features of particular patterns of social consensus. Nor does he distinguish efforts to explain the original development and presentation of a set of ideas from their subsequent diffusion and acceptance within particular sectors of the scientific community or the larger culture.

Bloor's use of Douglas' theory is unsatisfactory on at least two counts. First, it is insufficiently critical on those points where Douglas' work is too obscure for ready translation into the sociology of science. Second, it fails to read Douglas in the appropriate esoteric context: the on-going dispute among anthropologists concerning the appropriate theoretical framework of their own discipline (Sahlins 1976a, 1976b; Harris 1977, 1979; Douglas 1966, 1970, 1975).

I will deal first with the second criticism. The preface to *Implicit Meanings* (Douglas 1975) extends Durkheim's account of the sacred and the profane to "engulf fundamentalist theories of knowledge as well as fundamentalist religious doctrines."(p. xv). Douglas seeks to do this *without* engulfing cognitive claims which do *not* rest on fundamentalist theories of knowledge. She argues that Durkheim's failure to develop a sociological critique of science may be traced to his unexamined assumptions concerning "objective scientific truth", or of a "non-context-dependent" or non-culture-dependent sense of truth as correspondence to reality. She rejects Durkheim's assumption in favor of a theory of knowledge in which "the mind is admitted to be actively creating its universe."(*ibid*., p. xviii). She holds that all societies are based on an apprehension of some general pattern of what is right and necessary in social relations, and that "this apprehension generates whatever *a priori* or set of necessary causes is going to be found in nature." (*ibid*., p. 281). But she does not reduce all cognitive claims to their underlying social context. She holds that "anyone who would follow Durkheim must give up the comfort of stable anchorage for his cognitive efforts. His only security lies in the evolution of the cognitive scheme, unashamedly and openly culture-bound, and accepting all the challenges of that culture. It is part of our culture to recognize at last our cognitive precariousness." (ibid., p. xviii).

The "recognition of our cognitive precariousness", as she has it, must preclude any new "list of intuitions or innate ideas common to the human race." Bloor, however, uses the Douglas-Bernstein scheme of classification and control as if it were a graphic representation of such a list, as if all ideas and, by implication, every aspect of human culture, could be explained by a common human interest in social control.

Bloor might remain closer to the spirit of Douglas' position if he would take account of its location in current anthopological debate, particularly that between Douglas and Marvin Harris concerning the interpretation of the prohibitions of Leviticus. For Douglas (1966), the ancient Hebrew taboo concerning the flesh of the pig is explained by reference to a system of symbols: an animal which "parts the hoof but does not chew the cud" is a taxonomic anomaly (ibid., p. 54). Although she subsequently discusses the possibility that the relevant ethnozoology was itself related to concerns for group solidarity (1970, pp. 77-92), she first interpreted and explained it in terms of a comprehensive, coherent system of classification with its own distinctive symbolic meaning.

Harris' contrasting account explains the taboo on pork in terms of a direct cost/benefit analysis. Fundamental changes in the ecology of the region putatively made it extremely inefficient to invest scarce resources in the domestication of swine. The taboo efficiently redirected these resources to more productive channels (Harris 1977, 1979). Harris spends considerable energy criticizing Douglas on the grounds that she remains confined to a trivial circle of ideas and fails to explain the Leviticus ethnography in terms of its ecological costs and benefits. He would have little or no reason to complain of Bloor's version of Douglas' theory, since Bloor reads Douglas as explaining all cultural products in terms of their role in a struggle for social power. Bloor's use of Douglas falls under the same strictures M. Sahlins places on those who make a "fetish of utility" (Sahlins 1976a, p. x).

To return now to my first criticism of Bloor, Harris' interpretation of Douglas (as a non-utilitarian symbolist) is preferable to Bloor's (he makes her out to be a social utilitarian) because the former account relies less on Douglas' obscure claims concerning the anthropological significance of Bernstein's (1971) account of classification and control. Bloor fails to translate this scheme into the sociology of science in any systematic fashion. Consider Douglas' typical diagram of those dimensions of the individual's environment corresponding to "classification" and "control".

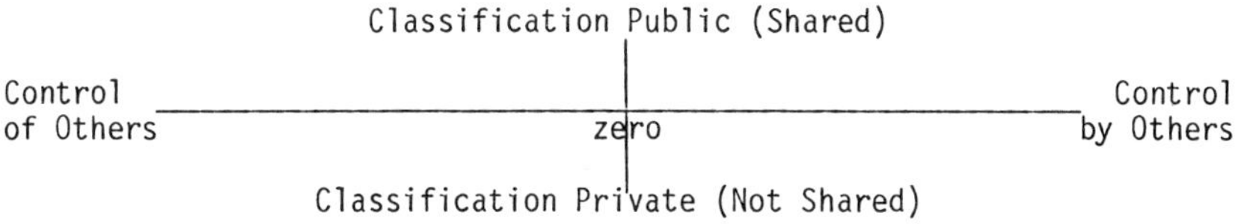

Douglas 1975, figure 14.1, p. 218, after Bernstein 1971; also see Douglas 1970, pp. 77-112.

As the scope of shared classifications increases (e.g., "Help me, I am a kinsman," or "Help me, I am indigent"), the possibilities for control through classification also increase. But as the scope of shared classification moves through zero and becomes negative, the nature of the control variable changes. Control not based on classification must depend either on strong personal bonds or on violence. The control axis in Douglas' diagram is a confused representation of two or three types of social relations, which may or may not vary concomitantly.

A second problem arises when the domain of reference of Douglas' diagram is shifted (as it must be for use in the sociology of science) from individuals to groups or to the internal relations which distinguish different types of groups. Several different types of problems are involved in attempts to analyze the social context of scientific discovery, e.g., that of the molecular configuration of DNA: (1) the analysis of the pair-wise relations connecting Watson and Crick to Pauling and Delbruck or to Wilkins and Franklin; (2) the person-to-person relations within each of these pairs; (3) institutional relations differentiating and connecting the Cavendish laboratory at Cambridge with the corresponding laboratories at the California Institute of Technology and King's College, London (Olby 1974; Judson 1979). Neither Bloor nor Douglas explores the difficulties involved in defining and measuring relations of classification and control in such diverse contexts. Until such issues are clarified, however, Bloor's hope of resting the strong programme on Douglas' account of classification and control is based on the doubtful premise that it has sufficient theoretical coherence and empirical significance to warrant its translation into the sociology of science.

2. Barnes' Modest Version of the "Strong Programme"

Barry Barnes wants to save the distinction of science and ideology even though both sorts of belief, on his view, are partly determined by social and cultural circumstances. Knowledge is ideologically determined, he argues, only insofar as it is created, accepted, or sustained by concealed, unacknowledged, or illegitimate interests (Barnes 1974, p. 128; 1977, p. 33). He denies that science and ideology are inextricably intertwined to constitute a pervasive "total ideology" (1974, pp. 138, 145). Knowledge does not have to be ideologically determined, and should not be. Social institutions wherein knowledge is generated and sustained entirely under the impetus of legitimate cognitive interests are both possible and desirable (1977, p. 43).

However, Barnes also holds that we lack a theory of natural rationality powerful enough to explain how the cognitive propensities of actors, in given cultural contexts, would or would not lead them to particular programs or theories simply from legitimate cognitive interests (1977, p. 34). Consequently, the mode of determination of particular beliefs cannot be used as indicators of their inherent scientific or ideological character.

Barnes also argues that since scientific knowledge is a resource for social-political action rather than a direct determinant of such action, it is inappropriate to search for laws linking the cognitive claims of science and the social order, or even to propose abstract instructions for the sociological investigation and explanation of scientific knowledge. The sociology of knowledge can only develop in immediate moves from one concrete historical case study to another. It cannot pretend to articulate and test an abstract theoretical structure (Barnes 1977, pp. 85-6).

MacKenzie and Barnes (1979) analyze the controversy over human heredity and the process of evolution which divided biometricians (Karl Pearson and others) and Mendelians (William Bateson and others) in the early years of this century. The main features of their study illustrate the methodology recommended by Barnes. They first attempt to identify the central scientific feature of the dispute, in this case: "Is natural selection or mutation (saltation) the basic mechanism of evolutionary change?" The first step is the assessment of the role played in the dispute by relevant "technical factors and esoteric professional interests." In this instance, the contending parties did not have access to different kinds of evidence and the differences in their training and skills did not account for their failure to interrupt their dispute to await further evidence or to look for ways in which their differing accounts might be applied to different kinds of evolutionary change.

MacKenzie and Barnes argue that the structure of the controversy becomes clearer when it is placed in the appropriate social context, that of the eugenics movement. Karl Pearson's interests were aligned with a rising professional class, concerned with the management of the industrialization of British society. If natural selection were the mechanism of evolution, human evolution might be managed and rationally directed in keeping with the goals of the new society. On the other hand, Bateson was aligned with the conservative opposition to modernization, particularly the "blighted atomistic individualism of the utilitarians." If mutation were the mechanism of evolution, the newer forms of rational management could not be applied to the direction of human evolution.

Barnes insists that that his intent is not to explain the behavior of individual scientists, not even that of Pearson and Bateson. His hypothesis draws on the evidence of individual utterances and publications but is not intended to explain the cognitive claims of individual authors. For example: "One would expect the controversy (over eugenics) to decline in significance if social structural changes weakened the particular interests sustaining it, or ... if social structural changes resulted in (the claims of the biometricians and the Mendelians) no longer being the most expedient means of furthering or legitimating those interests... . The central claim being made is that, in the absence of the social-structural factors referred to, the controversy would never have emerged, at least in the particular form observed." (Barnes 1977, p. 61). Barnes identifies a necessary condition for particular social-structural features of the eugenics controversy, but explicitly disavows the claim to have set forth general laws or necessary connections linking knowledge and the social order, as well as the claim to explain the "affiliations, actions, and utterances which make up individual biographies."(*ibid.*, pp. 63, 85).

3. Summary: Unnoted Differences and Common Shortcomings

There are at least four major but unnoted differences in the published work of David Bloor and Barry Barnes. (1) Barnes distinguishes the referents of 'science' and 'ideology'; Bloor does not. (2) Bloor claims to establish laws and test a general theory of causal connections between cognitive and social factors in the history of science. Barnes denies the possibility of such laws and theories, recommending a case-by-case analysis which always begins by attempting to identify the technical factors (evidence, instrumentation, skill, research training) relevant to the differentiation of legitimate cognitive interests. (3) Barnes insists on maintaining a structural level of analysis and explanation. He denies that social-cultural factors (in the absence of theories linking social and individual psychological events) explain the utterances or textual archives associated with individual scientists. Within his program, social factors explain social phenomena; e.g., explananda include such events as variations in intensity or in the social form of particular scientific controversies. Bloor's discussions of Popper, Kuhn, Seidel, Boyle and Newton, on the other hand, seem to rest uncritically on the assumption that social structure determines the behavior of individual human actors. (4) Barnes recognizes the possibility of social institutions within which scientific knowledge can be generated and sustained exclusively under the impetus of legitimate cognitive interests. Bloor is silent on this topic, but it does not seem unfair to read his case studies as implying that such institutions are either impossible, or, if they occur, inexplicable.

Barnes' modest version of the strong program exhibits an entirely different level of reflexivity than Bloors'. For the latter, detailed case studies function only to explain knowledge as a tool or weapon in a struggle for social power. As it is presently constituted, Bloor's program could not survive the reflexive application of its own methods. On Barnes' account (and in his practice), however, reflexivity requires the identification of social factors underlying the social-structural impact of a research program upon both the scientific community and its larger social context. Such analysis does not touch immediately on the properly scientific question of the legitimacy of the program's findings, but is necessary for the discovery of ideological (concealed or illegitimate) interests in its elaboration. Such reflexivity contributes to the program's use in the identification of social conditions favoring the development of knowledge under legitimate cognitive interests.

On the other hand, neither Barnes nor Bloor fully explores the symbolic uses of scientific language. As a result, neither provides an adequate foundation for the critical analysis of the interplay of scientific, ideological, and moral interests in the growth of scientific knowledge. For example, the figurative representation of a new scientific theory may express, reduce, or intensify the cultural conflict associated with its introduction (Gliserman 1975). Such representation also fulfills persuasive and polemical functions required for the formation of new specialty groups or the reorganization of groups associated with old theories (Manier 1978, pp. 168-171, 181-190). These broad social and cultural functions complement the cognitive functions Hesse (1966) has noted: (1) the redescription of puzzling phenomena, (2) the redefinition of central concepts, and (3) the extension of proven formalisms to new subjects. The meanings of a scientific theory are the outcomes of relationships between the theory and its antecedent and consequent cultural contexts. In complex and pluralistic modern cultures, these relationships are dialectical, equivocal, or analogical, and rarely if ever simple or deterministic (Manier 1980). A more thorough analysis of the various symbolic uses of scientific knowledge is necessary for the defense and concrete application of Barnes' distinction of science and ideology (Barnes 1974, pp. 53-59, 86-96).

Bloor's most recent work (1981, pp. 15-23) indicates an interest in combining the positions of Hesse and Douglas to show how a "concern with the prediction and control of nature is not automatically sacrificed by giving nature a moral employment." It is true that Hesse's network theory of scientific inference is compatible with the existence of causal connections between social forces and the logical structure and conceptual content of science, but Hesse (1970) is careful to avoid any implication that the meaning of scientific theories is to be found in their uses as weapons in the struggle for social power. At this point, Barnes' cautious use of case-study materials points to a path avoiding the eliminative reduction of cognitive claims to the categories of social control. Bloor's enthusiastic accounts of the causal links connecting a society's distinctive forms of social control, its uses of nature, its scientific knowledge, and the scientific activities of its individual historical actors, on the other hand, suggests a thoroughgoing sociological reduction of science, and, a *fortiori*, of Bloor's own program.

Notes

[1]I am very appreciative of David Bloor's friendly and generous responses to early versions of my criticism of his program. The development of my evaluation of Douglas' use of the scheme of "classification and control" owes much to collegial discussion with J. David Lewis, Department of Sociology and Anthropology, University of Notre Dame. The efforts of Bloor and Lewis isolate responsibility for errors remaining in the paper squarely on my shoulders. My work on this topic was initially facilitated by a conference grant from the Center for the Study of Man in Contemporary Society, University of Notre Dame; it was continued while I was on leave as a Fellow of the National Humanities Institute at the University of Chicago.

References

Barnes, B. (1974). Scientific Knowledge and Sociological Theory. London: Routledge & Kegan Paul.

---------. (1977). Interests and the Growth of Knowledge. London: Routledge & Kegan Paul.

Barnes, B. and Shapin, S. (eds.). (1979). Natural Order. San Francisco: Sage.

Bernstein, B. (1971). Class, Codes and Control. London: Routledge & Kegan Paul.

Bloor, D. (1976). Knowledge and Social Imagery. London: Routledge & Kegan Paul.

--------. (1978). "Polyhedra and the Abominations of Leviticus." British Journal for the History of Science. 11: 245-72.

--------. (1981). "Durkheim and Mauss Revisited: Classification and the Sociology of Knowledge." To be published in The Language of Sociology. Edited by John Law.

Boon, L. (1979). "Review of Knowledge and Social Imagery." British Journal for the Philosophy of Science. 30: 195-9.

Douglas, M. (1966). Purity and Danger. London: Routledge & Kegan Paul.

----------. (1970). Natural Symbols. New York: Pantheon. (Cited edition (1973) New York: Vintage.)

----------. (1975). Implicit Meanings. London: Routledge & Kegan Paul.

Freudenthal, G. (1979). "How Strong is Dr. Bloor's 'Strong Programme'?" Studies in History and Philosophy of Science. 10: 67-83.

Gliserman, S. (1975). "Early Victorian Science Writers and Tennyson's 'In Memoriam': A Study in Cultural Exchange." Victorian Studies. 18: 277-308, 437-59.

Harris, M. (1977). Cannibals and Kings. New York: Random House.

---------. (1979). Cultural Materialism. New York: Random House.

Heimann, P.M. (1978). "Review of The Newtonians and the English Revolution, 1689-1720." History of Science. 16: 143-151.

Hesse, M. (1966). Models and Analogies in Science. Notre Dame, IN: University of Notre Dame Press.

--------. (1970). "Hermeticism and Historiography: An Apology for the Internal History of Science." In Historical and Philosophical Perspectives of Science. (Minnesota Studies in Philosophy of Science, Volume V.) Edited by Roger Stuewer. Minneapolis: University of Minnesota Press. Pages 134-160.

--------. (1974). The Structure of Scientific Inference. Berkeley: University of California Press.

Jacob, J.R. (1972). "The Ideological Origins of Robert Boyle's Natural Philosophy." Journal of European Studies. 2: 1-21.

Jacob, M.C. (1976). The Newtonians and the English Revolution, 1689-1720. Ithaca: Cornell University Press.

Judson, H.F. (1979). The Eighth Day of Creation. New York: Simon and Shuster.

Lakatos, I. (1976). Proofs and Refutations. Edited by J. Worrall and E. Zahar. Cambridge: Cambridge University Press.

MacKenzie, D. and Barnes B. (1979). "Scientific Judgment: the Biometry-Mendelism Controversy." In Barnes and Shapin (1979). Pages 191-210.

McMullin, E. (1978). Newton on Matter and Activity. Notre Dame, IN: University of Notre Dame Press.

Manier, Edward. (1978). The Young Darwin and His Cultural Circle. Dordrecht: Reidel.

--------------. (1980). "History, Philosophy and Sociology of Biology: A Family Romance." Studies in the History and Philosophy of Science 11: 1-24.

Olby, R. (1974). The Path to the Double Helix. Seattle: University of Washington Press.

Sahlins, M. (1976a). Culture and Practical Reason. Chicago: University of Chicago Press.

----------. (1976b). The Use and Abuse of Biology. Ann Arbor, MI: University of Michigan Press.

Shapin, S. and Barnes B. (1979). "Darwin and Social Darwinism: Purity and History." In Barnes and Shapin (1979). Pages 125-42.

Worrall, J. (1979). "A Reply to David Bloor." British Journal for the History of Science. 12: 71-78.

Part VIII

History and the Metaphilosophy of Science

Toward a Historical Meta-Method for Assessing Normative Methodologies: Rationability, Serendipity, and the Robinson Crusoe Fallacy[1]

Stephen J. Wykstra

University of Tulsa

During the past two decades, much philosophy of science has been focused on issues about the norms and methods by which scientific theories are rationally appraised; and increasingly, philosophers have turned to history of science as a touchstone for assessing normative methodologies purporting to elucidate scientific rationality. But even among such historical methodologists, there is much disagreement and unclarity about how historical study of science can arbitrate between rival methodological theories; and until progress is made at this meta-methodological level, the very legitimacy of this role for history will remain controversial. (Keynotes in the controversy are sounded in Kuhn (1970b , pp. 235-41; Lakatos (1971); Giere (1973); McMullin (1976); Burian (1977); and Laudan (1977 , pp. 158-63).) This paper begins by arguing that the meta-method implicit in much historical methodology is different from the explicit meta-methodology most often touted. This implicit meta-method—involving the rationability principle—appears to lead almost inevitably to the methodological anarchism of Feyerabend, and (in mitigated forms) of Lakatos and Kuhn. Hence the main aim of this paper: to redeem the rationability principle by arguing that this specter of anarchism can be exorcized from it, provided that we avoid several misconceptions about the nature of rational norms. The most serious of these is a Robinson Crusoe fallacy which, having originally misled Kuhn to anarchistic conclusions, has more recently confounded a dispute between Grünbaum and Worrall.

1. Two Approaches to Historical Methodology

The methodologist's first cue is Einstein's advice that to understand the methods of physics, one must attend to what the physicist does, not to what he says he does. For in the large sense of "does"

PSA 1980, Vol. 1, pp. 211-222

which includes cognitive doings, what the scientist does when he theorizes about the world is only darkly seen in the glass of what he says he does when he pauses to reflect on his doings. But on entering the stage, the methodologist seldom restricts himself to mere description of the norms implicit in scientific activity: rather, he aspires to give an account of the norms that ought to govern scientific theorizing. Of course, this more interesting and ambitious enterprise of normative methodology itself cries out for higher-order reflection: how should the methodologist assess a theory which purports to prescribe norms of scientific rationality? This is our meta-methodological question; and one can hardly begin the business of normative methodology until one has at least implicitly answered it. But perhaps Einstein's maxim also applies to what methodologists say when they talk about their methods. If so, we need to pay attention to how methodologists actually reason about scientific norms, not just to how they say they reason about them.

Among historical methodologists, the dominant explicit meta-methodologies enshrine what can be called an "intuitionist" strategy. The methodologist tells us he begins with certain "pre-analytic intuitions" (or "basic value judgments"), leading him to suppose that if rationality is to be found anywhere, it is embodied in (most of) a given set of historical scientific episodes. He then assesses a candidate methodological theory by asking whether, by its lights, these episodes were rational episodes. A normative methodology, on this approach, does purport to describe the norms implicit in the cognitive doings of scientists: but a normative methodology which succeeds in this explication also has prescriptive force, because the explicanda are those cognitive doings about which the methodologist has firm "normative intuitions" to begin with. And a good normative methodology is useful (not just a rationalization) because the methodologist, having explicated the norms implicit in episodes about which he has clear intuitions, can apply those norms to cases about which our intuitions are unclear.

This intuitionist strategy has been most clearly and recently expressed by Laudan (1977, pp. 158-63); variations on it are also evident in Hall (1971), Lakatos (1971), Maull (1976), Millman (1976), and the Lakatos-inspired essays collected in Howson (1976). This type of approach to historical methodology is notably akin to one meta-methodology within the decidedly non-historical tradition of logical empiricism: Carnapian inductive logic also appeals to "pre-analytic intuitions", although these are of course about elementary confirmation situations and logical requirements, not about complex episodes in the history of science. (Cf., Creary (1971) and Cohen (1975).)

In the actual procedures of historical methodologists, however, one can find implicit commitment to a very different meta-method. Consider briefly four examples. Thomas Kuhn (1970b) rejects certain norms of scientific rationality because he judges that they would, if accepted, preclude a type of commitment to "dogma" which he considers essential to scientific progress. Paul Feyerabend (1975) rejects other normative methodological theories (including Kuhn's) because they would, if accepted, preclude a pluralism which he argues is crucial to progress. Larry Laudan (1976, 1979) criticizes various forms of the "cumulativity postulate" or "correspondence principle" (most recently the variation enshrined in the "convergent realism" of Putnam and Boyd) because such a principle would have precluded a number of theoretical episodes which are paradigms of scientific progress. And finally, Imre Lakatos and his followers—cf., Quinn (1972) and Musgrave (1976)—reject norms which derive "heuristic advice" from "methodological appraisal", on the grounds that if such norms had been used in the past, they would have precluded research which culminated in highly progressive theoretical advances.

Each of these arguments uses the history of science as a methodological touchstone; but the dominant historical meta-methodology does not begin to do justice to the pattern of argumentation implicit in them. For the above examples do not test candidate norms merely by asking whether these norms implicitly governed the cognitive behavior of past and presumedly rational scientists. Rather, the crucial question is whether these norms would have conduced to progress, if (hypothetically or even contrary-to-fact) they had been used. The idea behind the question differs from the intuitionist meta-methodology as much as Reichenbach's approach to inductive logic differs from Carnap's—and indeed in much the same way. (See Creary (1971) on this parallel contrast between Carnapian and Reichenbachian meta-methodologies.) For the idea is that one "vindicates" a set of candidate norms by showing that their employment optimally promotes the realization of scientific aims or values. Since "progressive" scientific episodes are (by definition) ones which embody a substantial realization of scientific aims, a candidate norm is "suspect" whenever it can be cogently argued that its employment would have precluded or inhibited such episodes.

And how can such a thing be "cogently argued"? Sketched briefly, one strategy comprises two phases. The first is to identify certain steps or commitments as "essential" to paradigmatically progressive episodes. Arguably, the advance to a unified heliocentric astronomy and physics could not have transpired unless scientists like Copernicus and Kepler had doggedly committed themselves to a weak infant rival to the mature Ptolemaic system. In the second phase, the methodologist's crucial question will be whether a candidate normative methodology would sanction these commitments as "rational". But the term "rational" needs special scrutiny here. The methodologist is not asking whether, by the lights of the candidate norms, any particular scientists were rational in their commitments: whether Copernicus or Kepler were motivated by good reasons, though crucial on an intuitionist meta-method, is quite irrelevant here. The question instead is whether, by the lights of the candidate norms, there were good reasons available for the commitments requisite to the progressive episode. To mark this distinction, obscured by an ambiguity in the term "rational", I hereby coin the new term "rationable": a commitment to a theory is "rationable" if and only if there were, in the historical context, sufficiently good reasons available to warrant the commitment.

We can thus see that while the meta-methodologies here contrasted both employ the history of science to arbitrate between rival normative methodologies, they do so in different ways and for different reasons. Unlike an intuitionist meta-method, the "rationability meta-method" need not begin with any prior convictions about the rationality of some, most, or all past scientists: for its aim is not to explicate the norms which governed the reasoning of presumedly rational scientists. It rather begins with presumptions about the progressiveness of a set of theory changes: and a presumption about whether the emergence of the heliocentric theory was a progressive theoretical advance is clearly and radically different from an intuition about whether Copernicus was, in a person-specific sense, rational in his commitments. Beginning with such progress-presumptions (which serve to identify the aims of science prior to having a philosophical account of those aims), this meta-methodology inquires whether a candidate norm would--if it had been employed--have promoted or inhibited the realization of scientific aims: would it have sanctioned as "rationable", or forbade as "irrationable", the commitments requisite to the progressive episodes? Hereafter, I shall refer to this as "the rationability principle", or "the rationability meta-methodology".

Elsewhere (Wykstra 1978), I have argued at some length that the rationability meta-methodology has some striking advantages over intuitionist approaches. It allows us to give a much more defensible account of why factual historical studies are germane (and indeed indispensable) to an ultimately normative enterprise; it permits a satisfying solution to the problem of how one can test norms of scientific rationality without taking for granted the very norms at issue; and it opens the prospect for a genuinely constructive methodological enterprise, which can develop new normative principles by which future science ought to be done, even if these principles have not been even implicitly operative in past scientific reasoning. But, space not permitting a defense of these claims here, I instead undertake a more modest task: to show that the rationability principle does not license some objectionable conclusions which its tacit (mis)application has led some methodologists to espouse.

2. Serendipity and the Specter of Methodological Anarchism

Tacit commitment to the rationability principle has led some philosophers to conclusions which are, for many of us, so odious as to constitute a reductio ad absurdum of any meta-methodology which licences them. I refer to the "methodological anarchism" of Feyerabend and (in more restrained forms) of Lakatos and Kuhn. But in defense of the rationability approach, I shall argue that these thinkers are led to "anarchistic" conclusions only because they contaminate the approach with defective auxilliary assumptions about norms of scientific rationality.

The road to methodological anarchism begins from a realization that even the best norm-governed judgments will be revocable: even the best norms of theory-rejection will, for example, sometimes lead scientists to reject ideas which later turn out to have unexpectedly great explanatory power. This characteristic revocability opens the door to what can be called "the serendipity factor": it allows that on some occasions, a scientist might dogmatically commit himself to a theory which by any plausible norms is indefensible, but which—as a result of his dogged labors on its behalf—fortuitously progresses until it outshines all its rivals in explanatory power, verisimilitude, problem-solving effectiveness, or whatever else it is that makes a good theory good. Quinn thus observes that even though some theory T gets a dismally low rating by Lakatosian norms, it is entirely possible that "irrationally pertinacious adherents (to T) might score spectacular triumphs denied to scientists who follow the directives Lakatos proposes." (1972, p. 147). This is because Lakatosian norms can warrant only revocable verdicts: and such norms can thus never guarantee that the rational scientist who follows them will fare better than the irrational scientist who flaunts them. As Quinn puts it: "Fortune may smile on the irrational gambler..., and even in science there will always be room for serendipity."(1972, p. 146).

Even a modicum of serendipity in scientific progress forces on us the following question: if the rationability approach is used to assess candidate norms of scientific rationality, will not the conclusion follow that no norms should be used to govern scientists' judgments, decisions, or commitments? This is just the anarchistic corner into which Feyerabend, Lakatos, and Kuhn have, in various degrees, painted themselves. Consider Feyerabend, who seems to like the corner: by adducing historical episodes in which science progressed through violations of the norms of this or that methodology, Feyerabend argues that "given any rule, however 'fundamental' or 'necessary' to science, there are always occasions when it is advisable not only to ignore the rule, but to adopt its opposite." (1975, p. 23). Hence Feyerabend's infamous "methodological anarchism": "It is clear, then, that the idea of fixed method, or a fixed theory of rationality, rests on too naive a view of man and his social surrounding. To those who look at the rich material provided by history, and who are not intent upon impoverishing it in order to please their lower instincts, their craving for intellectual security in the form of clarity, precision, 'objectivity', 'truth', it will become clear that there is only one principle that can be defended under all circumstances and in all stages of human development. It is the principle anything goes." (1975, p. 27-8).

Lakatos, in contrast to Feyerabend, did advocate normative principles for assessing whether theories are "progressive" or "degenerating"; but despite much criticism, Lakatos refused to specify any point beyond which it would be irrational for a scientist to stick to a degenerating programme: instead, Lakatos completely divorced "methodological appraisal" of a theory's merits (or liabilities) from "heuristic advice" about whether to pursue the theory. Many critics urge that Lakatosian norms are rendered impotent when they are so castrated of advisory implications, and that Lakato's methodology—since it then puts no constraints on scientific commitments—is as anarchistic as Feyerabend's sole "normative principle" that "anything goes." (See Musgrave 1976). But bracketing this issue, we need to note here only that both Feyerabend and Lakatos are led to eschew certain sorts of methodological norms because they use, implicitly, a rationability meta-method: a norm is objectionable if its employment would, on certain historical occasions, have precluded as "irrationable" commitments which led to scientific progress. Both positions thus underscore our question: insofar as we use a rationability meta-methodology, will we not (reckoning historically with the serendipity factor) put normative methodology out of business? In answer, I shall now argue that Feyerabend, Lakatos, and (as we shall see) Kuhn are led to their anarchistic conclusions about normative principles only because they contaminate the rationability approach with several auxiliary assumptions about rational norms. Showing what is wrong with these assumptions will exorcise the specter of anarchism from the rationability meta-method.

3. The Exorcism of Anarchism

The first and simplest step in the exorcism is to resist vague talk about "commitment to (or acceptance of) a theory," and instead begin to differentiate the varieties of "cognitive stances" that can be taken toward theories. To commit oneself to working on a theory is one sort of cognitive stance; to take the theory for granted in testing other theories is another; to require that any successor to the theory preserve its explanatory successes (or refer to the same entities) is yet a different cognitive stance (or two); and to use the theory to put men on the moon, yet something else. Any adequate theory of scientific rationality must provide us with a sensitive taxonomy of such cognitive stances, and thus allow us to exploit Laudan's insight that different cognitive stances involve different "modalities of appraisal", in which different normative principles are appropriate, and "very different sorts of questions are raised about the cognitive credentials of a theory." (1977, p. 108). Such differentiation allows us to see what is true in Feyerabend's premise: historically it would indeed have been scientifically disastrous to use any plausible norms of theory acceptance to govern certain kinds of theory commitment: infant (or embryonic) theories are often inferior to mature rivals precisely because they are so young, and they hence need to be given an incubation period during which their competitive potential can be nursed. But properly applied, the rationability approach will not lead us to conclude that in the interests of progress, we should reject any norms which deem such infant-theories "unacceptable": for this would fail to see that the cognitive stance appropriate to (some) infant-theories is not "acceptance" (however this problematic notion is unpacked) but pursuit, which requires its own "norms of rational pursuitability".

The preceding point reduces the magnitude of the problem, but it admittedly does not solve it completely, for the problem will again arise if we focus on any one modality of appraisal. Focusing on norms of rational pursuitability, it will still be the case that insofar as norm-governed judgments about "pursuit" are necessarily revocable, there again will be fortuitous occasions on which a theory-pursuit, though violating the best of such norms, serendipitously leads to strikingly progressive scientific advances. If we use the rationability approach, will we not then find ourselves—like Lakatos—rejecting any set of norms of rational pursuitability, on the grounds that their employment in the past would have precluded as irrationable commitments vital to progressive scientific episodes?

How anarchistically we answer this will depend, I now want to argue, upon how we regard the "universalizability" of rational norms; and here the second step of the exorcism is necessary. To raise the issue, consider the norm of rational pursuitability proposed by Laudan: "It is always rational to pursue any research tradition which has a higher rate of progress than its rivals." (1977, p. 111). What we need to scrutinize here is simply the sort of norm this is: it is one which, insofar as it gives advice, advises all scientists (in a given historical situation) to pursue the same theory. Is this a necessary feature of rational norms? That is: if cognitive stances are to be governed by rational norms, must it be the case that when ever it is rational for some scientists to take a given cognitive stance toward a theory, then it is rational for all scientists to do so (and if it is irrational for all scientists to commit themselves in a given way to a theory, then it is irrational for any scientist to do so)? If we think that norms of (say) rational pursuitability must be "universalizable" in this sense, it is likely that using the rationability approach will lead us to anarchistic conclusions.

The need for reflection here is evident from a recent dispute between John Worrall and Adolf Grünbaum (seconded by Alan Musgrave). Grünbaum claimed that "the heuristic rationality of further pursuit of research...does not conform to Kant's rule that the principle of one's action must be such that one could consistently universalize it." (1975, p. 89). Instead he argued that the rationality of pursuit must be "statistically relativized" in the following case: "Since scientific inquiry is conducted by a community of scientists, research strategies, policies, or practices that would be irrational if adopted by that community as a whole or by a majority of it need not necessarily be irrational when only a certain gifted minority engages in them." (1975, p. 89; quoted by Musgrave 1976 , p. 480).

Grünbaum's argument, however, has provoked some debate in Lakatosian quarters. Alan Musgrave, acknowledging a debt to Kuhn, endorses its basic conclusion that norms of rational pursuit apply only to the majority of scientists: X-rated theories may yet be pursued by some scientists—Grünbaum's "gifted minority"--who are, as it were, above the rules binding on the herd. (1976, p. 480 and passim.) John Worrall, on the other hand, rejects it with the adamant if somewhat terse claim that it "is really to give up the game." (1976, p. 163). Certainly it is not hard to sympathize with Worrall's misgivings: there is something unsettling about proposing a normative principle, and then adding the rider: "Of course, you as a (gifted?) individual may properly ignore this principle--but only so long as not too many of your peers also ignore it." If such riders are not giving up the "game" of normative methodology they are at least changing rather radically the spirit of it.

To resolve this dispute, we need not to take sides, but to draw a distinction overlooked by both sides. It is one thing to claim that truly rational action (and belief) must conform to norms which are binding on everyone; it is quite something else to claim that a proper norm of (say) rational pursuitability must entail--when applied to a given situation--that every scientist in the situation pursue the same theory. It is a mistake (perhaps committed by both sides of the dispute) to conflate these claims, or to think that the second follows from the first. This mistake appears implicitly in Grünbaum's remark, quoted above, concerning the non-Kantian character of rational pursuitability. And it appears that both Worrall and Grünbaum assume that claim 1 contains (or entails) claim 2: Worrall thus feels obliged to affirm claim 2, since he affirms claim 1; while Grünbaum, having denied claim 2, is led to deny claim 1.

Behind this mistake lies a seductive Robinson Crusoe conception of norm-governed activity which exclusively regards norms as guiding the commitments of scientist qua autonomous individuals, rather than as members of a community with common goals. As a corrective analogy, consider a group of people lost in a cavern: one can quite easily imagine circumstances in which it would be rationally necessary for some members of the group to explore one possible escape passage, but irrational for all members of the group to do so. The norms implicit in such decisions about "rational explorability" could here be accepted by all members of the group; but because the norms pertain to the efforts of the group, they need not advise all members to act in the same way. So also in science, norms of rational pursuitability are rules for rationally distributing (or focusing) the labors of a community: though binding on each member, such rules need not require all members to pursue the same theory. In pursuing some X-rated hypothesis, Grünbaum's "gifted minority" may thus yet be conforming to, rather than violating, norms which have the no-riders-attached universality Worrall desires!

If the Robinson Crusoe fallacy sets Grünbaum and Worrall at cross-purposes, it has much more perniciously confounded the Kuhnian arguments to which, ironically, Grünbaum and Musgrave acknowledge a debt. Seeing why will allow me to complete this defense of the rationability approach.

The "insights" to which Grünbaum and Musgrave refer lie in Kuhn's argument that "if a decision must be made under circumstances in which even the most deliberate and considered judgment may be wrong, it may be vitally important that different individuals decide in different ways. How else could the group as a whole hedge its bets?" (1970a, p. 186). From this need (limited, for Kuhn, to times of "crisis") for a heterodoxy by which the scientific community can hedge its bets, Kuhn draws a conclusion concerning scientific judgment about the severity of anomalies facing the reigning paradigm: "If everyone agreed in such judgments, no one would be left to show how existing theory could account for the apparent anomaly as it usually does. If, on the other hand, no one were willing to take the risk and seek an alternate theory, there would be none of the revolutionary transformations on which scientific development depends." (1970b, p. 241).

Without denying the wisdom of bet-hedging, we must surely sift Kuhn's "insights" about this more critically than do Grünbaum and Musgrave. For in claiming that scientific progress would be jeopardized "if everyone agreed in their judgments", Kuhn is of course arguing that everyone should <u>not</u> agree in their judgments; and from this he draws his larger and more infamous conclusion that such judgments must <u>not</u> be governed by rule-like "norms", but rather must be guided by "values" or "ideology". (See especially Kuhn (1970b), p. 262.) The pattern of argument is a tacit but recognizable use of the rationability principle: if using some (or any) set of norms would preclude as irrational certain steps essential to scientific progress, then the norms are objectionable norms. But the argument is also a perfect illustration of how "anarchistic conclusions" follow only because the rationability approach is corrupted by precisely the two mistakes this paper has tried to expose. First, the argument assumes that scientists would not try to square a paradigm with anomalies unless they <u>believed</u> this effort would be successful (and that they would not seek an alternative paradigm unless they believed the effort would fail). This sort of psychological presumption (which also lies behind Kuhn's analogous argument for monolithic dogmatism during "normal science") fails even to attempt the differentiation of "cognitive stances" that we have seen is crucial to normative methodology. Second, the argument assumes that a healthy "heterodoxy" in the context of pursuit requires that not all scientists "agree in their judgments." Of course, such heterogeneity requires that to the question "Should I pursue H?", scientist Smith will answer "Yes, I should.", while Jones will answer "No, I should not.". But this egregiously trades on the multiguity of the speaker-relative "I" which, like "now", has a floating reference. The illusion of "disagreement" evaporates the moment we refuse to couch the pertinent question in a first-person sentence: to the question "<u>Who</u> should pursue H?" our scientists can <u>agree</u> that Smith should while Jones should not. For all his emphasis on the "social dimension" of scientific inquiry, Kuhn's anarchistic deprecation of rational norms (during times of crisis) belies a Robinson Crusoe conception of norm-governed judgment. It is this (coupled with his failure to differentiate between the modalities of appraisal), and not the rationability approach itself, which leads Kuhn to his version of methodological anarchism.

4. Conclusion

By recognizing the need to differentiate the varied "modalities of scientific appraisal" and by avoiding a Robinson Crusoe conception of norm-governed judgments, we greatly increase the rational space for the serendipity factor, and hence exorcize the specter of anarchism from the rationability meta-methodology. For we can now grant the Feyerabendian premise that science sometimes (often?) has progressed because some scientists have committed themselves to theories which it would have been irrational for all of their peers to have pursued, or for any of their peers to have accepted. But this premise no longer pressures us to conclude that such minority-commitments must violate universal norms, whose employment would have precluded the progressive episodes at issue. Whether the rationability approach will enable methodologists to develop useful normative principles remains to be seen; and no doubt other problems confront the approach which need resolution.[2] But if the present exorcism has been successful, perhaps there is reason to hope that other demons can be cast out as well.

Notes

[1]I wish to thank Larry Laudan and Philip Quinn for their comments on the dissertation chapter from which this paper is abstracted.

[2]I have discussed some of these problems and refined the rationability approach to handle them in Wykstra (1978), Ch. 7.

REFERENCES

Buck, Roger C. and Cohen, Robert S. (eds.). (1971). PSA 1970. (Boston Studies in the Philosophy of Science, Vol. 8). Dordrecht: Reidel.

Burian, Richard. (1977). "More than a Marriage of Convenience: On the Inextricability of History and Philosophy of Science." Philosophy of Science 44: 1-42.

Cohen, L. J. (1975). "How Empirical is Contemporary Logical Empiricism?" Philosophia 5: 299-315.

Creary, Lewis. (1971). "Empiricism and Rationality." Synthese 23: 234-65.

Feyerabend, Paul. (1975). Against Method. London: New Left Books.

Giere, Ronald. (1973). "History and Philosophy of Science: Intimate Relationship or Marriage of Convenience?" British Journal for the Philosophy of Science 24: 282-97.

Grünbaum, Adolf. (1975). "Falsifiability and Rationality." Unpublished manuscript.

Hall, Richard J. (1971). "Can We Use the History of Science to Decide Between Competing Methodologies?" In Buck and Cohen (1971). Pages 151-9.

Howson, Colin (ed.). (1976). Method and Appraisal in the Physical Sciences. Cambridge: Cambridge University Press.

Kuhn, Thomas. (1970a). The Structure of Scientific Revolutions. Second Enlarged Edition. Chicago: University of Chicago Press.

- - - - - - (1970b). "Reflections on My Critics." In Criticism and the Growth of Knowledge. Edited by I. Lakatos and A. Musgrave Cambridge: Cambridge University Press. Pages 231-78.

Lakatos, Imre. (1971). "History of Science and Its Rational Reconstructions." In Buck and Cohen (1971). Pages 91-136.

Laudan, Larry. (1976). "Two Dogmas of Methodology." Philosophy of Science 43 : 585-96.

- - - - - - - (1977). Progress and Its Problems. Berkeley: University of California Press.

- - - - - - - (1979). "A Refutation of Convergent Realism." Unpublished manuscript.

Maull, Nancy. (1976). "Reconstructed Science as Philosophical Evidence." In Suppe and Asquith (1976). Pages 119-29.

McMullin, Ernan. (1976). "History and Philosophy of Science: A Marriage of Convenience?" In PSA 1974. (Boston Studies in the Philosophy of Science, vol. 32.) Edited by R. S. Cohen, et al. Dordrecht: Reidel. Pages 585-602.

Millman, Arthur B. (1976). "The Plausibility of Research Programmes." In Suppe and Asquith (1976). Pages 140-50.

Musgrave, Alan. (1976). "Method or Madness?" In Essays in Honor of Imre Lakatos. (Boston Studies in the Philosophy of Science, Vol. 34.) Edited by R. S. Cohen, P. K. Feyerabend, and M. W. Wartofsky. Dordrecht: Reidel. Pages 457-491.

Quinn, Philip. (1972). "Methodological Appraisal and Heuristic Advice" Studies in History and Philosophy of Science 3: 135-49.

Suppe, F. and Asquith, P. (eds.). (1976). PSA 1976, Vol 1. E. Lansing: Philosophy of Science Association.

Worrall, John (1976). "Thomas Young and the 'Refutation' of Newtonian Optics: A Case-Study in the Interaction of Philosophy of Science and History of Science." In Howson (1976). Pages 107-79.

Wykstra, Stephen. (1978). The Interdependence of History of Science and Philosophy of Science: Toward a Meta-Theory of Scientific Rationality. Unpublished Ph.D. Dissertation, University of Pittsburgh.

Moderate Historicism And The Empirical Sense of 'Good Science'

G. H. Merrill

Loyola University of Chicago

Much recent discussion has been devoted to questions concerning what role considerations of the history of science should play in developing and evaluating philosophical analyses of science.[1] It is generally, if not universally, conceded that the philosopher of science must appeal in *some* way to actual scientific practice and to the historical development of actual science if his analysis is to have any content, and it has been argued (Achinstein 1977) that even the most "ahistorical" of the positivists made, or presupposed, such an appeal.

It is now fairly common to distinguish two polar positions regarding the relation of history to philosophy of science: *radical logicism* is the doctrine that an appeal to actual scientific practice or to the history of science can never serve to refute any philosophical analysis of scientific concepts, while *radical historicism* is the view that really there is nothing more to the philosophy of science than historical description and analysis (i.e., that conformation to the history of science is *always* the determining factor in evaluating the adequacy of a philosophical analysis of science). It is agreed that these positions are somewhat perverted "idealizations", but it is possible to find philosophers who approach the "ideals" quite closely.

The recent discussions referred to above are concerned with discovering a position somewhere between these poles, and such a position is usually and naturally called 'moderate historicism'. The mark of a moderate historicism then is that neither purely historical nor purely *a priori* considerations alone determine the acceptability of a philosophical analysis of science.

PSA 1980, Volume 1, pp. 223-235

Before proceeding further, it is important to emphasize why moderate historicism is so attractive to philosophers of science. Both radical logicism and radical historicism are generally felt to be *prima facie* unacceptable -- the former because it does not require that philosophy of science bear any relation to science, and the latter because it results in an anarchistic view of science according to which "anything goes" and which prohibits us from making a non-arbitrary distinction between good and bad science. (For a fuller treatment of these issues see Burian 1977, section 1.) Moderate historicism promises us a philosophy of science which embodies both the "empirical content" missing from radical logicism and the true "understanding" of science absent from radical historicism. In order to achieve both of these desirable goals, it is necessary that the position provide us with clear criteria concerning when (*contra* radical logicism) historical considerations are to serve as grounds for rejecting a philosophical analysis and that it also provide us with such criteria concerning when (*contra* radical historicism) historical considerations may be ignored.

I believe it is not possible to formulate a coherent and non-trivial form of moderate historicism, but I shall not here offer a sustained argument in support of this rather sweeping claim though the discussion below will indicate how such an argument might proceed. Instead, I shall first describe a nasty dilemma that confronts any form of moderate historicism and then show that what is the most plausible way of escaping this dilemma is doomed to failure. Finally I shall offer some additional support for my arguments by examining one recent attempt to escape the dilemma in the manner I have mentioned.

1. The Historicist's Dilemma

Briefly, the primary problem facing historicism is this: The historicist tells us that if we wish to understand science and scientific methodology we need only look at the *history* of science. But when we go to look at the history of science, how do we know it is the history of *science* at which we are looking (as opposed to, say, the history of philosophy or the history of religion)? Of all the recorded episodes in history, how are we to tell which are the *scientific* ones? I do not see how there could be a purely historicist answer to these questions, and I believe that *any* historicist position (radical or moderate) ultimately will founder on the issue of distinguishing the history of science from the history of other disciplines. Indeed, the argument concerning moderate historicism that follows will be seen to turn upon a similar difficulty.

Radical historicism is (or implies) the view that any philosophical analysis of science that fails to accommodate _any_ episode in the history of science is to be rejected as inadequate: it fails to correspond _fully_ to the history of science. Another way of putting this is to say that for the radical historicist there is no distinction between actual science and good science, or that all actual science is good science. This is precisely what makes the position "radical".

Unlike radical historicism, moderate historicism is intelligible only if it can accommodate the _possibility_ that some actual science fails to be _good_ science, and we must ask what 'good' could mean here. If there are criteria available for classifying a methodology as good science even though it has never been employed in the actual history of science,[2] then it is these history-independent criteria -- and not historical considerations -- to which we must turn in evaluating the acceptability of any philosophical analysis of science.[3]

Seen in this way, then, the position of moderate historicism must presuppose a history-independent distinction between good and bad science. And this prohibits the moderate historicist from ever claiming that under certain kinds of circumstances (or in certain kinds of situations) historical considerations will force us to disregard those of a logical or philosophical nature since by 'the history of science' the _moderate_ historicist means the history of _good_ science, and so for him there _are_ _no_ historical considerations more fundamental than those _a priori_ ones upon which his crucial distinction is based.

Now once the distinction between good and bad science is made (_via_ the history-independent criterion), then the only reasonable requirement to impose is that a truly adequate philosophical analysis will conform to _all_ (and only) _good_ science. But it is just this distinction between good and bad science that is at issue among the moderate historicist and his more radical brethren, and to countenance a history-independent criterion for distinguishing good from bad science is to throw in with the radical logicists.

The dilemma faced by the moderate historicist is then this: In order to distinguish himself from the radical historicist he must allow for a distinction between good and bad science in the history of science and insist that an adequate philosophical analysis accommodate all of the good science. But then: (a) it is apparent that this distinction cannot be drawn in historicist terms; and (b) if it is drawn in logical or philosophical terms, this seems to preclude historical considerations from ever being relevant in the evaluation of a philosophical analysis of science (in which case the moderate historicist could not be distinguished from the radical logicist).

2. The Empirical Sense of 'Good Science'

It might be thought that there is a way out of the above dilemma for the moderate historicist. He need not, it might be argued, insist on any ahistorical sense to the description 'good science', for he need not hold that we must be able to say exactly which episodes in the history of science must be accommodated by our philosophical analysis. Rather it is sufficient to say only that in order for any such analysis to be acceptable it must treat most actual science as being good science (though there are no specific parts of the history of science which antecedently may be singled out as those that must be accommodated). Let us call this approach to the relation between actual science and good science the empirical approach to what is good science, and the sense of 'good science' involved here the empirical sense of this expression.[4]

Such a view preserves the distinction between moderate and radical historicism (according to which there is no distinction between actual and good science), and between moderate historicism and radical logicism (according to which an acceptable philosophical analysis need not conform to any actual science).[5] But a closer examination of this approach reveals it to be flawed in fundamental ways.

On what grounds do we justify the supposition that most actual science is good science? If this is construed as an empirical claim about the history of science then it is apparent that we are still faced with the problem of characterizing "good science" in some ahistorical way, and hence the dilemma is not to be escaped by taking 'most science is good science' to be an empirical assertion. But we may simply take this statement to be a meaning postulate: it is part of the meaning of 'science' (or of 'good science') that most science is good science. Such a meaning postulate is not as arbitrary as it may first appear, and although there are hidden dangers in adopting it, here is an argument in its favor.

There is a certain more or less well-defined discipline properly called 'science'. To be sure, there is some vagueness in the term and there will be some disagreement over whether certain "fringe" areas should fall under this description, and of course the sense of this term is not sufficiently clear to serve in a philosophical analysis of science. Nonetheless we all use the term 'science' to refer to basically the same events, programs, or methods, and thus this term has a common sense in that we agree upon what in fact (and in history) is its extension. Now obviously any discipline properly called 'philosophy of science' must bear some relation to what is generally regarded as science -- which is to say that the philosopher's sense of 'science' must overlap the common sense of the term. And this is to say that philosophy of science must have an historical dimension. Thus while it is not unreasonable to suppose that scientists have made some errors, it cannot be supposed that they always (or even usually) have been mistaken in their perception of what counts as good scientific practice. Consequently, if the phrase 'philosophy of science' is to have any non-arbitrary meaning, then we must regard it as part of the very meaning of 'science' that most science is good science.

In essence this argument offers a reason for adopting at least some kind of historicist position, and if it is correct then it appears reasonable and non-circular to adopt a position of moderate historicism according to which: (a) most actual science is in fact good science; and thus (b) any adequate philosophical analysis of science must conform to the history of science to the degree that it judges most actual science to be good science. Such a position avoids the difficulty seen in section 1 of presupposing an ahistorical concept of good science which renders historical considerations irrelevant in the evaluation of philosophical analyses of science.

The empirical approach to good science then has much to recommend it, and I believe it may be fruitful to take this view initially as a working hypothesis or guide that we use to draw our attention to various reasoning processes in science, for in this way the history of science may then be used to suggest methodologies to be analyzed by the philosopher of science. But the hypothesis must always be subject to rejection as a result of these analyses.

It should be noted that there is a latent ambiguity here regarding the phrase 'good science'. This could mean 'conforms to the practices of scientists' in which case the meaning hypothesis in question is trivial; or it could mean 'yields knowledge' or 'achieves the announced goals of science' in which case the principle that most science is good science may well be false. We may, for example, in one sense hold that most astrology is good astrology (i.e., conforms to the standards of professional astrologers) while in another sense deny that any astrology is good astrology (in that it does not yield the advertised knowledge or results). The situation is somewhat different with science since science is thought of as a paradigm epistemic enterprise (though we may wonder in passing on what grounds -- historical? philosophical? -- astrology is not felt to count as science). It is almost certain that we can show that most science is good science in the more significant (and ahistorical) sense of this term, but this will be a matter of analysis and demonstration rather than a matter of meaning postulates and dogma. Though I fear that this ambiguity of the phrase 'good science' is often not recognized and so leads to some avoidable confusions, I shall not dwell on it here since it is clear that if moderate historicism is to be distinguished from radical historicism, the sense of 'good science' employed must be the non-trivial one.

It now appears that by employing the empirical approach to what is good science we are able to formulate a position of moderate historicism that is clearly distinct from both radical historicism and radical logicism and which thus escapes the dilemma of Section 1. However, we shall presently see that the resulting position is so vague as to lack any content at all and that, in addition, it appears to possess the primary defect of radical historicism.

3. The Dilemma's Return

The empirical approach to what is good science has an interesting consequence: There may be philosophical analyses $\underline{A}_1$ and $\underline{A}_2$ such that $\underline{A}_1$ conforms to all of actual science _except_ episode $\underline{E}_1$ while $\underline{A}_2$ conforms to all of actual science _except_ episode $\underline{E}_2$. (Obviously this is a gross simplification, but the likely complications will not alter the force of my argument.) In such a case we may say that $\underline{A}_1$ and $\underline{A}_2$ are _historically comparable_ since each conforms to the history of science to the same degree.[6] Thus the fact that any number of (mutually incompatible) philosophical analyses may be historically comparable is a direct result of resorting to the empirical sense of 'good science', and this result may draw our attention to certain difficulties inherent in the empirical approach.

How are we to choose between the analyses $\underline{A}_1$ and $\underline{A}_2$? Certainly not on the grounds of the degree to which they conform to actual science. If we have two analyses which are _not_ historically comparable we _may_ follow the principle of rejecting that one which conforms less to actual science, but it is not obvious that we would do so if the alternative analysis were more "promising" in certain ways, were less _ad hoc_ , or were to account for episodes deemed "more significant".[7] It seems likely, for example, that we should prefer an analysis that could accommodate the advances of Newton, Lavoisier, Pasteur, and Einstein to one that could not, even though this latter one might accommodate _more_ of the history of science than the former. More important, however, is the fact that we have still said nothing more than that sometimes certain historical considerations will sway us in accepting an analysis and sometimes they will not, or that in accepting any such analysis a consideration of the history of science will play _some_ role. We have not yet seen any account of precisely _when_ or why historical considerations should carry the day; nor have we seen any account of how such historical considerations may be characterized independent of more basic logical and philosophical ones.

To say only that a philosophical analysis must conform to _most_ science, without specifying which episodes are to be included within the scope of 'most', is both to raise such questions as 'How _much_ conformation is required?' and to presume that all episodes in the history of science are comparable with regard to their importance or significance (for the only alternative is to suppose that we could have a criterion which would determine which episodes we might safely ignore). This question indicates the vagueness that is inherent in the empirical sense of 'good science' and manifested in the problem of resolving when two philosophical analyses are indeed historically comparable. As a consequence, moderate historicism becomes so vague as to be contentless since the moderate historicist cannot provide a criterion for distinguishing those cases when historical considerations will be the determining factor in assessing the adequacy of a philosophical analysis of science from those cases in which logical or philosophical considerations will play the determining role.

On the other hand, the presumption that all episodes in the history of science are comparable with regard to their significance prohibits any philosophical analysis of science from participating fully in what most take to be its peculiar task: the _evaluation_ of scientific reasoning processes. If it is first necessary to decide what 'good science' means in order to evaluate the adequacy of a philosophical analysis of science, then the most significant part of the game will already have been played before the philosopher is allowed onto the field. And if (in order to avoid the question-begging difficulties mentioned in Section 1) it is found necessary to presume that all episodes in the history of science are on a par with respect to their importance and significance, then the outcome of the game is determined prior to the philosopher's entry in precisely the same way that it is determined by the radical historicist: the philosopher's goal becomes that of developing an analysis which conforms to "most" of actual science, and he is prohibited from advancing any arguments that most of actual science has not, as a matter of _fact_, conformed to appropriate canons of reasoning. All such arguments _must_, from the point of view now being considered, be regarded as incorrect (since they conflict with the meaning postulate that most science is good science) and hence the "moderate" historicist is compelled to deny the _possibility_ that most actual science has been (or could be) defective in certain ways that may be identified and elucidated through philosophical analysis. It is the denial of this possibility, however, that was found to be the major fault of radical historicism: that in the end historical fact _must_ triumph over philosophical analysis and argument. And the position of moderate historicism as based upon the empirical sense of 'good science' shares this defect.

4. The Bankruptcy of the Empirical Approach: An Illustration

The difficulties I have been discussing may now be illustrated by considering briefly the views of one philosopher of science who favors a position of moderate historicism based on the empirical approach to what is good science. In contrasting the goals and methods of the historian with those of the philosopher, Frederick Suppe observes that:

> ...the historically oriented philosopher of science is concerned with generalization primarily as a prelude to critical analysis _and evaluation_. First he examines the patterns of reasoning involved in the advancement, evaluation, and justification of scientific hypotheses. Then, having identified the reasoning patterns used in a particular historical episode, he asks _whether the reasoning in question was good or not_. Do the reasons advanced really make the hypothesis plausible? Does the evidence adduced strongly favor the truth of the hypothesis as claimed? _If_ the philosophical examination yields affirmative answers to these questions, the philosopher can abstract the reasoning pattern and generalize that "when, or to the extent that, the following factors are present such and such a sort of theory is likely to be true," or that "it is reasonable to accept the theory as being established," and so on. And _if the philosophical evaluation is that the reasoning patterns are deficient_, the philosopher becomes concerned with what alternative reasoning patterns _could_ or _should_ have occurred given the available information in the scientific domain of the time and available background knowledge. In dealing with these issues, the philosopher is free to invent alternative reasoning patterns that were feasible at the

> time which can be evaluated as to their adequacy; and if such invented patterns of reasoning are both feasible and successful, the philosopher may generalize from them just as he does from actually exemplified patterns which pass the *test of philosophical scrutiny*. (Suppe 1977, p. 655, italics added).

This certainly emphasizes the *evaluative* role of the philosopher and explicitly suggests that he *might* find all, or a significant portion, of actual science to deviate from acceptable norms for reasoning patterns. The "test of philosophical scrutiny" is to play a major role in the assessment of actual science. But notice that in his formulation of reasoning patterns that scientists *should* use, the philosopher is always restricted to those that are *feasible at the time*. And later in the same paragraph Suppe asserts that:

> [the historically oriented philosopher of science] is claiming that *much* of what science characteristically does *is* rational and capable of yielding knowledge, although he also recognizes that *not all of science is rational* and thus that *not everything science does yields knowledge*. ...
>
> His aim is to establish a philosophical theory of rationality, reflective of *good* science *as actually practiced* wherein such actual science *is capable of yielding knowledge*... .(Suppe 1977, p. 656, italics added).

This is as explicit an endorsement of the empirical approach to what good science is as one is likely to find (the last quoted paragraph appears to echo Kuhn 1970, pp. 236-237), and it is further emphasized by his later remark that:

> ...in attempting to analyze such epistemic devices [as theories, explanations, reductions, etc.], it [the "new movement" in philosophy of science] demands that viable analyses characterize such devices as are actually employed in science, and so actual scientific practice, both historical and contemporary, plays an essential evidential role in the evaluation of such analyses. Ultimately, this reflects a deep conviction that philosophy of science *must show how science, as actually practiced*, is capable of yielding knowledge about the world. And an adequate philosophical analysis of epistemic devices such as theories *must display how they bear the epistemic burdens actual science imposes on them*. (Suppe 1977, pp. 657-658, italics added).

We are not to suppose that actual science might not, in general, really yield knowledge through the principles and processes to which it subscribes; that it might not support the epistemic burden imposed upon it.[8]

Given such a strong commitment to showing how actual science does yield knowledge -- i.e., it is assumed that it does, and the only question is how it does -- we must wonder, then, exactly how much latitude the philosopher is permitted in his negative evaluations of science. Not much, it would seem. In examining actual science the philosopher is permitted to ask "whether the reasoning in question was good or not" but he had better vote a resounding "Yes!" most of the time. And "the test of philosophical scrutiny" had better not result in the failure of much of actual scientific practice, or else this will serve as cause to reject the outcome of that test. But where are we left with regard to the relative weight of philosophical versus historical considerations when the philosopher abandons his objective evaluative role to become a propagandist for actual scientific practice?

Suppe clearly wishes to occupy a position of moderate historicism in which both philosophical and historical considerations play a role in our analysis and understanding of science. Upon closer examination, however, this position appears incoherent. Suppe originally makes a strong claim in support of the evaluative role of philosophy. But if the philosopher can determine that the pattern of reasoning in an arbitrary episode in the history of science is deficient, then logic dicates that it is possible for him to do the same for much (or even all) of the history of science. Thus a consequence of taking seriously the evaluative role that philosophy is to play is that much (or all) actual science might fail the test of philosophical scrutiny, and the philosopher is required to recognize this possibility in pursuing his evaluations of actual science. In addition, however, Suppe holds that the historically oriented philosopher of science claims that much of actual science does conform to appropriate criteria of rationality, and that such a philosopher has a deep conviction that the claim is true. But whence this claim and deep conviction? They do not arise as a result of philosophical analysis, for it is (according to Suppe) the job of the philosophical analyst to show that they are justified. Consequently they are a matter of dogma, and Suppe's "new" philosopher of science cannot recognize the possibility that much (or all) of actual science is deficient, for recognizing this is inconsistent with the deep dogmatic conviction that is guiding his analysis.

It is clear, then, how Suppe's treatment illustrates my earlier criticisms of moderate historicism as based on the empirical approach to what is good science, and the insurmountable difficulties in maintaining such a position should now be obvious. At best Suppe has given us an answer to the question "What role is philosophy to play in our understanding of the history of science?", but his fervent commitment to the view that most science is good science begs the question of what role history of science is to play in the philosophy of science. In the end it seems clear that philosophy is to be a handmaiden to history and that the only point of distinction between such a position of moderate historicism and radical historicism lies in the vague claim of the former that philosophy is to serve some (unspecified) evaluative role. And when we attempt to see just how the evaluative force of philosophy is to be exercised, we discover only that no evaluation is to be taken seriously if it conflicts with certain dogmatic views concerning the ultimate acceptability of actual scientific practice. Thus it is then when based upon the empirical approach to what is good science, moderate historicism shares the major defect of its radical relative.

Notes

1 For examples see the rather complete bibliography to Burian (1977).

2 Both Suppe, whose version of moderate historicism I shall consider in more detail in section 4, and Shapere acknowledge the possibility (or even the liklihood) of a reasoning pattern counting as good science even though it has never been employed in the history of actual science. (See p. 655 and note 116 on that page in Suppe 1977.)

3 To deny that there are (or could be) such criteria is to give up the moderate position for the radical one.

4 This is the approach of Suppe (1977) of which more will be said below.

5 But the radical logicist may easily hold that as a matter of fact the analysis does conform to actual science, or even that acceptable philosophical analyses are likely to conform in this way. He simply does not prejudge the issue by taking actual science to impose any constraints upon his analyses.

6 The problems here are similar to those in deciding among scientific theories which are empirically equivalent.

There is also a serious, and untouched, problem of individuating episodes in the history of science which is comparable to that of individuating scientific theories, but I shall not dwell on this particular difficulty.

7 The familiar difficulties involving factors which affect our acceptance or rejection of scientific theories are mirrored in the case of choosing among alternative philosophical analyses or rational reconstructions. In particular, that theory which accounts for the most phenomena is not necessarily the one to be preferred.

8 Note that one can agree that most science is *as a matter of fact* good science without adopting this as an article of faith or a methodological principle in the manner recommended by Suppe. For example, a radical logicist could hold such a view on the basis of his discovery that most science conforms to the reasoning patterns he has determined to be good. Suppe's assertion, however, is not an empirical claim whose truth we could discover. Rather it is analytic -- part of the meaning of 'science'.

References

Achinstein, P. (1977). "History and Philosophy of Science: A Reply to Cohen." In Suppe (1977). Pages 350-360.

Burian, R. (1977). "More Than A Marriage of Convenience: On the Inextricability of History and the Philosophy of Science." *Philosophy of Science* 44: 1-42.

Kuhn, T. (1970). *The Structure of Scientific Revolutions*. Second edition enlarged. Chicago: University of Chicago Press.

Suppe, F. (ed.) (1977). *The Structure of Scientific Theories*. Second edition. Urbana: University of Illinois Press.

History and the Norms of Science[1]

James Robert Brown

Dalhousie University

Most philosophers have a schizophrenic attitude toward the history of science. On the one hand, they want their philosophical accounts of how science ought to be done to do justice to typical scientific practice; but on the other hand, they want to avoid any confusion of historical *facts* with philosophical *norms*. A fine example of such a schizophrenic attitude is the one expressed by Wesley Salmon who says: "If a philosopher expounds a theory of the logical structure of science according to which almost all of modern physical science is methologically unsound, it would be far more reasonable to conclude that the philosophical reasoning had gone astray than to suppose that modern science is logically misconceived." (Salmon 1970, p. 73).

But after conceding so much, he takes it all back when he adds: "In spite of this the philosopher of science is properly concerned with issues of logical correctness which cannot finally be answered by appeal to the history of science... .[S]olutions, if they are possible at all, must be logical, not historical in character. The reason, ultimately, is that justification is a normative concept, while history provides only the facts." (Salmon 1970, p. 74).

In the first passage cited, Salmon grants the evidential role of history to normative methodology, but in the second passage he withdraws it. It would seem that historical facts can be used to refute a philosophical theory but they cannot be used to confirm one. Only *a priori* arguments could establish the likes of, say, Popper's falsificationism or Baysian inductivism, according to Salmon. However, this sort of asymmetry in the historical evidence is not typical of Salmon's thinking. He does not, for instance, think that experiments can only refute and not confirm scientific theories, nor do most other philosophers who share his attitude toward the history of science. Such an ambivalent attitude to history is certainly lamentable; but how could one avoid a schizophrenic state of mind on this subject?

PSA 1980, Volume 1, pp. 236-248

There are two ways: One way is to claim that it is entirely possible that all scientific practice, past and present, is totally irrational and "unscientific". The other way is to claim that history, contra Henry Ford, is not bunk, and, moreover, that it can be used as evidence for and against doctrines of how science ought to be done. In short, it is to claim that there is an evidential relationship between the history of science and the philosophy of science. I shall adopt this latter route.

I am certainly not alone in taking this course. The great William Whewell long ago wrote a *magnum opus* entitled *The Philosophy of the Inductive Sciences, Founded Upon Their History*, and more recently, Larry Laudan has made the same general assumption in his well-received *Progress and Its problems*. Indeed, to make the assumption that there is an evidential relationship between the history of science and the normative philosophy of science is to put myself into a growing minority of philosophers who maintain there is, generally, an intimate connection between history and philosophy, and who concur with Lakatos, who was so fond of paraphrasing Kant: "Philosophy of science without history of science is empty; history of science without philosophy of science is blind." (Lakatos 1971 , p. 91).

To maintain that there is a history-philosophy evidential relationship is one thing; to spell it out is quite another. The question we must face then is this: just *how* does history serve as evidence for normative methodology; how do the *facts* of the past connect with what *ought* to be done in the future? While there is a significant number of philosophers who agree that there is an intimate connection between history and philosophy, hardly any have made specific proposals as to just what the history-methodology evidential relationship is. I want to quickly examine one account which I think false and then propose an alternative which I take to be considerably more plausible.

Larry Laudan's account appears in his important new book, *Progress and Its Problems*. There he maintains that we have trustworthy intuitions about a number of episodes in the history of science and these can be used to test competing methodologies:

> ...[T]here is, I shall claim, a subclass of cases of theory-acceptance and theory-rejection about which most scientifically educated persons have strong (and similar) normative intuitions. This class would probably include within it many (perhaps even all) of the following: (1) it was rational to accept Newtonian mechanics and reject Aristotelian mechanics by, say, 1800; (2) it was rational for physicians to reject homeopathy and to accept the tradition of pharmacological medicine by, say, 1900; (3) it was rational by 1890 to reject the view that heat was a fluid; (4) it was irrational after 1920 to believe that the chemical atom had no parts; ...
>
> ... What I shall maintain is that there is a widely held set of normative judgments similar to the ones above. This set constitutes what I shall call *our preferred pre-analytic intuitions about*

> *scientific rationality* (or "PI" for short) ... [O]ur intuitions about such cases can function as decisive touchstones for appraising and evaluating different normative models of rationality ... [*T*]*he test of any putative model of rational choice is whether it can explicate the rationality assumed to be inherent in these developments* ... [*and*]*the degree of adequacy of any theory of scientific appraisal is proportional to how many of the PIs it can do justice to.* (Laudan 1977, pp. 160-161. His italics.)

Laudan's account of the evidential relationship between history and methodology has a number of serious difficulties, enough to render it most implausible. For one thing, his proposal is fundamentally at odds with his model of scientific rationality. There are a number of reasons for this which I shall get into shortly, but in the main it is because the PIs act as a foundation in an anti-foundationalist theory of method. Furthermore, his account is ill-motivated since it largely stems from his efforts to skirt the problem of "vicious circularity" which is, in fact, a mere pseudo-problem. I will elaborate on each of these difficulties.

The actual account of scientific rationality that Laudan proposes is to apply not only to science but to all activities which have cognitive aims. Thus, his model is to apply to itself, as I am sure he will readily agree. What I will argue is that his proposal for testing methodologies is in flagrant violation of that model itself. In order to show this I will first give a brief description of that methodology.

According to Laudan it is not individual theories which are tested directly, rather it is the *research tradition* which is the unit of appraisal. The aim of a research tradition is to solve problems, of which there are two kinds. *Empirical problems* arise for a theory when a competitor theory has explained some phenomenon. This point must be stressed. There is no such thing, says Laudan, as an un-theory-laden observation, or a neutral observation language. A particular phenomenom falls into the domain of a theory when a theory ought to explain it. A theory is required to explain it *only when* a competitor theory has managed to. Until then it is free to ignore the phenomenon in question. The other kind of problem, *conceptual problems*, arise for a theory when it clashes with other accepted theories (or research traditions), or when it is internally inconsistent.

Having unsolved problems, whether empirical or conceptual, counts against a tradition while solving them counts in a tradition's favor. Rationality, according to Laudan, consists in following these rules: *Accept* the tradition with the *greatest momentary adequacy*, that is, the one which has solved the most problems and has the fewest anomalies hanging over its head. *Pursue* the tradition which is the most *progressive*, that is, the one which is solving problems at the fastest rate.

When applying the model of rationality to the theory of how to test that model we need to know what the analogues of the conceptual and empirical problems are. It seems fairly easy to understand what a conceptual problem for a methodology might be. Internal inconsistency, of course, is one sort. Some commentators have expressed concern over Laudan's disavowal of "truth" and others have doubted Laudan's ability to ever be able to individuate his "solved problems". Whether these are legitimate worries or not is of no matter here. They illustrate the kinds of thing which could be conceptual problems for a methodology.

But what is the analogue of the empirical problem? The only candidate would seem to be capturing the PIs. A methodology solves empirical problems by making the individual episodes in the set of PIs turn out rational or irrational as the case may be. At any rate, I shall work on this assumption. Now to some difficulties.

According to Laudan's methodology, an empirical problem arises only when a competitor theory has explained a phenomenon. It follows that theory evaluation is basically *comparative*. Yet the PIs function as an absolute measure; they are put forward as a *sine qua non* for adequacy *before* any other theory has managed to explain them. Even if no account of scientific method could capture them, the PIs would still serve as an absolute standard. Thus, Laudan's methodology calls for evaluation by comparison while his account of testing methodologies is, contrary to this, quite anti-comparative.

A second difficulty concerns the fact that the PIs function as a sort of *foundation for methodological knowledge*. Laudan's model of rationality is highly anti-foundationalist. Nothing serves as a base, corrigible or not. He maintains a coherence view of justification. A consistent application of his model of rationality to the account of testing competing models would have to rule out any notion that the PIs were immune to revision. They might serve, like Popper's basic statements, as a convenient starting point, but they may all be tossed out as time goes by. By stressing conceptual matters and the theory-ladenness of observation in his model of rational scientific change, Laudan rightly downplays those spurious entities, *empirical facts*. By postulating the PIs, however, he has made himself a slave, when evaluating methodologies, to equally spurious entities, *normative historical facts*. The existence of an incorrigible set of PIs constitutes a type of foundationalism which is in conflict with his totally coherentist methodology.

Another difficulty for Laudan's account concerns intuition. The term "intuition" has two distinct senses. Which does Laudan mean when he says that there is "a subclass of cases of theory-acceptance and theory-rejection about which most scientifically educated persons have strong (and similar) normative intuitions "? (Laudan 1977, p. 160).

There is one sense in which to have an intuition is to have immediate knowledge of a concept where having this knowledge does not entail being able to define the concept. So, to say that some particular historical episode is intuitively rational is to say that we know that it *is* rational but we may still not be able to say *why* it is rational or even what rationality is. So-called "platonic intuitions" are often characterized this way.

If this is what Laudan means by "intuition" then the PIs must, once again, be viewed as foundational. They are pure, unadulterated, untheory-laden, normative historical facts. But, as already noted above, this is completely at odds with the spirit of his methodology.

The other sense of "intuition" has to do with common sense, or, more to the point, common belief or common prejudice. Many of our beliefs have this character. They seem obvious only because they have been ingrained since birth. It might well be the case that a particular example such as the rationality of adopting Newtonian mechanics by 1800, strikes us as intuitive simply because we have been told, over and over again since early childhood, that failing to adopt it constitutes the height of superstition and folly. If the PIs are intuitive in this sense then it might well be wondered why we should take them seriously at all. We might just as well pick them from a hat as select them on the basis of being intuitive.

This might seem too harsh a judgment because some philosophers have claimed a priority for common sense and maintained that we should trust common sense beliefs until they are shown to be false (e.g., Russell 1912, Ch. 2). Showing a common sense belief to be false usually amounts to overruling it by a theory for which there is good evidence. Ordinarily this is fine, but here the common sense beliefs *are the evidence* any methodological theory must account for; and so no methodological theory could overrule that evidence. Whichever episodes we pick for the PIs, the rationality of these episodes will thus become permanently entrenched common sense. Accordingly, Laudan's "intuitions" are certainly suspect, and they provide a poor criterion for selecting examples of good and bad scientific practice to test competing methodologies.

Laudan, quite rightly, stresses the great importance of history to the philosophy of science. The general aim of his whole program is one of making explicit the many complex and widespread inter-relationships between history and philosophy. In view of this it is surprising to find that *most of history will be evidentially neutral to normative philosophy of science*. The reason for this is very simple. The set of historical episodes is large while the set of PIs is rather small. Since "... the degree of adequacy of any theory of scientific appraisal is proportional to how many of the PIs it can do justice to " (Laudan 1977, p. 161) (that is, since it is only the PIs which matter), most of history is evidentially irrelevant.

This is so contrary to the spirit of Laudan's general ambitions that there is a strong temptation to think that, if queried on the matter, he would put things somewhat differently and add additional criteria to the requirement of capturing the PIs.

One possible response to this is to try to enlarge the set of PIs. Perhaps, as time passes, we could add more episodes. But, on what basis? A new criterion for indefinitely enlarging the set of PIs would be a criterion for distinguishing good science from bad science; that is, it would be a methodology. So, finding such a criterion is, in fact, equivalent to solving the initial problem.

I'll postpone until later my remarks on an alleged problem of circularity which seems to have motivated Laudan to propose the criterion that he did. It will be much more convenient to take it up below. I should note in passing, and I stress this, that my criticisms of Laudan's criterion are partly based on his own model of rational scientific change. It is a model which I find very appealing, and I think is more or less untouched by my criticisms of his "meta" claims.

A rival account of just how it is that history may serve as evidence for philosophical norms is the one proposed by Lakatos. I won't try to explicate and defend that account here. Let it be known that most of the philosophical world is divided into two camps on the matter. One group of philosophers thinks Lakatos' "rational reconstructions" are totally crazy; the other group has yet to read about them. I find myself in neither camp, and think that Lakatos makes more sense on this point than he is given credit for. But rather than engage in Lakatos scholarship I prefer to give my own account which I think right, and which is, I shall say, "inspired" by Lakatos.

I propose the following criterion as the best characterization of the evidential relationship between the history of science and the normative methodology of science. Just as theories of scientific method provide criteria by which scientific theories are to be evaluated, so the following is a meta-criterion of how these different normative methodologies of science are to be evaluated.

(MC) *That methodology is best which makes its theoretical reconstructions and normative reconstructions coincide for the greatest number of episodes in the history of science, and which best coheres with other accepted theories.*

Some explanations are in order: The term "rational reconstruction" can be used in two distinct ways. So, I drop it in favor of two new ones: "theoretical reconstruction" and "normative reconstruction".

By maintaining that the history of science without the philosophy of science is blind, or that history cannot be written without invoking norms, what is meant is that the history of science cannot be written without invoking the apparatus of some methodology of science. The concepts and categories of some philosophy of science are required in order to describe what has actually happened. This is very much akin to the thesis, now held by a great many philosophers of science, that our observations are theory-laden; we cannot describe the world in any neutral language. I shall simply make a similar claim about historical observations and descriptions.

A *theoretical reconstruction* is a description of some historical episode using the concepts of some methodology. For example, should we describe some episode using Popperian methodological concepts, the account will be in terms of "crucial experiments", "basic statements", "falsified theories", etc. If we were using the concepts of the so-called "methodology of scientific research programmes" then the history would be written in terms of "research programmes", "heuristics", "hard core", etc. These terms, "crucial experiment", "heuristic", and so on are theoretical terms. Any written historical episode must be a theoretical reconstruction; that is, it must invoke such concepts, concepts which owe their being to some methodology of science. This is because our histories are explanatory and they appeal to reasons as the causes of events. This requires an account of what good reasons are. Thus, it is *in principle* impossible to write the history of science without employing the conceptual apparatus of some philosophy or other. Such a history I call a "theoretical reconstruction".

Whereas a theoretical reconstruction is an attempted description of actual history, a *normative reconstruction* is quite another thing. Like a theoretical reconstruction it employs the conceptual apparatus of some methodology, but a normative reconstruction does not attempt to say how history actually went; rather, it declares how history *ought* to have gone according to that methodology.

For example, we might have the following theoretical reconstruction using Popperian methodology: Jones boldly conjectured a theory T and submitted it to a crucial test. T was falsified. Jones continued to maintain T and looked for ways to confirm it. Using the same Popperian methodology, the normative reconstruction of the historical episode would run: Jones boldly conjectured a theory T and submitted it to a crucial test. T was falsified. Jones abandoned T.

What I have called a "theoretical reconstruction" is an attempt to describe actual history, and what I have called a "normative reconstruction" might also be called "rational history". Without a clear distinction between these two types of reconstruction no sense can be made of the above proposal for evaluating a methodology. The metacriterion can be simply employed, for example, in the Jones case above. The normative and the theoretical reconstructions do not coincide here. So, this episode would have to count against Popperian methodology.

The condition in MC requiring coherence is important. For one thing it prevents the absurd consequence that the best methodology will be the one which makes *every* episode in the history of science rational. This we already know to be false; our knowledge comes from the prevailing psycho-social theories which tell us that scientists are only as God made them and some even worse. Thus, a methodology M will be obliged to make the likes of the Lysenko affair come out irrational; that is, the theoretical and the normative reconstructions of the Lysenko episode will have to diverge when M is applied to this event in the history of science, if the episode is to count in M's favor.

What can be said in favor of my proposal MC?

For one thing, it obviously captures the spirit of the initial demand that the history of science must be understood as significantly rational.

A second thing that can be said in its favor is that, formally, it parallels a commonly accepted methodological principle used in linguistics: the principle of charity. This principle says that when given a choice of translating a statement of a radically different language into either a true statement or a false statement of our language, we should choose to translate it into the true statement. The spirit of the two principles, MC and Charity, is the same. As Charity tries to maximize truth, MC tries to maximize rationality.

Most important is the fact that it overcomes all of the problems I pointed out in Laudan's criterion. It is not foundational; it does not rely on intuitions; no particular episodes in the history of science get entrenched; every episode in the history of science counts; and finally my proposal completely obviates a circularity problem.

Laudan, in discussing a number of difficulties which arise in any attempt to formulate an evidential relationship between the history of science and the philosophy of science, formulates the circularity problem this way:

> Foremost among these difficulties is the *vicious circularity* which it seemingly entails. If the writing of history of science presupposes a philosophy of science and if philosophy of science is then to be authenticated by its capacity to lay bare the rationality held to be implicit in the history of science, how can we avoid automatic self-authentication, since the history of science we write will presuppose the very philosophy which the written history will allegedly test? (Laudan 1977, p. 157).

Laudan is not alone in seeing this difficulty. Many others view it as the fundamental problem in establishing any sort of evidential-relationship between history and methodology (e.g., Giere 1973, p. 192). I think it is a pseudo-problem.

In order to lay the alleged circularity problem to rest I will reconstruct the arguments in a number of ways. The presuppositions which are said to hold between the history of science and the philosophy of science will be understood sometimes temporally and sometimes logically.

Here is one way of understanding the argument which makes it *seem* particularly devastating:

1. *Before (temporally) a philosophy of science can be chosen a history of science must be written.*
2. *Before (temporally) a history of science can be written a philosophy of science must be chosen.*

∴ *Before (temporally) a philosophy of science can be chosen a philosophy of science must be chosen.*

The first premise of this argument says that it is the history of science which provides the evidence for choosing among competing methodologies of science and that we must have that historical evidence before we can make a choice. The second premise says — to use the terminology developed above — that the written history of science is a theoretical reconstruction. Given that nothing can be chosen before it is chosen, the conclusion would appear to express an absurdity. Consequently, if the argument is valid then one or both of the premises is in trouble.

It would seem reasonable to interpret Laudan as understanding the argument this way. His reaction to the argument is to break up the history of science into two sets. He says we have clear "pre-analytic intuitions" about one set of these historical episodes. The competing philosophies of science are to be evaluated on the basis of these, not on the basis of a detailed, theory-laden, or theoretically reconstructed history of science. Laudan, however, may have taken the route he did because he *wrongly* accepted the argument as valid and the conclusion is absurd.

The argument is not cogent. There is an equivocation on "chosen". In the first premise "chosen" means something like "accepted" or "confirmed as the best going", while in the second premise it means something like "utilized" or "employed". If these terms are substituted into the argument in place of "chosen" we find that the conclusion is a relatively innocuous one:

1. *Before (temporally) a philosophy of science can be* confirmed *a history of science must be written.*
2. *Before (temporally) a history of science can be written a philosophy of science must be* utilized.

∴ *Before a philosophy of science can be* confirmed *a philosophy of science must be* utilized.

It is not obvious that this conclusion should lead to any further trouble. In fact, it is specifically required by the criterion stated above, MC. In order to confirm a philosophy of science as better than others, it and its competitors first have to be utilized in the reconstruction of the history of science.

These same premises seem to be used to draw a different conclusion. Laudan worries that the philosophy of science used to do the theoretical reconstruction cannot avoid "automatic self-authentication". But his fears are quite unjustified as can be seen by inspection of the argument when it is laid out.

1. *Before (temporally) a philosophy of science can be confirmed a history of science must be written.*
2. *Before (temporally) a history of science can be written a philosophy of science must be utilized*

∴ *The philosophy of science utilized will be confirmed.*

This conclusion simply does not follow from the premises. The argument is obviously invalid.

So far, temporal versions of the circularity argument have been given. Let us look at a logical version now.

1. *Philosophy of science (logically) presupposes history of science.*
2. *History of science (logically) presupposes philosophy of science.*

∴ *Philosophy of science (logically) presupposes philosophy of science.*

Laudan claims that we end up in a vicious circle. The conclusion here is of the form *p presupposes p* which is perfectly harmless. Vicious circles are of the form *p presupposes q* and *q presupposes ~p*. We do not have a vicious circle here at all. (Incidentally, I am using "presupposes" in the sense of material implication, i.e., "p presupposes q" means "If p then q".)

Let us fill out the premises of this argument with a bit more detail. Suppose H is some account of an episode in the history of science which is theoretically reconstructed using the conceptual apparatus of some normative philosophy of science M. And further, suppose H is the evidence for M.

1. *M is chosen (logically) presupposes H is written.*
2. *H is written (logically) presupposes M is chosen.*

∴ *M is chosen (logically) presupposes M is chosen.*

As in the argument immediately above, this is a perfectly harmless conclusion. After all, we are free to infer tautologies with impunity. If we substitute "confirm" and "utilize" for the two occurrences of "chosen" we get, once again, an innocuous conclusion:

M is confirmed (logically) presupposes M is utilized.

We clearly do *not* get the conclusion which would be harmful:

M is utilized presupposes M is confirmed.

In order to drive this point home I will return to the MC account of how it is that history serves as evidence for philosophy. MC would seem to imply all of the premises in all of the above arguments. Specifically, that

1. Before (temporally) a philosophy of science can be confirmed a history of science must be written.
2. In order for a philosophy of science to be confirmed a history of science is (logically) presupposed.
3. Before (temporally) a history of science can be written a philosophy of science must be utilized.
4. In order for a history of science to be written a philosophy of science is (logically) presupposed.

Assertions 1 and 2 mean that a methodology of normative philosophy of science is temporally and logically dependent on the historical facts for evidential support. Assertions 3 and 4 mean that all written history or historiography of science is theoretically reconstructed.

If the circularity argument is cogent it should work using these four premises. In order to show that the conclusion

M is utilized presupposes that M is confirmed.

does not follow from these premises, I will show that any such conclusion cannot be inferred except with the addition of extra premises saying just how history serves as evidence for methodological theories.

Recall how MC says a philosophy of science is confirmed. First, it must be utilized in two ways: to make a theoretical reconstruction (actual history) and a normative reconstruction (rational history). Second, the theoretical and the normative reconstructions are to be compared. The best methodology is the one in which theoretical and normative reconstructions of historical episodes more closely coincide than do the corresponding reconstructions of other philosophies of science.

Let us add this criterion to the premise of the argument. Recall that M is some normative philosophy of science, a methodology. H is a history, theoretically or normatively reconstructed according to the methodology M. I will distinguish between them by using the symbol "H(M-theoretical)" to mean the actual event described through the eyes of the methodology M, and the symbol "H(M-normative)" for the history as it *ought* to have gone according to the methodology M were everything done rationally.

1. *M is confirmed presupposes H(M-theoretical) is written.*
2. *H(M-theoretical) is written presupposes M is utilized.*
3. *M is confirmed presupposes H(M-theoretical) and H(M-normative) more closely coincide than H(P-theoretical) and H(P-normative) for any other methodology P.*

Now, does the problematic conclusion

M is utilized presupposes M is confirmed.

follow from these premises? Obviously not. Suppose we add another premise.

4. *H(Q-theoretical) and H(Q-normative), for some methodology Q, more closely coincide than H(M-theoretical) and H(M-normative).*

Then we could conclude that

M is not confirmed.

And this conclusion follows in spite of the fact that M is utilized. This would seem to show that the utilization of a methodology in the reconstruction of history does not automatically lead to its self-authentication.

Accordingly, MC avoids any problem of circularity. Indeed it shows the circularity problem which so many have fretted about to be a mere pseudo-problem. And this, I take it, is an additional piece of support for my proposed account of just what the history-methodology evidential relationship is.

Notes

1 For helpful discussions on related matters I thank R. Butts and K. Okruhlik of the University of Western Ontario. Of course I alone am responsible for blunders.

References

Giere, R. (1973). "History and Philosophy of Science: Intimate Relationship or Marriage of Convenience?" British Journal for the Philosophy of Science 24: 282-297.

Lakatos, I. (1970). "Falsification and the Methodology of Scientific Research Programmes." In Criticism and the Growth of Knowledge. Edited by I. Lakatos and A. Musgrave. Cambridge: Cambridge University Press. Pages 91-195.

----------. (1971). "History of Science and Its Rational Reconstruction." In PSA 1970. (Boston Studies in the Philosophy of Science, Vol. 8.) Edited by R. Buck and R. Cohen. Dordrecht: Reidel. Pages 91-136.

Laudan, L. (1977). Progress and Its Problems. Berkeley and Los Angeles: University of California Press.

Russell, B. (1912). The Problems of Philosophy. London: Oxford University Press.

Salmon, W. (1970). "Bayes's Theorem and the History of Science." In Historical and Philosophical Perspectives of Science. (Minnesota Studies in the Philosophy of Science, Volume 5.) Edited by R. Stuewer. Minneapolis: University of Minnesota Press. Pages 68-86.

Whewell, W. (1847). The Philosophy of the Inductive Sciences, Founded Upon Their History. 3 volumes. London: J.W. Parker.

Part IX

The Cat Paradox and Quantum Logic

A Formal Statement of Schrödinger's Cat Paradox[1]

James H. McGrath

Indiana University at South Bend

Schrödinger's argument about a "cat penned up in a steel chamber" is a timely challenge to those concerned with the philosophy of quantum theory; the argument prompts difficult decisions about correlated quantum mechanical systems, locality, reality and completeness. Here Schrödinger's own text is reproduced and, in view of the text, several plausible axioms and formal rules are chosen. Then, beginning with the axioms and using the rules, a contradiction is derived. This result establishes that Schrödinger's argument can be viewed as a paradox, a derivation of a contradiction from plausible assumptions. A final section of the paper refines the paradox and treats two possible resolutions as representative of a watershed issue in the foundations of quantum mechanics.

Comments concerning the method of formalization were provided when I applied the method to the Einstein-Podolsky-Rosen argument (1978).

1. Schrödinger's Argument

Schrödinger's cat argument is a small but well-known part of a three section paper published in 1935, a paper acknowledging the motivation of the Einstein-Podolsky-Rosen article which had appeared earlier the same year. These two papers, together with two others Schrödinger published one year later, have proven seminal for the foundations of quantum theory. English translations of the cat argument have been given by Jauch (1973, p. 106) and (1968, p. 185) and by Mehra (1974, p. 72). An English translation by J. D. Trimmer of the entire 1935 Schrödinger article is forthcoming. Reproduced here is Trimmer's translation of the argument.

> A cat is penned up in a steel chamber, along with the following diabolical device (which must be secured against direct interference by the cat): in a Geiger counter there is a tiny bit of radioactive substance, _so_ small, that perhaps in the course of

---

PSA 1980, Volume 1, pp. 251-263

one hour one of the atoms decays, but also, with equal probability, perhaps none; if it happens, the counter tube discharges and through a relay releases a hammer which shatters a small flask of hydrocyanic acid. If one has left this entire system to itself for an hour, one would say that the cat still lives if meanwhile no atom has decayed. The first atomic decay would have poisoned it. The ψ function of the entire system would express this by having it in the living and the dead cat (pardon the expression) mixed or smeared out in equal parts. It is typical of these cases that an indeterminacy originally restricted to the atomic domain becomes transformed into macroscopic indeterminacy, which can then be resolved by direct observation. That prevents us from so naively accepting as valid a "blurred model" for representing reality. In itself it would not embody anything unclear or contradictory. There is a difference between a shaky or out-of-focus photograph and a snapshot of clouds and fog banks.

2. Transition To a Formal Restatement

An idealized mathematical scheme adequate to describe the measurement process of Schrödinger's cat experiment can be initiated by distinguishing three physical systems, the mathematical representations of those systems and their relationships in time. The measuring system, M, consists of the entire counter-hammer-cyanide-cat sequence as described by Schrödinger. Here M will be represented by the Hilbert space vector ψ. The system to be measured, S, is the radioactive substance and will be represented by Φ. When S and M are considered together, the composite system, S+M, will be represented by Ψ. At the earliest time, t_0, S and M are physically isolated and without interaction; later, S and M interact; and finally, at t, S+M is observed. A time evolution operator, U, maps $\Psi(t_0)$ to $\Psi(t)$.

3. Axioms

Required assumptions of the argument will now be stated as axioms upon which the argument rests. By calling the assumptions of the argument axioms, it is not intended that they be accepted as established or self-evident principles. Quite the contrary, to state explicitly the assumptions as axioms of the argument is to invite scrutiny; as we shall see, to relieve the paradox one must deny at least one of the axioms (or rules introduced in the next section). Yet the axioms are plausible claims and here are defended against some of the most glaring misinterpretations and objections that might be leveled against them.

The first axiom makes explicit a statement from the previous section: M denotes the measuring system and is represented by ψ. To secure the distinction between physical systems and the mathematical representations of such systems, the abbreviation $\{Rep\}$ is introduced to indicate that the Hilbert space vector ψ stands in a relation of representation to the system M. (Here the relation must be taken as unanalyzed.) Accordingly the first axiom is written

$$\psi \{\text{Rep}\} \text{ M} \qquad (t_0) \qquad \text{(i)}$$

The axiom asserts that ψ represents M at t_0, a time when M is physically isolated from S. (Axioms are numbered with lower case Roman numerals.)

Right off, two possible objections to the axiom ought to be summarily discharged. To those who might point out that system M contains a (conscious, living) cat whose description is beyond the purview of physics, it is responded that, although offering éclat, the cat plays no essential role and could be replaced by an appropriate macroscopic recording device. Schrödinger himself, as well as Einstein (1959, p. 670) and Jauch (1968, p. 191), stressed this point. Nor ought it be objected that a Hilbert space vector is improperly ascribed to a macroscopic object for, at least within the context of the orthodox account of measurement, this is unavoidable.

Two other possible objections to the first axiom are more substantial. Some might urge that at t_0 M should be represented by a statistical mixture and not by a single vector. But to do so would lead to unnecessary complication without conceptual gain. See Park (1968a, p. 226). Others including Ballentine (1970) have objected that in general the Hilbert space vector can only represent an ensemble, never a single system. If so ψ should represent $\mathbf{M}$, an ensemble of individual M systems. Whatever merit this proposal has, it does not seem to relieve Schrödinger's paradox. Even if ensemble $\mathbf{M}$ is represented by density operator ρ_ψ, axiom i can be rewritten and, unless other modifications are demanded, the paradox can still be derived. So while the present argument is developed in terms of individual system statements, ensemble statements could have been used.

The second axiom establishes the representation of the system to be measured, system S, prior to the start of the experiment.

$$\Phi = c_1\phi_1 + c_2\phi_2 \{\text{Rep}\} \text{ S} \qquad (t_0) \qquad \text{(ii)}$$

Here Φ is expressed as a linear superposition of two vectors, ϕ_1 and ϕ_2, and c_1 and c_2 are Hilbert space scalors or, in general, complex numbers. Comments concerning the representation of physical systems by linear superpositions follow in subsequent sections.

The third axiom asserts the existence of a unitary time evolution operator to map an arbitrary Hilbert space vector Θ from Θ_0 at time t_0 to Θ_t at time t.

$$\mathcal{U}: \Theta_0 \rightarrow \Theta_t \qquad \text{(iii)}$$

The axiom is read as if prefaced by an existential quantifier ranging over operators: "There exists an operator $\mho$ such that... ." Uncontroversial on orthodox accounts, the axiom is challenged by Belinfante (1975, sec. 1.12 and Appendix D).

Finally as a fourth axiom it is asserted that

$$S+M \text{ is macroscopic.} \qquad \text{(iv)}$$

Commentary on issues related to this axiom constitutes much of Section 7.

4. Rules

The rules used in the formal statement of the cat argument are extralogical rules. They have the same form and function as logical transformation rules; once certain statements have been established they secure the permissibility of asserting further statements. Unlike their logical counterparts, but like the four axioms just introduced, the rules used here are justified, if at all, on extralogical, quantum-theoretic grounds. Unlike the axioms which made specific assertions about the "experiment" in question, the rules function as schemata; they are intended to be completely general. To maintain this generality, placeholders S (with subscripts) and Θ, α, β (with subscripts) will be used in an obvious manner.

The first rule is a prescription sanctioning the introduction of the tensor product representation of a composite system given the Hilbert space vector representation of each of two individual systems. If vectors Θ_1 and Θ_2 represent S_1 and S_2 then, the tensor product, $\otimes$, is formulated as follows:

Rule $\otimes$ introduction ($\otimes$INT)

$$\begin{array}{|l} \Theta_1 \{\text{Rep}\}\ S_1 \\ \Theta_2 \{\text{Rep}\}\ S_2 \\ \hline \Theta_1 \otimes \Theta_2 \{\text{Rep}\}\ S_1 + S_2 \end{array}$$

Excellent commentary relevant to this rule and the use of tensor products in general is found in work by Park (1968b, p. 222), Jauch (1968, p. 179) and van Fraassen (1972, p. 352). The rule as stated here leaves tacit a restriction which is fulfilled in the Schrödinger situation: Θ_1 and Θ_2 must represent pure states (not statistical mixtures). It is a specific variant of the rule which will be required by the argument: Θ_2 is a superposition: $\Theta_2 = c_a\Theta_a + c_b\Theta_b$. Accordingly the rule takes the following specific form

Rule $\otimes$ introduction (Superposition form)

$$
\begin{array}{|l}
\Theta_1 \ \{\text{Rep}\}\ S_1 \\
\Theta_2 = c_a\Theta_a + c_b\Theta_b \ \{\text{Rep}\}\ S_2 \\
\vdash c_a(\Theta_1 \otimes \Theta_a) + c_b(\Theta_1 \otimes \Theta_b) \ \{\text{Rep}\}\ S_{1+2}
\end{array}
$$

The second rule is intended to capture an essential feature of the measurement process. Two initially isolated systems, when physically interacting, evolve in such a way that the vector representation comes to exhibit a correlation.

Rule Evolution (EVL)

$$
\begin{array}{|lr}
\Theta_0 = (\alpha \otimes \beta_1) \ \{\text{Rep}\}\ S_1 + S_2 & (t_0) \\
U : \Theta_0 \rightarrow \Theta_t & \\
\vdash \Theta_t = (\alpha_1 \otimes \beta_1) \ \{\text{Rep}\}\ S_1 + S_2 & (t\ )
\end{array}
$$

Here again, as with the $\otimes$ introduction rule, a superposition version will be required for the statement of Schrödinger's argument.

Rule Evolution (Superposition form)

$$
\begin{array}{|lr}
\Theta_0 = c_1(\alpha \otimes \beta_1) + c_2(\alpha \otimes \beta_2) \ \{\text{Rep}\}\ S_{1+2} & (t_0) \\
U : \Theta_0 \rightarrow \Theta_t & \\
\vdash \Theta_t = c_1(\alpha_1 \otimes \beta_1) + c_2(\alpha_2 \otimes \beta_2) \ \{\text{Rep}\}\ S_{1+2} & (t\ )
\end{array}
$$

Vectors such as Θ_t cannot be considered, in all cases, to be factorizable or reducible to vectors representing (a pure state for) S and a vector representing (a pure state for) M. Commentary relevant to this rule is given by Amai (1963, p. 550) and d'Espagnat (1971, p. 289).

Two final rules motivate the discussion of Section 7. The first asserts that observables of superposition states have no well-defined values. The restriction that Θ_a and Θ_b are eigenstates of observable $\mathcal{O}$ is tacitly assumed for the rule; the rule is thereby restricted to observables for the superposition states.

Rule Valuation (VAL)

$$
\begin{array}{|l}
\Theta = c_a\Theta_a + c_b\Theta_b \ \{\text{Rep}\}\ S \\
\vdash \mathcal{O} \text{ has no well-defined value in } S
\end{array}
$$

If this rule is accepted, there is no value o_i to satisfy the propositional form '$\mathcal{O} = o_i$'. Commentary follows in Section 7. The last rule formulates a position which may be called macroscopic realism.

Rule Macroscopic Realism (MR)

| S_1 is macroscopic
|———
| $\mathcal{O}$ has a well-defined value in S_1

In Section 7 refinements and qualifications are discussed. Here we assume that $\mathcal{O}$ is any observable of S_1.

5. The Formal Statement of Schrödinger's Argument

The set of four axioms and four rules is an inconsistent set. From the axioms a contradiction can be derived in accord with the rules. It is the purpose of the formal statement to make graphic this relationship.

i	ψ {Rep} M (t_0)	
ii	$\Phi = c_1\phi_1 + c_2\phi_2$ {Rep} S (t_0)	
iii	$U: \Theta_0 \rightarrow \Theta_t$	
iv	S+M is macroscopic.	
1	$\Psi = \Phi \otimes \psi = c_1(\phi_1 \otimes \psi) + c_2(\phi_2 \otimes \psi)$ {Rep} S+M (t_0)	i, ii, $\otimes$ INT
2	$\Psi = \Phi \otimes \psi = c_1(\phi_1 \otimes \psi_1) + c_2(\phi_2 \otimes \psi_2)$ {Rep} S+M (t)	1, iii, EVL
3	$\mathcal{O}$ has no well-defined value in S+M	2, VAL
4	$\mathcal{O}$ has a well-defined value in S+M	iv, MR

6. Four Distinctions

Three important distinctions are implicit in line 2 of the formal statement; the first two of these concern Ψ.

As written, line 2 must be distinguished from another different claim

$$\Psi' = |c_1|^2 \hat{P}_{[\phi_1 \otimes \psi_1]} + |c_2|^2 \hat{P}_{[\phi_2 \otimes \psi_2]} \quad \{\text{Rep}\}\ \text{S+M}\ (t) \qquad (2M)$$

where $\hat{P}_{[\chi]}$ is the projection operator onto χ. The distinction is the well-known one between superpositions (2) and statistical mixtures (2M). It has been discussed in a context relevant to Schrödinger's argument by Wigner (1963), Cartwright (1974) and Grossman (1972). It may be desirable to replace (2) by (2M); but, according to quantum theory, the correct assertion is (2). So proposals to reduce (2) to (2M) are obligated to provide justification for the reduction.

A second distinction is based on a contrast between (2) and (2I).

$$\text{Either } \hat{P}_{[\phi_1 \otimes \psi_1]} \text{ or } \hat{P}_{[\phi_2 \otimes \psi_2]} \quad \{\text{Rep}\} \text{ S+M} \quad (t) \qquad (2I)$$

(even if it is unknown which).

The problem Schrödinger presents is not that one is unable to say "S+M is in state $\hat{P}_{[\phi_1 \otimes \psi_1]}$" and also unable to say "S+M is in state $\hat{P}_{[\phi_2 \otimes \psi_2]}$." The problem is that one cannot even say "S+M is in state $\hat{P}_{[\phi_1 \otimes \psi_1]}$ or $\hat{P}_{[\phi_2 \otimes \psi_2]}$." This well-known but subtle and crucial distinction figures prominently in the discussion between Süssman (1957) and Bohm (1957) and is clearly presented by van Fraassen (1972, pp. 325-31).

The third distinction implicit in line 2 concerns S+M. Schrödinger explicitly stated that Ψ (of line 2) refers to "the entire system." It is

$$\Psi \text{ (a superposition) } \{\text{Rep}\} \text{ S+M} \qquad (2)$$

not

$$\Sigma_1 \text{ (some other superposition) } \{\text{Rep}\} \text{ M}$$

that figures in the paradox. Specifically, the claim

$$c_1\psi_1 + c_2\psi_2 \quad \{\text{Rep}\} \text{ M} \qquad (t) \qquad (2C)$$

plays no part in the argument. Since the correctly stated argument does not prove (2C), it is no resolution of the correctly stated argument to point out that (2C) should be replaced by

$$|c_1|^2 \hat{P}_{[\psi_1]} + |c_2|^2 \hat{P}_{[\psi_2]} \{\text{Rep}\} \text{ M.} \qquad (t) \qquad .$$

To accept a result which represents M by a mixture when S+M is represented by a superposition is to make no progress against Schrödinger's argument. To say that the paradox arises because, at the end of the experiment, "the cat is in a superposition" or "the cat is neither dead nor alive" reveals a failure to distinguish (2) and (2C).

The fourth and final distinction concerns Rule MR. Often in foundational discussions (Schrödinger's cat article itself is a good example), speculation is entertained about "observables" or unobserved properties of microscopic systems. Such speculation might culminate with the assertion or denial of a rule (which may or may not be related to MR) called microscopic realism (Rule MiR)

| S_1 is microscopic
|---
| $\mathcal{O}$ has a well-defined value in S_1

Whatever the eventual merits of Rule MiR, notice that it plays absolutely no part in the formal statement of Schrödinger's cat argument. The paradox appeals only to Rule MR not to Rule MiR.

7. A Final Perspective

Recent work suggests that the plausibility of Schrödinger's argument cannot be evaluated until the formal statement is refined and elaborated. For example, Moldauer (1972, p. 45) observes that while the cat can be "measured" by the veterinarian's stethoscope and while the radioactive substance by a geiger counter, it was not clear how observables of the combined system are to be measured. See also Krips (1976, especially p. 656). Following this observation, we will distinguish $\mathcal{O}^{M}$ of system M (with values: dead and alive) and $\mathcal{O}^{S+M}$ of S+M (with some other values). Rules MR and VAL can now be refined. Previously each was ambiguous between two possibilities now distinguishable as follows:

Rule VAL'	Rule VAL"
Some superposition {Rep} M	Some (other) superposition {Rep} S+M
$\mathcal{O}^{M}$ of M has no well-defined value in M	$\mathcal{O}^{S+M}$ of S+M has no well-defined value in S+M

Here, for clarity, the rules are made specific to the relevant systems and left vague in the exact form of the superposition. The forms of Rule MR are:

Rule MR'	Rule MR"
M is macroscopic	S+M is macroscopic
$\mathcal{O}^{M}$ has a well-defined value in M	$\mathcal{O}^{S+M}$ has a well-defined value in S+M

Once qualified in this way, the evaluation of the rules takes the following course.

Regardless of its plausibility, Rule VAL' can be eliminated from consideration on grounds of inapplicability to Schrödinger's argument. The paradox makes no claim about the representation of system M (recall Section 6, the third distinction). Since Rule VAL' applies to system M, the rule is inapplicable. Also when a combined system is represented by a superposition, no acceptable reduction formula specifies that component systems are represented by (some other) superpositions. Therefore, it is unlikely the paradox could be reformulated in a way which would appeal to Rule VAL'.

Despite a motif of orthodox quantum mechanics which would establish measurement as a precondition for an observable (sometimes even a macroscopic one) to have a value, Rule MR' seems unexceptionable. It seems reasonable to require that the cat is either dead or alive at all times. Yet, unexceptionable as Rule MR' may be, it can play no direct part in the paradox. The reason is formal. The paradox requires either Rules VAL' and MR' or Rules VAL" and MR". If one single-primed rule is used with one double-primed rule, no contradiction can be derived. For example, from Rules VAL" and MR' (and an adjustment of axiom iv) one could only derive the pair

$\mathcal{O}^{S+M}$ has no well-defined value in S+M

$\mathcal{O}^{M}$ has a well-defined value in M.

Unlike lines 3 and 4 of the actual paradox, these statements are not contradictory. Since Rule VAL' cannot be used, the single primed pair of rules cannot be used either and Rule MR' therefore plays no part in the paradox. Incidentally, many of those who dismiss the argument as fanciful often seem to have the correct intuition about these rules. The argument may seem to gain plausibility from the aliveness or deadness of the cat (Rule MR') but that state of affairs is not what the paradox describes.

If the paradox is to be maintained, both Rule VAL" and Rule MR" must be acceptable. It is that possibility we now discuss.

Without specifically mentioning Schrödinger's argument, Arthur Fine (1976) has recently developed a view which is applicable to the two rules in question. Fine urges that quantum-theoretic states underdetermine the values of observables. In Schrödinger's case, when a superposition represents S+M, the possible values of the observable $\mathcal{O}^{S+M}$ are determined but the actual values are not. Fine's program is one which attempts to show that $\mathcal{O}^{S+M}$ can still be regarded as having (an actual) value. But because quantum theory fails to specify what that value is, quantum theory is an incomplete theory. From this point of view Rule MR" is reasonable; it establishes the permissibility, in some cases, of just what Fine is confident is true in any case: $\mathcal{O}^{S+M}$ has a value. But because Rule VAL" allows us to establish that $\mathcal{O}^{S+M}$ has no value, Fine rejects that rule. Fine has persistently argued (see his [1970] for example) that a theoretical framework which sanctions Rule VAL" leads to contradictions similar to the one proven in sec. 5. Without Rule VAL" however, Schrödinger's paradox cannot be derived. Therefore Fine's rejection of the rule constitutes a resolution of the cat paradox which is at least formally adequate.

Yet as a physical fact, Rule VAL" appears plausible: we never observe the value $\mathcal{O}^{S+M}$ has. Anticipating such an objection, Fine (1976, p. 259) contends that while there may be an "antecedent physical specification" for the measurement of $\mathcal{O}^{S+M}$, there is none for $\mathcal{O}^{S+M}$. This contention could be buttressed with a recent argument of L. L. van Zandt. As a technical fact it seems impossible to make a measurement of $\mathcal{O}^{S+M}$ and thereby determine its value: "The gedanken experimenter is not prohibited in principle from observing an interference, but to do so, he must measure an operator $[\hat{O}^{S+M}]$ which translates the particles of the system from their positions in the branch Ψ_d to their positions in the branch Ψ_a. Such an operator can only be an intellectual construct." (1977, p. 55). In principle, $\mathcal{O}^{S+M}$ could have a value and Rule VAL" would fail. But from the perspective of "convenient" laboratory procedures, and while isolated from in-principle-speculation, $\mathcal{O}^{S+M}$ cannot be regarded as having a value and Rule VAL" therefore appears plausible.

While van Zandt's physical arguments are consistent with Fine's incompleteness perspective, it would be hasty to conclude that the arguments prove Fine's case. For it would also be consistent with van Zandt's position to arrive at a dramatically different conclusion. Adopting a principle which establishes the practical possibility of measurement as a prerequisite for the existence of values of an observable, one might attempt to use van Zandt's no-possible-measurement-of-$\mathcal{O}^{S+M}$ result to establish that $\mathcal{O}^{S+M}$ has no value. In this case, exactly contrary to Fine, one might conclude that Rule MR" fails while Rule VAL" is acceptable. Here again a formally adequate resolution of Schrödinger's paradox would be achieved; for Rule MR" is a necessary rule.

A watershed issue in the foundations of quantum mechanics is at stake. Schrödinger's paradox can be "resolved" in at least two ways. Following Fine, one can sacrifice Rule VAL", embrace incompleteness and preserve the possibility of realism. Alternatively, one can deny that quantum theory "represents reality" and give up Rule MR" while allowing the possibility of Rule VAL". This situation is often stated as the conclusion that quantum theory cannot give a contradiction-free, complete description of physical reality. It is well-documented that this is one of Einstein's reaction to quantum theory in general and to Schrödinger's argument in particular. See Einstein's 1950 letter to Schrödinger (reprinted by Mehra (1974, p. 73) or his reply to critics in the Schilpp volume (1959, p. 670).

While I am fully confident that the formal statement conforms to Schrödinger's statement of the cat argument and while I am equally confident that the Einstein perspective on the argument has been valuable, I think the complete-real context is a procrustean bed for Schrödinger's cat argument. As the translation of the cat article by Trimmer will soon make well-known, Schrödinger had much else on his mind when he wrote the cat argument. Schrödinger's thought resides not in the two paragraph argument but in the dozens of preceding and following pages. And from the formal point of view the paradox indicts an entire set of eight axioms and rules; not just two.

There is one conclusion with which Schrödinger would undoubtedly agree. The cat paradox is by no means fanciful and the options it forces are difficult ones. The issues woven into the argument motivated Schrödinger's much neglected 1935 and 1936 articles. We would do well to retrace these footsteps.

Note

[1]In its earliest stages, this paper was encouraged by Bas C. van

Fraassen and Roger T. Simonds and financial support was contributed by the Borden P. Bowne Foundation. Later, discussions with John D. Trimmer, Arthur Fine and Lonnie van Zandt influenced the course of research. Versions were read at physics colloquia at Purdue University and at the University of Notre Dame. For all this support it is a pleasure to acknowledge gratitude.

References

Amai, S. (1963). "Theory of Measurement in Quantum Mechanics." Progress of Theoretical Physics 30: 550-62.

Ballentine, L. (1970). "The Statistical Interpretation of Quantum Mechanics." Reviews of Modern Physics 42: 358-381.

Belinfante, F. (1975). Measurement and Time Reversal in Objective Quantum Theory. Oxford: Pergamon Press.

Bohm, D. (1957). "Discussion." In Observation and Interpretation: A Symposium of Philosophers and Physicists. Edited by S. Korner. New York: Academic Press. Pages 46-61. (Reprinted in Observation and Interpretation in the Philosophy of Physics. Edited by S. Korner. New York: Dover Publications, Inc., 1962.)

Cartwright, N. (1974). "Superposition and Macroscopic Observation." Synthese 29: 229-242.

d'Espagnat, B. (1971). Conceptual Foundations of Quantum Mechanics. Menlo Park: Benjamin. Pages 287-303.

Einstein, A. (1959). "Remarks to the Essays Appearing in this Collective Volume." In Albert Einstein: Philosopher-Scientist. Edited by P. Schilpp. New York: Harper and Row. Pages 665-83.

------------, Podolsky, B. and Rosen, N. (1935). "Can Quantum Mechanical Description of Physical Reality Be Considered Complete?" Physical Review 47: 777-80. (Reprinted in S. Toulmin (ed.). 1970. Physical Reality. New York: Harper. Pages 122-130.)

Fine, A. (1970)."Insolubility of the Quantum Measurement Problem." Physical Review D 2: 2783-87.

--------. (1976). "On the Completeness of Quantum Theory." In Logic and Probability in Quantum Mechanics. Edited by P. Suppes. Dordrecht: Reidel. Pages 249-281.

Grossman, N. (1972). "Quantum Mechanics and Interpretation of Probability Theory." Philosophy of Science 39: 451-60.

Jauch, J. (1968). Foundations of Quantum Mechanics. Reading: Addison-Wesley.

--------. (1973). Are Quanta Real? Bloomington: Indiana University Press.

Krips, H. (1976). "Foundations of Quantum Theory." Foundations of Physics 6: 639-659.

McGrath, J. (1978). "A Formal Statement of the Einstein-Podolsky-Rosen Argument." International Journal of Theoretical Physics 17: 557-71.

Mehra, J. (1974). The Quantum Principle: Its Interpretation and Epistemology. Dordrecht: Reidel.

Moldauer, P. (1972). "A Reinterpretation of von Neumann's Theory of Measurement." Foundations of Physics 2: 41-47.

Park, J. (1968a). "Quantum Theoretical Concepts of Measurement." Philosophy of Science 35: 225-31.

-------. (1968b). "The Nature of Quantum States." American Journal of Physics 37: 217-22.

Schrödinger, E. (1935). "Die Gegenwartige Situation in der Quantenmechanik." Die Naturwissenschaften 23: 807-12, 823-28, 844-49.

--------------. (1936a). "Discussion of Probability Relations Between Separated Systems." Proceedings of the Cambridge Philosophical Society 32: 555-63.

--------------. (1936b). "Probability Relations Between Separated Systems." Proceedings of the Cambridge Philosophical Society 32: 446-52.

Süssman, G. (1957). "An Analysis of Measurement." In Observation and Interpretation: A Symposium of Philosophers and Physicists. Edited by S. Korner. New York: Academic Press. Pages 131-136. (Reprinted in Observation and Interpretation in the Philosophy of Physics. Edited by S. Korner. New York: Dover Publications, Inc., 1962. Pages 131-136.)

Trimmer, J. D. "Translation of Schrödinger's 'Cat Paradox' Paper." Unpublished manuscript.

van Fraassen, B. (1972). "A Formal Approach to the Philosophy of Science." In Paradigms and Paradoxes: The Philosophical Challenge of the Quantum Domain. (University of Pittsburgh Series in Philosophy of Science, Vol. 5.) Edited by R. Colodny. Pittsburgh: University of Pittsburgh Press. Pages 303-366.

van Zandt, L. (1977). "A Separation of the Microscopic and Macroscopic Domains." American Journal of Physics 45: 52-55.

Wigner, E. (1963). "The Problem of Measurement." American Journal of Physics 31: 6-15.

Only if 'Acrobatic Logic' is Non-Boolean [1]

Slawomir Bugajski

Institute of Physics, Silesian University
Katowice, Poland

The purpose of this paper is to defend 'quantum logic' against a vigorous attack of J.H. McGrath (1978). The case is serious, as he allegedly discredits quantum propositional logic showing it unmotivated and useless. The object of his criticism is the procedure of "taking a well-known theory and distilling its logic" (Finkelstein 1969), or "reading off quantum logic from quantum mechanics" (McGrath 1978). This procedure is indeed the corner stone of the quantum logic philosophy, so if it were be "discredited in a most general way" and shown to be "fundamentally misconceived" (McGrath 1978) then the idea of quantum logic has broken down. I am going to show, on the contrary, that McGrath's account of the 'reading off' procedure is not sufficiently precise resulting in the failure of his 'acrobatic' counterexample. It is easy to demonstrate that 'acrobatic logic', if carefully 'read off', is Boolean as it should be, hence does not provide any argument against the 'read off' procedure.

Before I begin, let me stress that I do not pretend to universally represent views of 'quantum logicians', but only my own views. There are a few more or less defined schools in the field of quantum logic, and adherents of some of them would disagree with my arguments. The 'read off' procedure I describe is close to the original ideas of Birkhoff and von Neumann (1936) which have been further developed by van Fraassen (1973). A different variant of it has been elaborated by Mittelstaedt and his co-workers (Mittelstaedt 1978a).

I start by briefly describing the 'read off' procedure as applied to an abstract physical theory. At least three working physical theories fit into this scheme, namely: classical mechanics (CM), standard quantum mechanics (SQM), and operational quantum mechanics (OQM). They are described in Mackey (1963), Jauch (1968), and Davies (1976). From now on the terms 'theory' and 'physical theory' will refer to any fixed one of the three just mentioned theories, whereas 'system' or 'physical system' will mean a class of 'similar' single physical objects

---

PSA 1980, Volume 1, pp. 264-271

describable by the theory in question.

Our 'read off' procedure is anchored to physics at least at two points: it takes into account the set of all measurable quantities (observables) considered by the theory, as well the set S of all states (preparations) of the physical system. Both sets have to be clearly defined if we attempt to apply the procedure to a given physical theory.

The basic category of experimentally meaningful statements concerning the physical system in question consists of statements of the following kind: 'A measurement of a physical quantity A gives a value in a (Borel) subset X of the real line R', in short (A,X), for any measurable quantity A and any set X. For instance: 'A measurement of the z-component of the momentum of a free electron gives a value in a given interval of R'. I consider it natural to take the set $\mathcal{E}$ of all such sentences as the set of elementary sentences of the object language 'read off' the physical theory.

It should be mentioned that there are other possibilities for defining the set of all sentences. Thus we can consider all sentences of the kind (α, A, X) (read: 'Quantity A measured on state α gives result in set X'), or (A, X, λ) (read: 'Quantity A gives result in X with probability λ), or even (α, A, X, λ). An objection can be raised to the first possibility on the ground that states and observables have different statuses in the theory as preparation and measurement procedures respectively, so they should not be mixed together in the 'logical' description. Moreover, the striking and fruitful interpretation of physical states as logically possible worlds (van Fraassen 1973) would break down if we considered such sentences. Nevertheless there is no decisive argument against this possibility. The other cases evidently fall into a meta-language (Mittelstaedt 1978b) and thus do not challenge the 'orthodox' position formulated above.

Given a state $\alpha \varepsilon S$, the theory provides us with a mapping $v_\alpha : \mathcal{E} \to [0,1]$ (the real unit interval), with the following interpretation: $v_\alpha(A,X)$ is the probability of obtaining a result in X if we measure A on the system prepared according to α. (Measurement is considered here as a sequence of single acts of measurement, thus resulting in a probability instead of providing a single numerical value of the measured observable. We abandom completely, in a way common in physics, the troubles with the frequency-versus-probability question.) It is plausible to assume that: $v_\alpha(A,X) = 1$ means that the elementary sentence (A,X) is true (more precisely: it is contingently true at α); $v_\alpha(A,X) = 0$ means that (A,X) is false (is contingently false at α); $v_\alpha(A,X) \neq 0,1$ means that (A,X) is neither true nor false (at α). I will consider, at least for convenience, the values of $v_\alpha(A,X)$ different from 0,1 as additional 'probabilistic' truth values. It should be stressed that in the case of OQM we cannot dispense with these intermediate truth values (see Bugajski 1980). Then the functions $v_\alpha : \mathcal{E} \to [0,1]$, $\alpha \varepsilon$ S, are just truth-value assignments on $\mathcal{E}$, so S is the valuation space (van Fraassen 1971).

The functions $v_\alpha, \alpha \in S$, define a mapping, V, which to any elementary sentence (A,X) attaches a function V(A,X): $S \to [0,1]$, given by $V(A,X)(\alpha) = v_\alpha(A,X)$. The mapping V is evidently the semantical valuation over $\mathcal{E}$, whereas its image P is the corresponding set of propositions. Our propositions generalize the characteristic functions of subsets of S which represent propositions in the two-valued semantics.

The set of P of propositions carries a natural algebraic structure defined by the physical theory. For example, it is ordered: $a \leq b$ iff $a(\alpha) \leq b(\alpha)$ for all $\alpha \in S$, a, $b \in P$. We identify P as the semantical algebra of the arising linguistic structure. In the case of CM P is a Boolean algebra. For SQM P is an ortho-modular semi-modular complete atomic lattice, the notorious 'quantum logic'. For OQM P is not a lattice. A detailed discussion of this case can be found in Bugajski (1980).

The operations of the semantical algebra are semantical traces of syntactical connectives. This makes it possible to build up a syntax on the basis of the semantics obtained. A well defined, although not necessarily classical, multi-valued language is the result. This language is called the inner language (IL) of the physical theory. The inner language of classical mechanics, IL(CM), has an entirely classical syntax. That is not the case for IL(SQM) and IL(OQM). The former has the syntax of the well-known non-Boolean type, whereas the latter has no conjunction and disjunction, but some modal connectives and a huge family of non-standard conditions (Bugajski 1980).

The process of reconstructing the inner language from its fragments scattered on the physical theory proceeds from the theory to a semantics, and from the semantics to a syntax. This sequence can be continued, because a language (syntax plus semantics) enables us to reconstruct a logical system (syntax plus a consequence operator) to which the language serves as an interpretation. It is rather clear that the semantical algebra introduced above should then be identified with the Lindenbaum-Tarski algebra of the corresponding logical system. Some ambiguity arises about a consequence operator. The simplest choice is to take the semantical entailment (represented by the order over P) to play this role (Goldblatt 1974). There are also other possibilities, see for instance Kalmbach (1974), Dalla Chiara (1977), and Mittelstaedt (1978a).

The 'read off' procedure goes as follows:

(i) Two basic sets inside the considered theoretical description are distinguished: the set of physical quantities and the set S of states.

(ii) The set $\mathcal{E}$ of all elementary sentences is identified with the set of all pairs (A,X), where A is a physical quantity and X is a (Borel) subset of the real line.

(iii) Find in the theory the mappings v_α, $\alpha \in S$, of $\mathcal{E}$ into the unit interval (0,1), and identify them as all admissible truth-value assignments over $\mathcal{E}$.

(iv) Define the semantical valuation V which maps $\mathcal{E}$ into the set of all real functions on S, $V(A,X) = v_\alpha(A,X)$ for any $\alpha \varepsilon S$ and $(A,X) \varepsilon \mathcal{E}$. We identify the set $P = V(\mathcal{E})$ as the set of all propositions.

(v) Find the natural algebraic structure defined on P by the physical theory, and identify it as the semantical algebra.

(vi) Reconstruct the syntax corresponding to the obtained semantical structure. The inner language of the physical theory in question is constructed.

(vii) Identify the semantical algebra of IL with the Lindenbaum-Tarski algebra of an appropriate logical system.

(viii) Look for a consequence operator generating this Lindenbaum-Tarski algebra. We obtain an inner logic of the considered physical theory.

Note that the physical and logical objects mentioned in (i) though (vii) are clearly defined by the physical theory, so there is no ambiguity in identifying the inner language, as well as the Lindenbaum-Tarski algebra, of the theory. It is by no means irrelevant which of the ordered sets appearing in the formalism of the theory is chosen as its 'logic'. A less than careful treatment of this point presumably causes McGrath's fallacy: it is not the case that one looks for some class of entities inside, say, SQM and finding it sufficiently regular baptises it 'quantum logic'. On the contrary, the set of propositions together with its algebraic structure is defined by the theory.

There is indeed a good amount of debate about and unsolved problems connected with the described procedure. This is not surprising, as this is probably the first case in which the human intellect is confronted with empirically motivated non-classical languages. Viewed in this way, 'quantum logic', i.e., the result of applying our 'read off' procedure to SQM, is not a philosopher's invention aimed to remove some paradoxes or 'paradoxes' of quantum theories. It emerges from the ocean of experimental facts as a basic structure of the set of empirical sentences about microsystems. In a similar way classical logic, i.e., the result of applying the above procedure to CM, is related to empirical sentences about the macro-world. (The commonly used notions of micro-world and macro-world are to be understood rather as 'quantum world' and 'classical world' respectively.) This does not mean that "classical logic has been discovered by reading it off classical physics" (here we agree with McGrath). Our conclusion serves merely as an encouragement for those who believe that classical logic has been 'read off' the macro-world in a way similar to classical physics. In the case of quantum logic this is not the case: quantum logic must be 'read off' from the quantum mechanics because we are not able to read it directly off the quantum world.

There is no problem about the lattice structure of the semantical algebra 'read off' SQM. The set P just carries a lattice structure as a matter of fact. There is a problem of interpretation of this lattice structure by means of simple experimental manipulations, but this is another story. For a, b, c $\in$ P with a = b$\wedge$c (the lattice meet), the problem can be formulated as follows: how can the experimental procedure corresponding to a be realized by means of the experimental procedures corresponding to b and c? This question is of importance if we try to 'read off' quantum logic from 'physical intuitions' about measurements instead reading it off SQM. Troubles with generating the a-procedure from b- and c-procedures should be viewed merely as a symptom of the 'non truth-functional' character of logical connectives in quantum logic. Quantum logic shares troubles of this kind with many other logical systems. Thus I do not find any "serious mistake" (McGrath 1978) in the 'read off' procedure as described above.

Finally our procedure can be applied to McGrath's acrobatic friend (1978). First it is necessary to find the set of states and the set of observables of the acrobat. It is rather evident that the distinguished positions of the acrobat (8 in number) correspond to the set of her pure states. The mixed states are not pictured in (McGrath 1978) as to do this an (infinite) ensemble of acrobats is needed (or specific misleading circumstances of observing the single one). From the formal point of view if the whole set of states were at our disposal, P could be reconstructed from it (Mielnik 1969). Only the set of pure states is known so the basis for reconstructing the 'acrobatic logic' is far from complete.

There is, however, a way out, so obvious that it can be easily overlooked. The eight pure states of the acrobat can be distinguished so eight corresponding measurable quantities can be defined. Thus the observable corresponding to state α is defined operationally by the following instruction: look at the acrobat many times: if she is in the position corresponding to α every time, the result of measurement is 1; if she is in a position different from α every time, the result is 0; if she is sometimes in position α and sometimes not, the result equals the relative frequency of observing her in the position α. As a necessary condition of a correct measurement it must be assured that during the measurement the acrobat does not change her state. It is evident now that we can observe only the cases with values 1 or 0. The typical quantum behaviour described in the third case is excluded by the macroscopic nature of the observed object, as well by the intentions of the author of the discussed example. This kind of behaviour appears also on the macro-level for mixed states, but here we are dealing with pure states only.

Now we are able to introduce the notion of orthogonality for the acrobatic states. Two states are called orthogonal if there exists a measurable quantity which measured on one of them gives 1 whereas on the second gives 0 (Mielnik 1969). We find easily that all 8 pure states of the acrobat are mutually orthogonal. This is enough: if all pure states of a given physical system are mutually orthogonal, then the system is classical. This trivial fact belongs presumably to the folklore of the 'quantum logic' approach. The 'acrobatic logic' P appears to be Boolean; in fact isomorphic to 2^8. There is nothing implausible in this result; moreover we expected it, so the 'read off' procedure does not lead to absurdities in this case. To be quite fair: we have assumed that the acrobat does not exhibit quantal behavior, the Boolean structure of her logic is then only a simple consequence of this assumption. It seems, however, that McGrath's intention was to describe a classical macroscopic system with a queer logic. Only in this case could his counterexample be sufficiently paradoxical. If so, then our assumption does not go beyond the scope of the example.

The 'acrobatic logic' of McGrath consists of different elements than ours. Instead of P as defined in (iv) above, he considers a non-commutative group of what he calls 'operations',i.e., acrobatic flips and rotations from one position to another, suggested presumably by Finkelstein's analysis (1969). It should be stressed that the acrobatic 'operations' do not correspond to the physical tests (filtering operations) considered by Finkelstein: the latter do not connect mutually orthogonal states! The filtering operations serve as a kind of physical semantics for P as was shown by Finkelstein (1969), whereas the acrobatic 'operations' have nothing in common with the experimental propositions concerning the acrobat. By the way, we also see that the 'acrobatic' counterexample does not work against "logics of (physical) operations" (Stachow 1980).

Thus quite surprisingly the 'acrobatic' counterexample of McGrath proves itself unsound and misconceived, which invalidates his whole attempt to discredit the quantum logic. Our discussion indicates that 'quantum logicians' can still safely study the logic of the quantum world. Of course, the fundamental question about the empirical nature of logic is not solved by the above remarks, as the philosophical status of the inner language and the inner logic of a physical theory is far from clear. Our precise account of the 'read off' process can contribute to the debate as an indication that the quantum logic is not a fancy, but it does emerge from physics in a way which cannot be easily discredited.

Notes

[1]The first draft of this paper has been written during my stay at the Institute of Physics of the Cologne University. I am greatly indebted to Professor P. Mittelstaedt and Dr. E.-W. Stachow for their warm hospitality and stimulating discussions, as well to the Alexander von Humboldt Foundation for a financial support.

References

Birkhoff, Garret and von Neumann, John (1936). "The Logic of Quantum Mechanics." _Annals of Mathematics_ 37: 823-843.

Bugajski, Slawomir (1980). "The Inner Language of Operational Quantum Mechanics." To appear in proceedings of the Workshop on Quantum Logic held at the Ettore Majorana Centre for Scientific Culture, Erice-Trapani (Sicily), 2-9 December 1979. Edited by E. Beltrametti. New York: Plenum Press.

Dalla Chiara, Maria L. (1977). "Quantum Logic and Physical Modalities." _Journal of Philosophical Logic_ 6: 391-404.

Davies, E.B. (1976). _Quantum Theory of Open Systems._ London: Academic Press.

Finkelstein, David (1969). "Matter, Space and Logic." In _Proceedings of the Boston Colloquium for the Philosophy of Science,_ 1966/1968. _(Boston Studies in the Philosophy of Science,_ Vol. 5.) Edited by R.S. Cohen and M. Wartofsky. Dordrecht: D. Reidel Publishing Co. Pages 199-215.

Goldblatt, R.I. (1974). "Semantic Analysis of Orthologic." _Journal of Philosophical Logic_ 3: 19-35.

Jauch, Josef M. (1968). _Foundations of Quantum Mechanics._ Reading, Mass.: Addison-Wesley.

Kalmbach, Gudrun. (1974). "Orthomodular Logic." _Zeitschrift fur Mathematische Logik und Grundlagen der Mathematik_ 20: 395-406.

Mackey, George W. (1963). _The Mathematical Foundations of Quantum Mechanics. (Mathematical Physics Monograph Series.)_ New York: W.A. Benjamin, Inc.

McGrath, J.H. (1978). "Only if Quanta Had Logic." In _PSA 1978,_ Vol. 1. Edited by P.D. Asquith and Ian Hacking. East Lansing, Mich.: Philosophy of Science Association. Pages 268-276.

Mielnik, Bogdan. (1969). "Theory of Filters." _Communications in Mathematical Physics_ 15: 1-46.

Mittelstaedt, Peter. (1978a). _Quantum Logic._ Dordrecht: D. Reidel Publishing Co.

-------------------. (1978b). "The Metalogic of Quantum Logic." In _PSA 1978,_ Vol. 1. Edited by P.D. Asquith and Ian Hacking. East Lansing, Mich: Philosophy of Science Association. Pages 249-256.

Stachow, Ernst-Walter. (1980). "On the Logical Foundations of Quantum Mechanics." _International Journal of Theoretical Physics._ Forthcoming.

van Fraassen, Bas C. (1971). Formal Semantics and Logic. New York: The Macmillan Company.

-------------------. (1973). "Semantic Analysis of Quantum Logic." In Contemporary Research in the Foundations and Philosophy of Quantum Theory. (University of Western Ontario Series in Philosophy of Science, Vol. 2.) Edited by C.A. Hooker. Dordrecht: D. Reidel Publishing Co. Pages 80-113.

A Model Theoretic Semantics for Quantum Logic

E.-W. Stachow

Institut für Theoretische Physik

Universität zu Köln, Köln, W.-Germany

1. Introduction

The purpose of this contribution is to consider a model theoretic semantics of quantum logic. The quantum logic to which we refer here comprises the system of sequential quantum logic (Stachow 1980b). This system establishes the implication relation with respect to logically connected propositions and also with respect to the more general class of sequentially connected propositions of the language of quantum physics.

A dialogic semantics for logically connected sentences is developed in Stachow (1976, p. 203), where it is shown that a logical system, usually called the system of quantum logic, is complete and sound with respect to the dialogic semantics. The dialogic semantics is extended to sequentially connected sentences in Stachow (1980b) and to modal sentences in Mittelstaedt (1979, p. 479).

In comparison with other approaches to a language for science, the dialogic approach is developed from a pragmatic and methodological point of view (Stachow 1980a). This approach consists in a systematic reconstruction of a scientific language which is based on the pragmatic possibilities of argumentation with respect to propositions. The possibilities of argumentation determine the frame of the syntactic as well as of the semantic concepts of the language. They can be formulated as rules in a dialog-game in order to establish a proof procedure for propositions. Truth and falsity of propositions are defined by means of win and loss of the dialog-game with respect to the propositions. In this way the dialogic approach establishes a proof theoretic semantics of a language for science.

Usually, a semantics of a language for science is presented not as a proof theoretic semantics but as a model theoretic semantics.

PSA 1980, Volume 1, pp. 272-280

Such a semantics is distinguished if we regard a scientific language from the outset as a language for the description of the structure of a certain reality. The central notion of this semantics is that of a model. A model comprises a set, the set of "possible worlds", and, in addition, certain structural entities which are defined with respect to the set of possible worlds. This structure determines the semantical concepts of the language. Propositions are defined to be certain sets of possible worlds. A proposition is said to be true in a possible world if and only if this possible world is an element of the set which corresponds to the proposition.

If a mapping (determined by a particular possible world) associates each proposition of the language with one of the two truth-values truth and falsity, we call it a bivaluation. The formal semantics of the normal propositional modal logic associates each possible world with a bivaluation (e.g., see van Fraassen 1971). This classical semantics can also be used for establishing a formal language of classical physics. In this case the set of possible worlds is given by the set of states of a classical system. The set of states with respect to an individual classical system forms a model of the language. Concerning a language of quantum physics it is well known that the classical semantics cannot be taken over without an additional pragmatic restriction of the valuations of propositions to those propositions which are pairwise commensurable. In this way an object language of quantum physics which takes into account the classical semantic concepts comprises a semantic as well as a pragmatic component.

In order to incorporate the pragmatic component into a modified model theoretic semantics of the language of quantum physics, one may consider the possibility to specify the set of possible worlds as the set of quantum states of a quantum physical object. However, this leads to the following difficulties. If mixed states are included in this set they do not present an adequate description of a possible world of a quantum physical object since subjective ignorance may be prevalent in a mixed state. If, on the other hand, it is assumed that a possible world can always be described by a pure state, one runs into the difficulties connected with the ignorance interpretation of mixed states. Therefore such a model theoretic semantics of the language of quantum physics seems to be unsatisfactory.

In this contribution we consider a model theoretic semantics which arises from the dialogic semantics if the pragmatic concept of the game is replaced by a "metaphysical" concept of a game. Because of this connection between the two kinds of semantics we at first make a few remarks on the dialogic semantics.

2. The Dialogic Semantics

We consider propositions about an individual quantum physical system S (like an atom, a nucleus, an elementary particle). Truth and falsity of such propositions are established by means of dialogs about the propositions. It is assumed that before a dialog the system is prepared in a certain state W which may also be established by means of preceding dialogs. The system is then characterized by S(W). We distinguish between elementary propositions, which assert particular properties of a quantum physical system, and compound propositions, which are linguistic constructions composed of elementary propositions.

Elementary propositions are proven within elementary dialogs. If a participant asserts an elementary proposition a, he is obliged, upon an attack by an opponent, to justify his assertion by means of a measurement of the corresponding property â. Let the system S be respresented in the frame work of quantum theory by the Hilbert space $\mathcal{H}(S)$, and let the property â be represented by the projection P_a defined on $\mathcal{H}(S)$ with the range M_a. If the system is prepared in a certain state, represented by the statistical operator W, a measurement of the property â which proves a results in the state $W'=P_a \cdot W \cdot P_a$. In case P_a satisfies $P_a \cdot W \cdot P_a = W$, the proposition a will certainly be proven by the measurement. In case $(1-P_a) \cdot W \cdot (1-P_a) = W$ holds, the proposition a will certainly be disproven by the measurement. In these two cases the truth and falsity respectively of an elementary proposition a can already be inferred from the preparation of the system. However, this is not possible for all elementary propositions. If $P_a \circ W \circ P_a \neq W$ and $(1-P_a) \circ W \circ (1-P_a) \neq W$, the result of the measurement is not determined but contingent.

Let W be the state of the system before the measurement and let W' be the state of the system after the measurement of a property â. Then we have the following proof and disproof conditions for the elementary proposition a:

(2.1) Definition:

(a) The elementary proposition a is true, given the system S(W),
$\overset{def}{\leftrightarrows}$ a is proven by means of the measurement M (denoted by $S(W) \vdash_M a$)
$\leftrightarrows$ $P_a \circ W' \circ P_a = W'$;

(b) the elementary proposition a is false, given the system (S(W),
$\overset{def}{\leftrightarrows}$ a is disproven by means of the measurement M (denoted by $S(W) \vdash_{\neg M} a$)
$\leftrightarrows$ $(1-P_a) \circ W' \circ (1-P_a) = W'$.

Compound propositions are checked within the material dialog-game D_m the rules of which are given in Stachow (1976, p. 203, 1980a, 1980b). The connectives are defined by means of the different possibilities to attack and to defend compound propositions. For instance, if a and b are elementary propositions, the sequential conjunction $a \sqcap b$ is defined by the following scheme:

sequential conjunction	attacks	defences
$a \sqcap b$	1. 1?	2. a
"a and then b"	3. 2?	4. b

As an example let us assume that an electron S is given in the state W. Two elementary propositions with respect to S are: "S has spin-up in the direction x" and "S has spin down in the direction y", denoted by $\underline{a}$ and $\underline{b}$ respectively. The sequential conjunction "S has spin-up in the direction x <u>and then</u> S has spin-down in the direction y" is checked within a dialog which is represented by the following game tree:

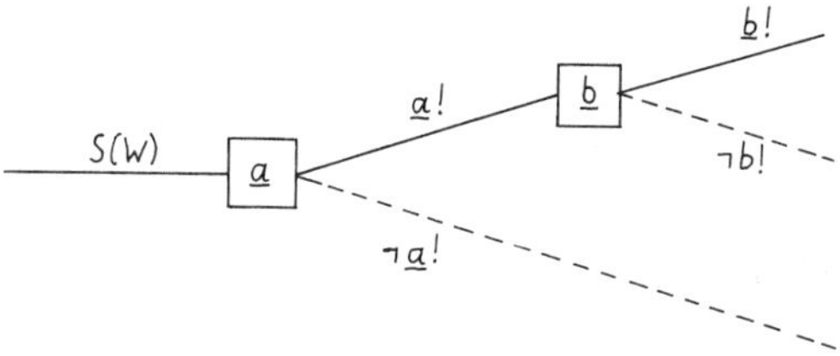

Each possible run of the dialog about $\underline{a} \sqcap \underline{b}$ is indicated by a particular branch of the above game tree which depends on the proof results with respect to the elementary propositions. After the first defence of $\underline{a} \sqcap \underline{b}$ by $\underline{a}$ the elementary proposition $\underline{a}$ is checked by means of a measurement. If $\underline{a}$ is false, denoted by $\neg\underline{a}!$, the dialog about $\underline{a} \sqcap \underline{b}$ is lost and, therefore, the proposition $\underline{a} \sqcap \underline{b}$ is false. If $\underline{a}$ is true, denoted by $\underline{a}!$, the dialog about $\underline{a} \sqcap \underline{b}$ is continued, and after the second defence of $\underline{a} \sqcap \underline{b}$ by $\underline{b}$ the elementary proposition $\underline{b}$ is checked by means of a subsequent measurement. In case $\underline{b}$ is false, the proposition $\underline{a} \sqcap \underline{b}$ is false, in case $\underline{b}$ is true, the proposition $\underline{a} \sqcap \underline{b}$ is true. The branches of the game tree are established by individual dialogs about $\underline{a} \sqcap \underline{b}$, given S(W). Let W' denote the state of the system S after the measurement of the property $\hat{\underline{a}}$ corresponding to the elementary proposition $\underline{a}$, and let W" denote the state of the system S after the subsequent measurement of the property $\hat{\underline{b}}$ corresponding to the elementary proposition $\underline{b}$. Then we have for the above dialog, in case $\underline{a} \sqcap \underline{b}$ is true: $P_a \circ W' \circ P_a = W'$, $P_b \circ W'' \circ P_b = W''$. In case $\underline{a} \sqcap \underline{b}$ is false, denoted by the dotted lines in the above game tree, we have $(1-P_a) \circ W' \circ (1-P_a) = W'$ or $P_a \circ W' \circ P_a = W'$, $(1-P_b) \circ W'' \circ (1-P_b) = W''$ respectively.

The proof of the sequential conjunction $\underline{a} \sqcap \underline{b}$, however, does not establish the truth of the logical conjunction $\underline{a} \wedge \underline{b}$. In order to prove $\underline{a} \wedge \underline{b}$ it must be checked whether the two subpropositions $\underline{a}$ and $\underline{b}$ are <u>commensurable</u>. The commensurability of $\underline{a}$ and $\underline{b}$ means that in an arbitrary sequence of proofs of $\underline{a}$ and $\underline{b}$ the proof results of both of the propositions do not change (Stachow 1980a). In the object language of quantum physics <u>commensurability propositions</u>, denoted by $\underline{k}$ ($\underline{a}$, $\underline{b}$), can be defined which assert the commensurability of $\underline{a}$ and $\underline{b}$ as a property of a quantum physical system. Commensurability propositions are proven in the same way as elementary propositions by means of measurements. With the aid of the commensurability proposition $\underline{k}$ ($\underline{a}$, $\underline{b}$) the <u>logical conjunction</u> can be proven by means of the following possibilities of attack and defence within a dialog-game:

logical conjunction	attacks	defences
a ∧ b	1. 1?	2. a
"a and b"	3. 2?	4. b
	5. k(a,b)?	6. k(a,b)

The dialog-game about a ∧ b is represented by the game tree:

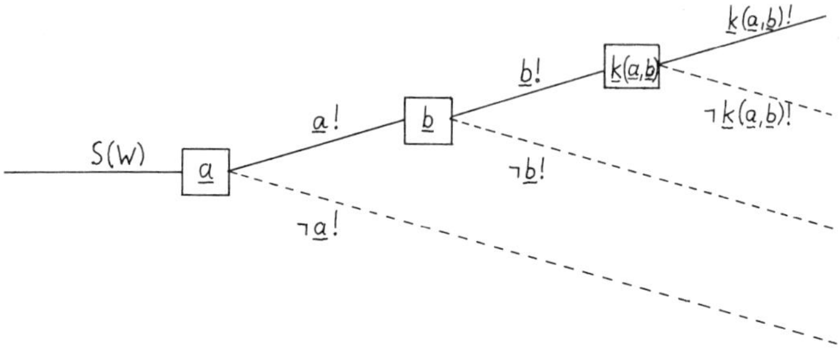

In our above example of elementary propositions a and b the commensurability proposition is disproven in each dialog-game. Therefore the branch with k(a,b)! does not exist in our example and the proposition a ∧ b is always false.

By means of the dialog-game truth and falsity of compound propositions are defined by:

(2.2) Definition:

(a) The compound proposition $\mathcal{A}$ is true, given S(W), $\overset{def.}{\Leftrightarrow}$ the proponent wins the dialog-game D_m about $\mathcal{A}$ (denoted by $S(W) \vdash_{D_m} \mathcal{A}$) ;

(b) the compound proposition $\mathcal{A}$ is false, given S(W), $\overset{def.}{\Leftrightarrow}$ the proponent loses the dialog-game D_m about $\mathcal{A}$ (denoted by $S(W) \vdash_{\neg D_m} \mathcal{A}$).

In the framework of the dialogic semantics it is assumed that the system S is prepared in a certain state W before a dialog about a proposition $\mathcal{A}$ is performed. Hence, in general, the quantum physical system under consideration has already experienced a certain history before a dialog-game.

However, one may also consider this history as a particular branch of an extended game tree which describes the whole history of the quantum physical system. The idea is that the extended game completely describes the actual history of the system and all possible runs of the history in the past and in the future as branches of its game tree. This "metaphysical" concept of a game with respect to a quantum physical system leads to a model theoretic semantics of the language of quantum physics.

3. The Model Theoretic Semantics

At first we explain what we mean by a possible material process. Let us consider a particular sequence of proofs and disproofs of elementary and commensurability propositions as they occur for instance in the above examples of game trees for $\underline{a} \sqcap \underline{b}$ and $\underline{a} \wedge \underline{b}$. If such a proposition, let us say $\underline{a}$, is proven, the corresponding property $\hat{\underline{a}}$ pertains to the system under consideration. We assume that, in case $\underline{a}$ is disproven, a certain property, denoted by $\neg\hat{\underline{a}}$, pertains to the system. A sequence of properties pertaining to the system is called a possible material process. For instance the sequence $\langle \hat{\underline{a}}, \hat{\underline{b}}, \hat{\underline{k}}\ (\underline{a},\underline{b})\rangle$ is a possible material process established within the above dialog-game about $\underline{a} \wedge \underline{b}$.

Now we regard possible material processes from the outset as potential entities of a certain reality. We want to characterize an individual physical system S by a particular set of possible material processes such that each process is a possible history of the same individual system S. A game tree may now describe this set as the set of its branches. A particular dialog about the system S, like the above dialog about $\underline{a} \wedge \underline{b}$, given S(W), occurs as a part of the "metaphysical" game tree corresponding to S, as is indicated in the illustration:

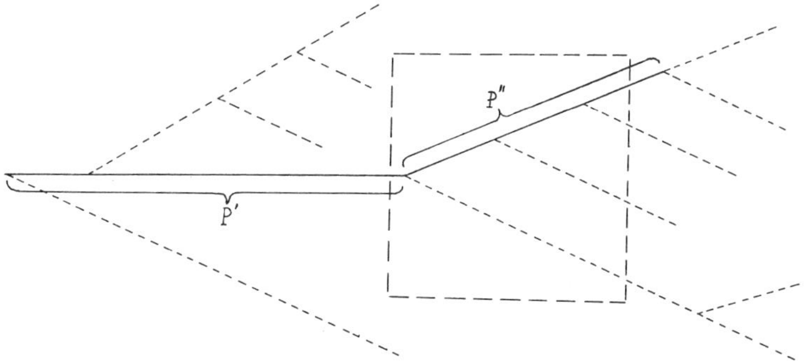

All possible material processes have the same beginning at the left side of the illustration. The material process p' specifies how the system S was prepared in the state W before the indicated dialog. p" denotes the partial process $\langle \hat{\underline{a}}, \hat{\underline{b}}, \hat{\underline{k}}\ (\underline{a},\underline{b})\rangle$. The join of p' and p", denoted by p' ⊎ p" in the following, again is a possible material process.

We assume that all possible material processes p which in this way correspond to an individual quantum physical system S form a set, denoted by P(S). A model of the formal language of quantum physics comprises a set P(S) and, in addition, certain relations which are defined later. For the reason of simplicity we do not take into account in this contribution that certain possible material processes are identical and that equivalence classes of processes should be used instead of processes.

In the framework of this semantics propositions $\mathcal{A}$ are defined as particular sets of possible material processes. These processes are either processes $q'' \in P(S)$ such that q'' determines a branch of the dialog-game about $\mathcal{A}$ in which the proposition $\mathcal{A}$ is proven. Or they are processes $q' \uplus q'' \in P(S)$ such that the partial process q'' again determines a branch of the dialog-game about $\mathcal{A}$ which proves $\mathcal{A}$. For instance the conjunction $\underline{a} \wedge \underline{b}$ is defined as the set which consists of $p'' := \langle \underline{\hat{a}}, \underline{\hat{B}}, \underline{R}(\underline{a},\underline{b}) \rangle$ if $p'' \in P(S)$, and which consists of all $p = p' \uplus p'' \in P(S)$.

Truth and falsity of propositions are defined by:

(3.1) Definition:

(a) The proposition $\mathcal{A}$ is true in the possible material process p $\overset{def.}{\leftrightsquigarrow}$ $p \in \mathcal{A}$;

(b) the proposition $\mathcal{A}$ is false in the possible material process p $\overset{def.}{\leftrightsquigarrow}$ $p \in \neg\mathcal{A}$.

Finally we want to extend our model theoretic semantics to modal propositions of the object language of quantum physics. In order to define the modal propositions we make use of the following relations.

(3.2) Definition:

For each proposition $\mathcal{A}$ the relation $R_{\mathcal{A}} \subseteq P(S) \times P(S)$ is given by $(p,q) \in R_{\mathcal{A}} \overset{def.}{\leftrightsquigarrow} p = p' \uplus p''$ and $q = p' \uplus q''$, where $p' \in P(S)$ and the partial processes p'' and q'' determine branches of the dialog-game about $\mathcal{A}$.

Obviously the relation $R_{\mathcal{A}}$ is an equivalence relation.

The modal proposition "$\mathcal{A}$ is necessary", denoted by $N\mathcal{A}$, is defined as the set $\{p \in \mathcal{A} : (p,q) \in R_{\mathcal{A}} \rightsquigarrow q \in \mathcal{A}\}$. The negation $\neg N\mathcal{A}$ is defined as the set $\neg\mathcal{A} \cup \{p \in \mathcal{A} : \exists_q((p,q) \in R_{\mathcal{A}} \,\&\, q \in \neg\mathcal{A})\}$. The modal proposition "$\mathcal{A}$ is possible", denoted by $M\mathcal{A}$, is defined to be the proposition $\neg N \neg\mathcal{A}$. Truth and falsity of modal propositions are defined analogously to (3.1). We have:

(3.3) (a) $N\mathcal{A}$ is true in p $\leftrightsquigarrow \forall_q((p,q) \in R_{\mathcal{A}} \rightsquigarrow q \in \mathcal{A})$;

(b) $M\mathcal{A}$ is true in p $\leftrightsquigarrow \exists_q((p,q) \in R_{\mathcal{A}} \,\&\, q \in \mathcal{A})$.

Propositions which consist of iterations of the connective and modal signs can also be defined as sets of possible material processes. If $\mathcal{S}_e$ is the set of elementary propositions, the set of logical propositions is given as the least set $\mathcal{L}$ such that: (i) If $a \in \mathcal{S}_e$, then $a \in \mathcal{L}$, (ii) if $A, B \in \mathcal{L}$, then $A \wedge B$, $\neg A$, $\underline{k}(A,B)$, $NA \in \mathcal{L}$. The total set of sequential propositions considered here is given as the least set $\mathcal{S}$ such that: (i) If $A \in \mathcal{L}$, then $A \in \mathcal{S}$; (ii) if $\mathcal{A}, \mathcal{B} \in \mathcal{S}$, then $\mathcal{A} \sqcap \mathcal{B}$, $\neg\mathcal{A}$, $N\mathcal{A} \in \mathcal{S}$.

(3.4) Definition:

(a) The valuation over the model structure $\langle P(S), \{R_{\mathcal{A}}\}_{\mathcal{A}}\rangle$ is the function
$v: M \subseteq P(S) \times \mathcal{S} \longrightarrow \{T,F\}$ with
$M := \{(p,\mathcal{A}) : \mathcal{A} \in \mathcal{S} \;\&\; p \in \mathcal{A} \cup \neg\mathcal{A}\}$

defined by

(i) $v((p,\mathcal{A})) = T \overset{def.}{\leftrightarrow} p \in \mathcal{A}$;

(ii) $v((p,\mathcal{A})) = F \overset{def.}{\leftrightarrow} p \in \neg\mathcal{A}$.

(b) An admissible valuation of $\mathcal{S}$ is the function
$v_p: \Delta_p \subseteq \mathcal{S} \longrightarrow \{T,F\}$ with
$\Delta_p := \{\mathcal{A} \in \mathcal{S} : p \in \mathcal{A} \cup \neg\mathcal{A}\}$

defined by $v_p(\mathcal{A}) := v((p,\mathcal{A}))$.

The set of admissible valuations of $\mathcal{S}$ is the set $\{v_p : p \in P(S)\}$.

A more careful analysis which uses equivalence classes for p instead of possible material processes leads to the following characterization of admissible valuations of the set $\mathcal{L}$ of logical propositions:

(3.5) A function $v_p: \Delta'_p \subseteq \mathcal{L} \longrightarrow \{T,F\}$ with
$\Delta'_p := \{A \in \mathcal{L} : p \in A \cup \neg A\}$

which satisfies

(i) $v_p(A \wedge B) = T \leftrightarrow v_p(A) = v_p(B) = v_p(\underline{k}(A,B)) = T$

(ii) $v_p(\neg A) = T \leftrightarrow v_p(A) = F$,

(iii) $v_p(NA) = T \leftrightarrow v_q(A) = T$ for all q with $(p,q) \in R_A$

is an admissible valuation of $\mathcal{L}$.

By means of these semantic concepts the other semantic concepts (like validity) of the language of quantum physics can be established as usual (e.g., see van Fraassen 1971).

References

Mittelstaedt, P. (1979). "The Modal Logic of Quantum Logic." _Journal of Philosophical Logic_ 8: 479-504.

Stachow, E.-W. (1976). "Completeness of Quantum Logic." _Journal of Philosophical Logic_ 5: 237-280. (As reprinted in C.A. Hooker (ed.). (1979). _Physical Theory as Logico-Operational Structure._ Dordrecht: D. Reidel. Pages 203-243.)

--------------. (1980a). "On a Game-Theoretical Approach to a Scientific Language." In _PSA 1978,_ Vol. 2. Edited by P.D. Asquith & Ian Hacking. East Lansing: Philosophy of Science Association. Forthcoming.

--------------. (1980b). "On the Logical Foundation of Quantum Mechanics." _International Journal of Theoretical Physics._ Forthcoming.

van Fraassen, B.C. (1971). _Formal Semantics and Logic._ New York: The Macmillan Company.

Part X

Time, Causation and Matter

Is Temporality Mind-Dependent?

Paul Fitzgerald

Fordham University

The question is oversimple. We are more likely to find the right set of answers if we focus sharply a distinction which is invariably blurred by those who discuss the issue, especially those who argue that "temporal becoming" is a mind-dependent feature of reality. We must distinguish the "indexicality theme" from the "elapsive theme". Failure to do so has bred some confusion, I think.

A bit of background first. Those who discuss the mind-dependence of temporal becoming think of that issue as analogous to the traditional one about the mind-dependence of secondary qualities. Do colors, felt hotness, and salty taste exist in the mind-independent world? Or only within experience? "Roses are red, just like some of our after-images." "No they aren't. Roses only have a property of selectively reflecting light in the 7200 Å region. That makes them *look* red to us. But they aren't really red." "You're both wrong. Of course roses are red. But to be red *just is* to be the sort of thing which would look red to normal observers under standard conditions. Similarly, to be soluble is simply to have the dispositional property of dissolving if placed in a solvent."

How about experienced temporality? Analogous questions arise for it. Our efforts to answer them are being hobbled by irrelevant issues twined around the essential one. The irrelevant issues have to do with *nowness, pastness, futurity* and other properties allegedly expressed by indexical expressions such as "is occurring now" and "happened yesterday". These properties are sometimes called "A-determinations", after McTaggart's A-series. I take the essential issue to be that of finding out to what extent the durational character, the "timeyness", "transiency", or "elapsive" quality of our experience, is found mind-independently in the physical world as well. This *elapsive theme* has almost nothing to do with the *indexical theme,* the question of whether nowness and other A-determinations are found outside of mind.

The indexicality theme runs this way. Temporal indexical expressions

---

PSA 1980, Volume 1, pp. 283-291

express features essential to "temporal becoming". Events are first future. They become ever less future, approaching us in time, and eventually become present. That's temporal becoming. They shed futurity and acquire another distinctive property, *presentness* or *nowness*. (I call this idea of time a *property acquisition model.*) This transient nowness is held briefly (how long?). Then events lose it, picking up *pastness* as a consolation prize. Futurity, presentness, and pastness are held to be irreducibly indexical determinations, in that they can not be expressed by such non-indexical predicates as the tenseless "*occur at time t*". Nor can statements attributing them be reductively translated into non-indexical statements.

Several related ideas are at work here. There is the general question of universals, of course. Do *any* expressions, indexical or not, express properties? Are there such things as properties to be expressed? If there are, can the identity conditions of these properties be "read off" the meanings of the predicates which express them? Are two universals distinct iff the predicates expressing them differ in meaning? Such a view sometimes underlies belief in irreducibly indexical properties, and the property-acquisition model of time. For we sometimes see arguments for this model which run as follows (Gale 1968).

First, statements containing temporal indexicals, such as "Event E is occurring now", resist translation into statements lacking them, such as "Event E *occur at time t* ". Second, these indexical statements convey information which cannot be conveyed by any non-indexical means. To say that a battle *occur at time t* (fill in a proper name of the present time for "*t*"), is not to say that the battle is occurring now. For a person might know that the battle has that date without thereby having enough information to figure out that it is occurring now, and *vice versa*.

Third, temporal indexicals communicate their peculiar kind of information by predicating of their subjects irreducibly indexical properties. This step is the mistaken one, I think. Rather than predicating an irreducibly indexical property, indexical predicates, I have argued, play their peculiar role by a kind of linguistic pointing (Fitzgerald 1974a, 1974b). But let us continue the argument.

Fourth, since some statements predicating temporal indexical properties are true, the properties are genuinely instantiated. Fifth, these A-determinations are neither mental nor mind-dependent properties. For some statements predicating them report facts which would obtain even if there were no minds. For example, even if there were no minds the galaxy would still be rotating on its axis right now.

Note that this argument is independent of the fact that the indexicals in question are *temporal*. If it works at all, it works for spatial and other "indexical determinations" as well. So all attempts, such as Nicholas Rescher's (1973, pps. 124-5) to argue that "full-blooded temporality" is mind-dependent would, if successful, show the same thing about "full-blooded spatiality". And attempts such as those of Wesley Salmon (1974) and Adolf Grünbaum (1971) to construe nowness as mind-dependent should be matched by attempts to do the same thing for *hereness*.

Step three is questionable. There are no A-determinations, within or outside of mind. Temporal indexical statements do not play their peculiar linguistic role by predicating irreducibly indexical properties but by a kind of linguistic pointing, analogous to literal pointing gestures. When we say that something is occurring *now* we simply mean that it is occurring *at this time*. To say that is to predicate the relational non-indexical property of occurring at a time, and, by means of "this", to *point linguistically* at the time in question. We don't predicate *thisness* in addition. Similarly, when we say that something is occurring *here* we mean that it is occurring at *this* place. We predicate occurrence at a place, and point to the place. We don't predicate *thisness* of the place. There is no need to raise questions about whether some supposed *hereness* is mind-independently instantiated or only mind-dependently so, or whether "full-blooded spatiality" is found only in the mind.

Some might claim that arguments from translatability, functioning of indexicals, and the like, are beside the point, because we directly experience *nowness*. Lynn Rudder Baker, for instance, says:

> A better reason for claiming, as Grünbaum does, that "E is occurring now" (said at t) and "E occurs at t" predicate different properties is that the *becoming present* of an event, unlike its occurrence at a clock time, is not a physical property. But failing to be a mind-*in*dependent property of events, nowness can adequately be construed as a mind-dependent property. The reason to take nowness to be a mind-dependent attribute of events is that transiency plays an undeniably significant role in our experience. Thus...in order to "save the phenomena", an account of becoming should allow for the ongoing and transient quality of experience:... . (Baker 1979, p.356).

Transiency, yes; indexicality or nowness, NO! That is the conflation against which I have been protesting. I find no irreducibly indexical herenesses and nownesses, thisnesses and thatnesses, within my experience. Nor is there *you-ness*, *vous-té*, *itness*, *that-away-ness* nor *down-yonderness*. But some will claim that though these last are suspect, we surely have to acknowledge *presentness* within experience. Isn't there a kind of felt transition from futurity to presentness to pastness which is an immediate datum? So aren't there A-determinations within experience?

No, there are not. But we must do justice to the facts which lead people to think there are. There *is* felt temporality within experience, a directly experienced transient or elapsive character. But it is not essentially indexical. Phenomenal temporal features of experience are fully expressible in tenseless language, as in "my feelings of nausea *surge* up quickly and *pass* as suddenly," or "The after-image on the left *lasts* longer than the jumpy green one to its right." (Cumbersome references to times have been omitted for brevity.) The dawdle and whoosh of experienced time is perfectly well expressed by tenseless talk. To think otherwise is like thinking that you can't correctly translate

"I love you" into Arabic because Arabic doesn't *sound* like a loving language (compare: tenseless locutions *suggest* timelessness, not transiency). It is a serious question whether these phenomenal temporal features found within experience characterize the physical world mind-independently. *That* question is the analogue for time of the issue concerning the objectivity or subjectivity of secondary qualities.

Why do people think there is an irreducibly indexical transiency within experience, that is, an irreducibly indexical property of presentness? The reason, I suggest, is this. When you say that some particular event is "happening now", that use of words is only temporarily appropriate for that event. People think this is because the words predicate a property, nowness, which is only temporarily possessed. Not so, however. "Is happening now" is only temporarily predicable not because the particular of which it's predicated has an evanescent property of nowness, but because it is evanescent. Later acts of predication are too late to be simultaneous with the short-lived particular.

To firm up my claim, let's take the analogous case of *hereness*. Someone might say that it is an undeniable fact that his green afterimage is *here* (mental pointing), and the red one is *there*. The beating of my heart is *here* (hand over heart); the beating of your heart is *there* (I point my finger to your chest). The wrong step would be to infer that the entities in question are thereby shown to have spatial indexical properties. "Here" and "there" *pick out* places in an indexical way. They are not used to *predicate* indexical properties of those places -- same for "now" and "then".

Several thinkers, such as Adolf Grünbaum (1971), Nicholas Rescher (1973), Wesley Salmon (1974), and Lynn Rudder Baker (1974-5, 1979) have held that there is an irreducibly indexical kind of nowness within experience which is essentially mind-dependent or at least mind-related. They differ from one another in their analyses of this nowness and, to a lesser degree, in their arguments for its mind-dependence. Salmon, for instance, criticizes details of Grünbaum's account and gives a different analysis of nowness. Lynn Rudder Baker partly defends Grünbaum against Salmon and partly offers reasons for modifying Grünbaum's account. Baker's thesis could better be described as asserting the "mind-centricity" of temporal becoming, and the same may hold for Rescher's. On Baker's view an event *acquires nowness* at the time of its occurrence even if no minds exist then. The meaning of this nowness-acquisition, however, is cashed in terms of what would be experienced by an observer who met certain requirements, if there were such an observer.

I have distinguished indexical nowness from non-indexical felt transiency. Felt transiency provides a closer analogue to the standard secondary qualities than does nowness, in that considerations relevant to deciding if felt transiency is mind-dependent are more like those which bear on secondary qualities. We should be asking whether non-indexical transiency, elapsiveness, is mind-dependent.

But we might ask the same question about the alleged property of indexical nowness. Is it essentially mind-dependent? You know that I think there is no such property in any conceivable world. But let's pretend that there is such a property, and that its identity conditions are given by the meaning of such predicates as "is occurring now" and "exists now". Is that property genuinely instantiated? If so, is it instantiated independently of minds? Does its analysis make essential reference to mind? We can examine which of the arguments for its mind-dependence would also, *mutatis mutandis*, be effective arguments for the mind-dependence of non-indexical transiency. Both nowness-acquisition and felt elapsiveness have often been regarded as essential to "temporal becoming". So we can use "temporal becoming" to speak in a general way about either or both phenomena.

The first argument for the mind-dependence of temporal becoming is that physics finds no need to posit nowness to explain happenings in the mind-independent world (Grünbaum 1967, Smart 1963). Similarly, we have no reason to believe that colors are found in the mind-independent world, since we can account for all the appearances without positing colors in things. Properties of selectively reflecting light will do instead.

The analogy creaks a bit. It is at best a conceptually contingent fact that colors aren't in things. We know what evidence would drive us to attribute to the mind-independent world colors like those found within experience. Suppose light didn't come in different wavelengths, and that the only way to account for why we see different objects as having different colors is that they *are* differently colored, independently of minds. But no one has ever successfully explained what it would be like for physics to require A-determinations and their spatial analogues, if it doesn't require them as things stand. (I think that Grünbaum (1967, 1969) has refuted the arguments of Hans Reichenbach (1956) and H.A.C. Dobbs (1969) that today's physics requires positing an indexical nowness.) But *if* I believed that indexical expressions expressed indexical properties with semantic identity conditions, I would argue that they *are* actually instantiated, and mind-independently, even though physics has no need of them. Physics focusses on those properties required to differentiate the actual world from other conceivable spatio-temporal worlds. But presentness, pastness and futurity are found in any conceivable temporal world. That's why physics need not mention them. Similarly, it does not bother to mention such properties as *being self-identical*.

Attackers of mind-independent nowness say that this move trivializes temporal indexical properties (Grünbaum 1971). So it does, in the sense of rendering them impotent to explain traits which are (conceptually) contingently associated with time in our world. But how could it conceivably be otherwise? No one has ever explained that. Moreover, psychology does not require indexical properties to explain felt transiency, or anything else in our experience. So *if* the argument from physics showed that indexical properties are absent from the mind-independent world, a parallel argument from psychology would show that they are not instantiated within consciousness either.

Will the argument from physics work against the mind-independence of felt transiency? Not without much more labor. For physics does find it necessary to attribute to physical things a mind-independent temporal extension which we call *duration*. Is this kind of duration the same that we find instantiated within consciousness, by after-images, for example? Act-object analyses of after-imaging, and sensation, with sense-data as objects, have been powerfully defended by Jackson (1977, see also Fitzgerald 1979). These will introduce further subtleties into the analysis of experiential time.

Here we meet the analogues of the classical questions about secondary qualities. To explain directly experienced duration are we required to posit the same quality in the mind-independent world? This is harder to answer than the analogous questions for standard secondary qualities, because so little is known about time perception. That's partly because we can't ordinarily block it out of consciousness, as we can close our eyes and stop seeing colors. But there are abnormal mystical-type experiences of which subjects say that during a certain period of clock time they were in a state of awareness in which time seemed to stand still, or be absent. Perhaps what was absent was some kind of felt transiency, though physical duration was present. And perhaps the conditions for the presence, absence, and variation of this kind of transiency would show that it is mind-dependent. These are just hints.

A second argument for the mind-dependence of temporal becoming would suggest that such mind-dependence is an *a priori* truth, insensitive to empirical findings. Here's a quotation which sums up the argument.

> We turn now to presentness, the second key aspect of full-blooded time. As observed above, A-series time requires the specifying of a particular 'now' to supplement the general machinery of B-series precedence. ... The 'now-of-the-present' requires an explicit *act of awareness* for its identification. ... The presentness of the now is of its very nature an *experiential* presentness, and this fact renders it overtly mind-involving. ... Full-blooded time (unlike bare temporality) is not merely conceptually mind-invoking, but is outright mind-dependent, because the essential factor of presentness introduces an overt reference to the experiencing mind into an analysis of what is at issue . (Rescher 1973, pps.124-5).

The argument can be summarized, I hope without caricature, this way.
1) No (full-blooded) time without a *now*.
2) No *now* without an act of specifying one.
3) No act of specifying a *now*, without a mind, and therefore
4) No (full-blooded) time without a mind.

A defender of mind-independent nowness would simply reject premiss 2. A *now* is simply a time which has nowness. Every time has nowness when it is the present time. To be sure, picking out a *now* requires a mind, because picking out anything is a mental act. But the entity thereby selected, a time, and the nowness of that time, are not shown by this argument to be in any way mind-dependent. Needless to say, the argument has no relevance to the status of non-indexical transiency. Yet some such directly experienced elapsiveness might well be taken to be what differentiates "full-blooded temporality" from bare physical duration or clock time. Not that there's any need to quibble about "full-blooded".

A third argument for the mind-dependence of temporal becoming has been advanced by Adolf Grünbaum (1967, pp. 45-56). This says that temporal becoming, and subjective time itself, comes in successive *nows* of an atomic kind. But the instants of the physical world are dense and continuous. Therefore the "atomic-now" structure of temporal becoming is not found in the physical world.

I would deny that the durations present in our experience come as atomic, individual beads of duration, one after the other. The *nows* which we sometimes set ourselves to select deliberately are another matter. Acts of now-selection necessarily come one after the other, bead-like. No matter how quickly we pick out successive *nows*, each has finite duration and an immediate successor. But this shows more about acts of now-selection than about the phenomenal temporality of ordinary experience sliding by when we're tying our shoes and not indulging in the artificial exercise of (sheepishly) counting *nows*.

But an argument like this one may show that phenomenal temporal duration is distinct from physical duration. Take an after-image, for example. It doesn't seem to be composed of instantaneous temporal stages having the structure of the continuum, especially if, as seems probable, it is not literally in physical spacetime (Fitzgerald 1979). So its phenomenal temporal structure would seem to be different from that which physics attributes to matter. That doesn't mean that its phenomenal duration has an *atomic* character. Maybe it does, but I'm not convinced. Its phenomenal duration might instead have some indefinite potential divisibility, as Aristotle thought physical time did.

A fourth argument for the mind-dependence of temporal becoming is derived from relativity theory. Say event E is occurring now. "Now" means "at this time". But what frame-time, exactly? Relativity of simultaneity entails that even instantaneous events occur not just at one time, but at an infinity of frame-times. When we say something's occurring *now*, is this so riddled with false implications of frame-time uniqueness that it is necessarily false? Are all nowness-attributions to physical events false, for this reason? Is the uniqueness condition satisfied only for intramental events? If so, that's a reason to regard nowness as mental.

If we regard ordinary tensed attributions of present occurrence as false on these relativistic grounds, we'll have to say the same thing about their usual tenseless substitutes. "Event E *occur* on Oct. 16, 1980" equally fails to pick out a single time, through failing to specify a frame. I would suggest that neither ordinary tensed nor tenseless claims that things are happening now should be regarded as so fatally flawed by failure to specify a frame as to be automatically false. Why not see them as vague enough to be true? If so, relativity provides no ground to regard nowness as mind-dependent. Its implications for non-indexical transiency must be discussed another time.

References

Baker, Lynn Rudder. (1974-5). "Temporal Becoming: The Argument from Physics." The Philosophical Forum 6: 218-236.

------------------. (1979). "On the Mind-Dependence of Temporal Becoming." Philosophy and Phenomenological Research 39: 341-357.

Cohen, R.S. and Schaffner, K.E. (eds.). (1974). PSA 1972. (Boston Studies in the Philosophy of Science, Vol. 20.) Dordrecht: Reidel Publishing Co.

Dobbs, H.A.C. (1969). "The 'Present' in Physics." British Journal for the Philosophy of Science 19: 317-324.

Fitzgerald, Paul. (1974a). "Nowness and the Understanding of Time." In Cohen and Schaffner (1974). Pages 259-281.

----------------. (1974b). "Critical Study of The Language of Time by Richard M. Gale." The Philosophical Forum 5: 424-440.

----------------. (1979). "Feature Book Review: Perception by Frank Jackson." International Philosophical Quarterly 19: 103-113.

Freeman, E. and Sellars, W. (eds.). (1971). The Meaning of Time. LaSalle, IL: Open Court.

Gale, Richard. (1968). The Language of Time. New York: The Humanities Press.

Grünbaum, Adolf. (1967). Modern Science and Zeno's Paradoxes. Middletown, CT: Wesleyan University Press.

---------------. (1969). "Are Physical Events Themselves Transiently Past, Present and Future? A Reply to H.A.C. Dobbs." British Journal for the Philosophy of Science 20: 148-153.

---------------. (1971). "The Meaning of Time." In Freeman and Sellars (1971). Pages 195-228.

Jackson, Frank. (1977). Perception: A Representative Theory. New York: Cambridge University Press.

Nakhnikian, George. (ed.). (1974). Bertrand Russell's Philosophy. London: Gerald Duckworth and Co.

Reichenbach, Hans. (1956). The Direction of Time. Berkeley: University of California Press.

Rescher, Nicholas. (1973). Conceptual Idealism. Oxford: Basil Blackwell.

Salmon, Wesley. (1974). "Memory and Perception in Human Knowledge."

In Nakhnikian (1974). Pages 139-167.

Smart, J.C.C. (1963). Philosophy and Scientific Realism. New York: The Humanities Press.

Are There Cases of Simultaneous Causation?

A. David Kline

Iowa State University

1. Introduction

The problem of the direction of causation has received considerable press. Justifiably so, since it is a major bottleneck in accounts of causation and explanation. But there is another, not unrelated, yet unexamined, temporal issue for those same accounts. Numerous writers have uncritically assumed that there are instances of simultaneous causation. (A necessary condition for *e* and *e*', two causally related events, to be in a *simultaneous* causal relation is that there be no time-difference between *e* and *e*' or that *e* and *e*' occur at the same time.)

The assumption is not impotent. Alleged instances of simultaneous causation have been used to argue against various views of causation/explanation. For example, Richard Taylor (1966), Douglas Gasking (1955) and Baruch Brody (1972) have used such cases to critique regularity or nomic-subsumption accounts. Myles Brand (1979) in a very recent paper uses simultaneous causation to challenge Wesley Salmon's statistical relevance analysis (1970).[1]

In a more positive vein the belief in simultaneous causation has fostered alternative accounts of causation/explanation. If there are instances of simultaneous causation and if the causal relation is not symmetrical, i.e., if one and only one of the causally related events is the cause, then clearly temporal order can not be the key to understanding the asymmetry. Gasking's (1955) manipulability theory of causation and Brody's (1972) Aristotelian theory of explanation are striking and well discussed efforts to deal with this "problem".

The task of the paper is to subject the widespread belief in simultaneous causation to criticism. I am not under the illusion that my arguments prove that simultaneous causation is *impossible* (in any sense of that term). The arguments give one reason for being suspicious of

---

PSA 1980, Volume 1, pp. 292-301

what has heretofore been seen as uncontestable.

2. The Examples

Each of the before mentioned authors has his favorite "examples" of simultaneous causation. Richard Taylor finds two cases persuasive.

> Consider . . . a locomotive that is pulling a caboose and to make it simple, suppose this is all it is pulling. Now here the motion of the locomotive is sufficient for the motion of the caboose, the two being connected in such a way that the former cannot move without the latter moving with it. But so also, the motion of the caboose is sufficient for the motion of the locomotive, given that the two are connected as they are, it would be impossible for the caboose to be moving without the locomotive moving with it. From this it logically follows that, conditions being such as they are - both objects are in motion, there are no other moves present, no obstructions to motion, and so on - the motion of each object is also necessary for the motion of the other. But is there any temporal gap between the motion of one and the motion of the other? Clearly there is not. They move together, and in no sense is the motion of one temporally followed by the motion of the other.
>
> . . . consider the relationship between one's hand and a pencil he is holding while writing... . It is surely true . . . that the motion of the pencil is caused by the motion of the hand. This means, first, that conditions are such that the motion of the hand is sufficient for the motion of the pencil. Given precisely those conditions, however, the motion of the pencil is sufficient for the motion of the hand; neither can move under the conditions assumed - that the fingers are grasping the pencil - without the other moving with it. It follows, then, that under these conditions the motion of either is also necessary for the motion of the other. And, manifestly, both motions are contemporaneous; the motion of neither is followed by the motion of the other. (1966, p. 35).

Douglas Gasking provides examples that at least on the surface appear to be of a rather different sort.

> . . . current from a battery is flowing through a variable resistance, and we have a voltmeter connected to the two poles of the battery to measure the potential difference. Its reading is steady. We now turn the knob of our variable resistance and immediately the voltmeter shows that the potential difference has increased. If someone asks: "What caused the increase?", we reply: "The increase of the resistance in the circuit." But here again the effect was not something subsequent to the cause, but simultaneous.
>
> Now, if someone saw a bar of iron glowing and, being quite

> ignorant of the physical facts, asked: "What makes the iron glow? What causes it to glow?" we should answer: "It is glowing because it is at a temperature of 1,000° C or more." The glowing, _B_ is caused by the high temperature, _A_. And here the _B_ that is caused is not an event subsequent to the cause, _A_. (1955, p. 479).

In the course of criticizing the Hempelian or covering law model of explanation Brody gives the following example. From the context it is clear that he thinks it is an example of simultaneous causation.

(1) if the temperature of a gas is constant, then its pressure is inversely proportional to its volume.
(2) at time t_1, the volume of the container _C_ was V_1 and the pressure of the gas in it was p_1.
(3) the temperature of the gas in _C_ did not change from t_1 to t_2.
(4) by t_2, I had so compressed the container by pushing on it from all sides that its volume was 1/2 _V_.

---

(5) the pressure of _C_ at t_2 is $2p_1$. (1972, p

Brody believes that the volume of _C_ becoming $1/2V_1$ at t_2 causes the pressure of _C_ to be $2p_1$ at t_2. Furthermore he believes that _as soon as_ the volume of _C_ changes the pressure of _C_ changes.

Brand also appeals to the volume/pressure case and gives an additional example reminiscent of Taylor's cases. Jack's going down on the seesaw causes Mary to go up but there is no time-lag. As soon as Jack starts down Mary starts up (Brand 1979, pp. 272-273).

Enough for the examples, these are certainly representative and will serve for my purposes. Before sketching the central argument, a word on causation is needed. I know of no fully adequate account of the causal relation but certain modest assumptions will be being made about the causal relation - assumptions that ought not be controversial.

If _e_ and _e_', two events, are causally related then, (a) _e_ and _e_' must be _distinct_ events and (b) the relata must have spacio-temporal location. The emphasis in (a) is on 'distinct' not 'event'. I do not want to beg the question among philosophical preferences for events, states, occasions, etc. but merely claim that the relata must be individuatable. Conditions (a) and (b) must be met if causal laws are to be confirm

3. The Critique

There are two general strategies one can take in arguing against the stated examples. (i) One can challenge the relation as being a genuine causal relation, for example, by arguing that the relata are not of the appropriate ontological type.[2] (ii) One can admit that the relation is a causal relation but show that there is a temporal separation between the relata.

The basic strategy taken in this paper is (ii). With respect to the stated examples of "simultaneous causation", I am willing to grant that the respective events are distinct and causally related, i.e., they pass challenge (i). As a matter of fact in each case the respective events are spatially separated. (This is part of the reason why it is clear that they are distinct.) But, as shall be shown, given that the events are spatially separated and given the truth of the special theory of relativity (STR), they cannot be temporally simultaneous, i.e., they fail challenge (ii).

All of the examples do fail in the above way. But it is instructive to attempt to patch one of them up - construct the case so that the relevant events are not spatially separated. When this is done, it is not evident that challenge (i) is avoided.

The details of the argument must be worked out in the context of the specific examples. Taylor's choice of a train example is appropriately relativistic, but as we shall see little else is. Taylor, of course, wants to exclude play between the coupling of the locomotive and caboose. Referring to Figure 1, for simplicity let us suppose there is a solid car with the forward set of wheels as drive wheels.

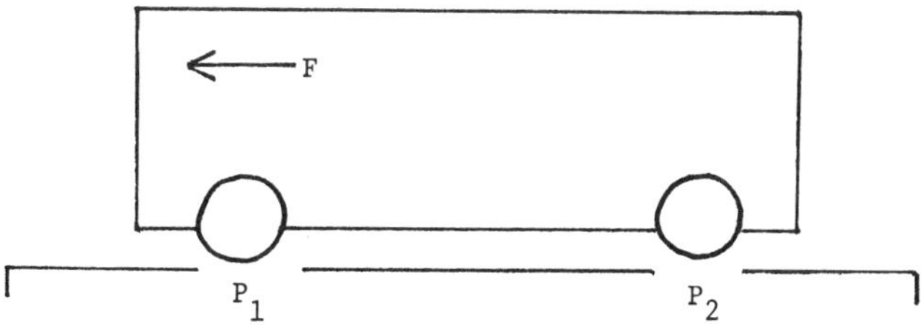

Figure 1

A force, $\underline{F}$, sufficient to move the train is exerted as point $\underline{p}_1$ at time $\underline{t}_1$. Taylor believes that the movement of the real wheel at $\underline{p}_2$ also occurs at $\underline{t}_1$ and is caused by the force applied at $\underline{p}_1$.

Cases like Taylor's, in which a causal effect is propagated at a speed greater than the speed of light, in his case instantaneously, have classically been rejected for two reasons: (I) Causal signals with velocities greater than the speed of light are alleged to generate causal anomalies. Since I regard this point as controversial I shall not rely on it here.[3]

(II) For two spatially separated points, $\underline{p}_1$ and $\underline{p}_2$, if event $\underline{a}$ at $\underline{p}_1$ is the cause of $\underline{B}$ at $\underline{p}_2$ then physical information is transmitted from $\underline{p}_1$ to $\underline{p}_2$. Transmitting information involves the transmission of energy, i.e., mass. The formula for relativistic mass,

$$(1) \quad m = m_o / \sqrt{1-V^2/C^2},$$

shows that if an object has a rest mass, $\underline{m}_o$, greater than zero, then as the object's velocity, $\underline{V}$, approaches $\underline{C}$, the speed of light, the relativistic mass approaches infinity. In Taylor's case $\underline{V}$ goes from 0 to ∞, hence $\underline{m}$ would be infinite and the causal signal to be so accelerated would require an infinite force, which is impossible.

Taylor is treating the train as a rigid rod through which causal information is instantaneously transmitted. Given STR rigid rods are not possible. Actually the causal signal from $\underline{p}_1$ to $\underline{p}_2$ is an elastic propagation that travels at powers of magnitude less than $\underline{C}$.

It makes no difference to the case if we let the train be traveling at a constant speed. It is still not the case that $\underline{a}$ at $\underline{t}_1$ is the cause of $\underline{b}$ at $\underline{t}_1$. The above argument still applies since simultaneous causation would require that mass be accelerated from some speed less than $\underline{C}$ to a speed greater than $\underline{C}$.

It is not difficult to see that Taylor's second example and Brand's seesaw case are defective for identical reasons. They implicitly assume the existence of rigid rods - a type of entity that is physically impossible.

The remaining examples do not obviously succumb to the same difficulty. The gas laws have been perplexing to philosophers of science in a number of ways but especially with respect to causation. The problem is quite simple. The gas law,

$$(2) \quad PV = NrT,$$

holding $\underline{NrT}$ constant, allows us to solve for $\underline{P}$ or $\underline{V}$ given the other quantity. But we have strong causal intuitions that changes in pressure are caused by changes in volume and not vice versa. Furthermore, we tend to think that the relations between $\underline{V}$ and $\underline{P}$ are instantaneous. Accordingly, no appeal to formal features of the law or to temporality will help account for our intuitions.

At the outset a certain confusion in Brody's example needs to be cleared up. The gas laws are intended as descriptions of gases in equilibrium. They are not to be taken as describing processes. Notice that time does not occur as a variable in the equation. Strictly speaking Brody's use of the law is illegitimate. Once a change in the volume occurs it is some time until the gas reaches equilibrium and is correctly described by (2). One might try to restate Brody's point though. During changes of V, let us grant that (2) does not describe the appropriate relation between P and V but nevertheless doesn't P instantaneously increase?

We can begin to see our way clear of this problem if we consider the behavior of gases from a molecular point of view, in particular, an elementary derivation of the pressure formula. Pressure is defined as force per unit area. Given the kinetic-molecular model, the supposition is that the pressure on the walls of the container is caused by the molecules impinging on those surfaces.

Assume a strong cubical container containing molecules of a pure ideal gas. Each molecule has the same mass and moves rapidly about colliding with the six walls. The average effect on the walls is as though 1/3 of the molecules were traveling in the x-direction, 1/3 in the y-direction, and 1/3 in the z-direction. Consider a single molecule moving in the x-direction.

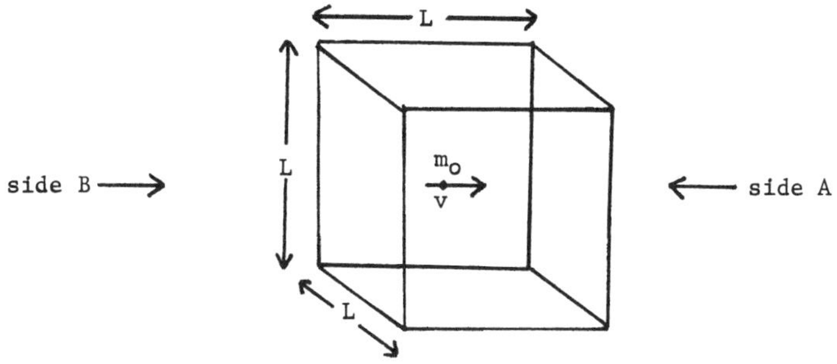

Figure 2

Let the speed of the molecule by v; its momentum is vm_o. When it strikes A, supposing an elastic collusion, it rebounds with momentum $-vm_o$. Therefore for each impact the change of momentum is $vm_o - (-vm_o) = 2vm_o$. In a round trip (A to B to A) the molecule travels a distance 2L in time 2L/v. So the molecule will make v/2L impacts at A per unit time. Thus the change of momentum per unit time is $2vm_o \times v/2L = m_o v^2/L$.

Supposing that there are N molecules in the cube; the average effect on A, given our simplifying assumption that 1/3 of the molecules are moving in the x-direction, will be the total change of momentum per unit time owing to impacts, i.e., $1/3\ N \times m_o\bar{V}^2/L$. ($\bar{V}^2$ represents the average of the squares of the molecules speeds. This must be taken into account given intermolecular collisions.) The force exerted by A on the molecules is the time-rate of change of momentum, $\overline{F} = (1/3\ Nm_o v^2)/L$. But by Newton's third law and the fact that the impacts are so frequent that for practical purposes they can be regarded as constant we have, $F = (1/3\ Nm_o v^2)/L$, for the force exerted on the wall by the molecules. Recalling that our interest is in pressure, therefore $F/L^2 \equiv P = (1/3\ Nm_o v^2)\ /L^3$. Since L^3 is the volume, V, of the container then $P = (1/3\ Nm_o v^2)/V$, i.e., the pressure formula.[4]

The molecular model makes it clear that the pressure is the result of impulsive forces due to the collisions of molecules on the walls of the container. Changes in volume alter the number of impacts per unit time and hence affect the pressure. Our question is whether changes in volume instantaneously introduce changes in pressure. Consider a cylinder, filled with an ideal gas, one end of which is a piston with surface, A.

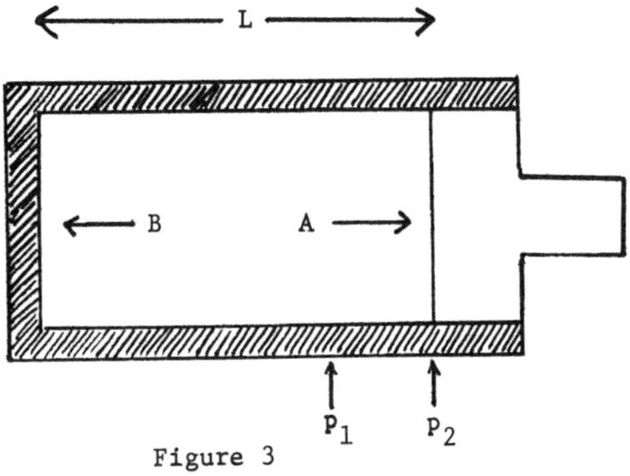

Figure 3

Now if the cylinder at $\underline{p}_1$ (time T_o) moves to $\underline{p}_2$ at (time T_1) does the pressure increase as soon as the piston begins to move, i.e., at T_o? First consider B; does the rate of impacts on B increase at T_o? Molecules moving to the left will strike B at the same time they would have struck B even if A hadn't moved.[5] Molecules moving to the right will strike B at an earlier time than they would have if A hadn't been moved. But notice that the earliest of these molecules to reach B, and hence increase the bombardment of B, must cover part of L before it reaches B. This journey takes time. Therefore some time will elapse after A is moved and the rate of impacts on B increases.

In fact the pressure will increase at different points on the container at different times. This is one reason why the gas laws are restricted to equilibrium states - a point after the change when the pressure, temperature, etc. are uniform.

The critic will probably respond as follows: What about the rate of impacts on A? At least it goes up as soon as A moves. There is a certain background rate of impacts on A. As soon as A moves there are molecules that strike A which wouldn't have struck it until a later time. By considering molecules that are closer and closer to A before A moves, we can make the distance A must move, before the rate of impact increases, arbitrarily small.

I am not sure that this argument is correct but let us assume that as soon as A moves the rate of impacts on it increases. Now we have a further question - does the force on A increase as soon as the number of bombardments on A increases? It will help to study a simplified version of this problem.

Referring to figure 4, consider an elastic collision between two bodies of equal mass. I consider their relationships at three temporal instants.

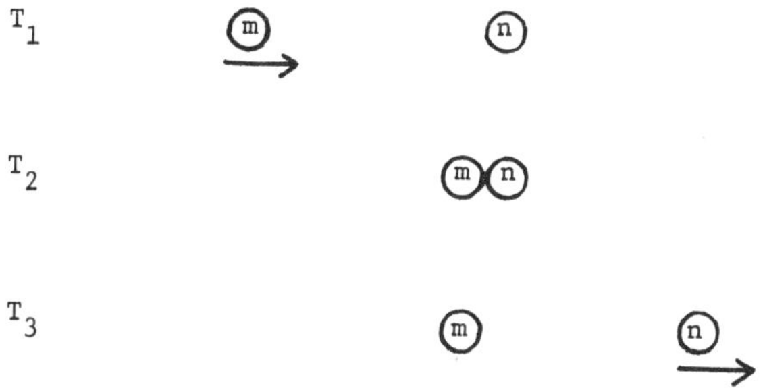

Figure 4

It is non-controversial that the momentum of m at T_{-1} is the cause of the momentum of n at T_3.[6] But is m at T_2, when they collide, the cause of n at T_2? Earlier it was proved that this is impossible if the momenta of m and n are spatially separated. Are the moments of m and n spatially separated when m and n collide? Where is the momentum of an object located - in the center of the mass, at a point on the surface, or is the question nonsensical?

If the momentum of a body is located at the body's center of mass then simultaneous causation is always impossible since the centers of mass of two bodies will always be spatially separated. To think simultaneous causation possible, under these circumstances, is to regard the bodies as rigid rods.

If the momentum of an object is located at a point on its surface or if a body can be regarded as a mathematical point then perhaps one could speak of the momenta of the objects having the same spatial location. The necessary abstractions seem rather extreme but even if they could be carried out the old problem of individuating the momentum of n arises.

The defender of the gas laws as an instance of simultaneous causation will have to face these questions. There is not a simultaneous increase in pressure throughout the gas. If there is any simultaneous increase it is on surface A. To argue this requires arguing that the molecular force is instantaneously transferred to A. I have been trying to show that such a claim is at the very least problematical.

Gasking's Ohm's law case can be handled in a way similiar to my treatment of the gas law. One needs to examine the microscopic version of Ohm's law, $j = \sigma E$, (Weidner and Sells 1973, pp. 540-546). The physics of the glowing poker case is quite complex. But again if examined at the micro-level one will discover that in so far as a convincing case is made for a causal relation that relevant events are spatially separated.

I have been arguing against the existence of simultaneous causation by challenging examples that are prominent in the literature. If one thinks there are better examples, state them, and we shall see. From my arguments one general point is clear: for a purported case of simultaneous causation to have a chance of being a genuine case the relevant events must be distinct, causally related, and not spatially separated.

Notes

1 There are temporal problems with Salmon's account but they are not the ones Brand points out (See Kline 1980).

2 Little has been written along these lines, but see Hedman (1972).

3 For a description of the supposed anomaly see Salmon (1975, pp. 122-124). For an effort to render tachyons banal see Kline(1977).

4 This derivation follows very closely that given in Holton and Roller(1958, pp. 437-439).

5 This assumes that the molecules are moving to the left faster than the piston. If this were not true the argument could be recast so that my central point will still stand. Furthermore, I consider motion only in the x-direction and disregard intermolecular collisions. These complications will not effect the spirit of the argument.

6 It is also the case that n at T_1 is the cause of m at T_3 given Gallilean relativity, but I shall disregard this kind of symmetry in the discussion.

References

Brand, Myles. (1979). "Causality." In Current Research in Philosophy of Science. Edited by Peter D. Asquith and Henry E. Kyburg. East Lansing, Michigan: Philosophy of Science Association. Pages 251-281.

Brody, B. A. (1972). "Toward an Aristotelian Theory of Explanation." Philosophy of Science 39: 20-31. (As Reprinted in Klemke, Hollinger and Kline (1980), pp. 112-123.).

Gasking , Douglas. (1955). "Causation and Recipes." Mind 64: 479-487.

Hedman, Carl G. (1972). "On When There Must Be a Time - Difference Between Cause and Effect." Philosophy of Science 39: 507-511.

Holton, Gerald and Roller, Duane. (1958). Foundations of Modern Physical Science. Reading, Massachusetts: Addison-Wesley.

Klemke, E. D., Hollinger, Robert and Kline, A. David. (eds.). (1980). Introductory Readings in the Philosophy of Science. Buffalo, New York: Prometheus Books.

Kline, A. David. (1977). The Direction of Causation. Unpublished Ph.D. Dissertation, University of Wisconsin. Xerox University Microfilms Publication Number 77-6615.

---------------. (1980). "Screening Off and the Temporal Asymmetry of Explanation." Analysis: forthcoming.

Salmon, Wesley C. (1970). "Statistical Explanation." In Nature and Function of Scientific Theories. (University of Pittsburgh Series in Philosophy of Science, Volume 4.) Edited by Robert G. Colodny. Pittsburgh: University of Pittsburgh Press. Pages 173-231. (As Reprinted in Salmon, Wesley et al. (1971). Statistical Explanation and Statistical Relevance. Pittsburgh: Press. Pages 29-87.).

----------------. (1975). Space, Time and Motion. Encino, California: Dickenson.

Taylor, Richard. (1966). Action and Purpose. Englewood Cliffs, New Jersey: Prentice-Hall.

Weidner, Richard T. and Sells, Robert L. (1973). Elementary Classical Physics 2nd ed. Boston: Allen and Bacon.

Recent Changes in the Concept of Matter: How Does 'Elementary Particle' Mean?

K. S. Shrader-Frechette

University of Louisville

1. How Versus What

"How does a poem mean?" asked John Ciardi, noted poet and critic (1959, p. 663). Because he was convinced that "there will never be a complete system for understanding poetry," Ciardi urged his Harvard and Rutgers students to discover "the total experience" of a poem and never to look merely for what it means (1959, p. 666).

Like a good poem, the total meaning of the concept of matter always eludes us.[1] As Heisenberg put it: what we know is not nature itself, but "nature exposed to our method of questioning." (Cited in Capra 1975, p. 126). For this reason, and because high-energy physics is in a state of rapid development, it would be presumptuous to attempt to say what matter or elementary particles are (See Woodruff 1963, p. 578). Moreover this "what" question is more properly a scientific one, about the correct referents of a specific theory of matter. "How does 'matter' mean?" or "how does 'elementary particle' mean?" is more properly a philosophical question. It is an epistemological query about the concept of matter.[2]

Let us leave the specific scientific question of what matter means to Nobel Prize-winning physicists and instead tackle the epistemological question of how matter means. Let us ask what criteria we use for employing the term, 'matter', as opposed to those that might formerly have been used. Is there a difference between the two? And what does our current employment of criteria reveal about a contemporary change (if any) in the concept of matter? These are important questions, both since correct answers to the more difficult "what" queries depend on them,[3] and because they might reveal how the role of the criteria has evolved through time and thus contributed to the evolution of the concept of matter and to that of explanation. Probing the "how" of matter also appears likely to lead one to discover the contemporary concept of matter, even though obviously the term 'matter' is no

PSA 1980, Volume 1, pp. 302-316

longer widely used by scientists, except in a somewhat popular way or as part of a prescientific language. If such a concept can be discovered, it might suggest ways in which a "natural philosophy" still guides scientists, even though it is not explicitly revealed either in their finished products or in the language of these products (See McMullin 1978b, pp. 295-98). In more modern jargon, this investigation might reveal how "themata" (See Holton 1978) influence science.

2. Matter as Elementary Particles

Before one can discover how a particular concept of matter guides physicists, however, one must identify some entities or events in scientific theory which are the analogue of the philosopher's notion of matter. The search for some ultimate substratum, begun at least as early as the sixth century BC, continues to the present day in the form of the atomistic, field, and symmetry theories currently used to describe microphysical phenomena. The history of the concept of matter, however, all the way from Democritus and Leucippus to Leibniz and Newton and beyond, is a fascinating tale whose scope is too broad to be treated here (See Feinberg 1977, pp. 1-18; McMullin 1963; 1978a; Gale 1971a, 1971b, 1973, 1974). What is significant, for the purposes of this paper, is that most physicists currently subscribe to atomistic views; they believe that matter is composed of subatomic particles.[4] In fact, "the study of these subatomic particles has been the central theme of physics in the second half of the twentieth century." (Feinberg 1977, pp. 165-66). Within the last three years, the "elementary constituents of matter" have been said to be identified; "the building blocks of nature", according to most physicists, are quarks and leptons (Greenberg 1978, p. 336; Turner and Schramm 1979, p. 42; Novozhilov 1975, p. 25; Frauenfelder and Henley 1974, p. 357; Sandorfi et al. 1978, pp. 756-57; Feinberg 1977, pp. 209, 220). As several scientists recently put it: "after the events of the last three years [the success of various quark predictions, including the discovery of the ψ particle], there are no longer any sceptics." (Applequist et al. 1978, p. 388).

Since the class of leptons includes only the muon, the electron, and two types of neutrino, whereas the vast majority of subatomic particles are hadrons, all of which are said to be composed of quarks, the quarks are, by far, the most important of the two types of particles said to be 'elementary'. Nevertheless, at least at first glance, certain obvious criteria for terming quarks 'elementary particles' seem questionable. As Sidney Drell put it, "during the past two decades we have come to the point of accepting into the exclusive family of elementary particles a guest [the quark] who would not have made the Social Register of an earlier generation." (1978, p. 23).

Apart from whether his claim is correct, the interesting epistemological issue, raised by Drell's comment, is how criteria for the term 'elementary particle' were used in the case of quarks. Understanding the criteria employed ought to give one some interesting insights, not only about how exclusive the Social Register for elementary particles has been and is, but also about how the concept of matter may have changed.

3. Observability as a Criterion for Particles

Although numerous criteria for 'elementary particle' and 'particle' have been employed, there are two whose association with the quark case seems to be of special philosophical import. The first, compositeness, has been one of those widely used to distinguish which particles are elementary, the other, observability, to determine which phenomena are particles of matter. Let us discuss the latter first.

Throughout the history of science, one criterion often used for affirming the existence of a postulated entity, such as a particle, has been experimental evidence. As Helmholtz expressed it in paying tribute to Faraday: the primary aim of the modern scientific method is "to purify science from the last remnants of metaphysics." (Cited in Drell 1978, p. 28). The suggestion behind his statement is that, to the extent that entities are said to exist without their being validated by "observation", then to that same degree has the scientific theory in question not been purified of possibly unwarranted assumptions.

Despite the fact that our actual concept of observation has been problematic and has undergone extreme changes through time (in contemporary microphysics, it is tied to photographs of ionization paths in bubble chambers or spark chambers), the criterion of in-principle-observability of postulated entities has held a central place in the practice of science. Moreover, despite the observational limits imposed by quantum theory, "there is no uncertainty in what we mean when we say that we <u>observe</u> an electron," for example, by means of contemporary detection techniques (Drell 1978, p. 26). Quarks, however, have not been "observed", in spite of many efforts to find them (Chupka <u>et al</u>. 1976, pp. 716-27; Avre <u>et al</u>. 1976, pp. 474-77; Nambu 1976, pp. 48, 55; Galik <u>et al</u>. 1974, pp. 1856-63; Nash and Yamanouchi 1974; Lubkin 1977b; Ellis and Kislinger 1974; Feinberg 1977, pp. 213-14). More importantly, they are said to be in-principle-unobservable. Let us see why they are thought to be so, determine what this belief does to the status of the apparent criterion of observability, and finally, discuss how the impossibility of ever seeing quarks affects our concept of particle and therefore matter.

Quarks are said to be in-principle-unobservable on the grounds that, were they observable, their alleged properties of low mass and fractional charge could easily have been detected in current chambers, at the energies now available (Nambu 1976; Sandorfi <u>et al</u>. 1978; Feinberg 1977, pp. 213-14; Glashow 1975; Greenberg 1978; 't Hooft 1978, p. 2). Hence, reason physicists, there must be some type of strong interaction acting among them, perhaps accounted for by (in principle) unobserved "virtual particles"[5] whose exchange produces the force needed to bind quarks together. Many mechanisms for binding quarks have been proposed, "but none of them has proven completely convincing, and the problem remains an open one." (Feinberg 1977, p. 215). Even though it is widely agreed that quarks can never be isolated and detected, and even though their confinement has not and may never be proved (Wilczek 1978, p. 32), however, they are almost universally acknowledged to be the basic building blocks of matter.

The unobservability of the quark leads one to wonder whether it ought to be likened to the ether ('t Hooft 1978). Many middle-seventeenth century writers, including Robert Boyle, objected that since the ether was not observed and its existence unverifiable, it ought not to appear as an ingredient in an experimental science (Hesse 1967, p. 67). Ought we to follow Boyle, reject the judgment of a majority of high-energy physicists, and claim that the quark likewise ought not to appear as an ingredient in an experimental science? Or, on the other hand, ought we to liken the quark to the neutrino, which was not observed until twenty-six years after its postulation? Originally proposed by Pauli in 1930, called a "neutron", and wrongly thought to be a constituent of the nucleus, the neutrino was at first met with almost universal scepticism by scientists. Eventually, however, theory concerning it was clarified, and in 1956 two physicists were able to "observe" its creation (Brown 1978).

The observational status of the quark is probably completely analogous neither to that of the field-theoretic ether nor to that of the atomistic neutrino. Although neither the ether nor the quark has been observed, the former did not yield the great variety of specific, successful predictions to which the quark has led; moreover the ether was not clearly believed (at least by Newton) to be in-principle-unobservable, as is the quark.[6] Likewise the neutrino was not believed to be in-principle-unobservable; practically speaking, Cowan and Reines needed merely a powerful nuclear reactor and an intense source to observe effects of this particle (Glashow 1978, p. 26; Brown 1978).

What is epistemologically interesting about the alleged neutrino-quark analogue, however, is the differing perspective, presented in the two cases, on the relative importance of the criterion of observability. Both particles are said to be elementary and both serve extremely useful functions in explaining microphysical phenomena. The neutrino was able to save the "exchange theorem" of statistics,[7] and the quark has been able to describe quantum numbers and other properties of individual hadrons. The histories of the acceptance of the quark and the neutrino, however, show quite different values assigned to the criterion of observability, especially as compared to that of saving conservation laws. In addition to preserving the exchange theorem, the neutrino was postulated by Pauli as a means of saving the conservation of energy and momentum in beta decay (Drell 1978, p. 26; Brown 1978). On the other hand, although the quark is able to account for the observed spectrum of the hadrons, "no conservation laws _require_ quarks." (Drell 1978, p. 30). Moreover even though the neutrinos were the only conceivable ways of saving energy and momentum conservation laws, they were not universally accepted as bona-fide particles until they had been "observed" twenty-six years after their postulation. Quarks, however, have apparently been almost universally accepted as particles, less than fifteen years after their postulation, even though they are required by no known conservation laws.

The differing emphases on observability in the two cases suggest that, in the last fifty years, observability may have become less important as a criterion for particles. The story is not quite so simple as this, however. What has been happening is not merely a devaluation of the worth of observability but rather an inflation of the value of theoretical simplicity as a criterion for accounts of particulate matter. The quarks have survived as fundamental particles because they have continued to fulfill the goal that motivated their postulation. They have provided a simple basis for the explanation of the observed multiplet structure and properties of subatomic phenomena, say high-energy physicists, and they have unified and simplified diverse observations (Flanagan 1977; Feinberg 1977, p. 205; Schwitters 1977, p. 56; Pati 1976, p. 38; Drell 1978, pp. 23, 26, 27). And, as Feinberg points out, it is not surprising that "belief in some entity or phenomenon is the result of detailed study of its properties and its relations to other phenomena, rather than the result of some critical experiment that demonstrates its existence directly." (1977, p. 19; Chandrasekhar 1979, pp. 25-27).

If the neutrino-quark case is typical, there might be several reasons why theoretical simplicity now seems to be gaining in importance over observability as a criterion for matter. First, "observation" is difficult in the high-energy realm. Although we can explain what we mean when we say how we "observe" many particles, often experimental results can be interpreted in a number of different ways (Shrader-Frechette 1977, pp. 417-21; 434; Pati 1976, pp. 58, 65-66; Lubkin 1977a, 1977b; Leon 1973, p. 133; Prince et al. 1975; Heisenberg 1976; Gaston 1973, p. 77). Secondly, alternative microphysical explanations often are deficient in predictive power, and therefore it is difficult to distinguish among them on the grounds of their satisfying criteria concerning observational effects (Lubkin 1977a, p. 20; Cox and Yildiz 1977). V.F. Weisskopf explained the problem of attempting to use more empirical criteria, such as observability, rather than theoretical ones, such as simplicity, for determing the structure of matter. He compared collisions among hadrons to collisions among fine Swiss watches. Whereas the products of watch collisions are present before impact, he noted, the products of hadron collisions are usually formed after impact. Precisely for this reason, says Weisskopf, "one should not try to learn about the structure of matter by smashing watches on each other, or even smashing atoms on each other." (Cited by Goldhaber and Heckman 1978, p. 200). Following the Weisskopf insight, contemporary scientists, with few exceptions, acknowledge "the hegemony of theory over experiment in particle physics." (Feinberg 1977, pp. 262, 166).

If the criterion of theoretical simplicity has in fact superseded that of observability in particle physics, then an interesting transition in the concept of matter has taken place, perhaps a "third phase" of its dematerialization. Russ Hanson pointed out the first two such phases when he noted that one of the central problems of philosophy has been to define the objective, observable qualities of matter. This was the rationale behind the primary quality/secondary quality distinction. Prior to Berkeley, said Hanson, the concept of matter was defined in part by the criterion of objective observability. Berkeley ushered in the first stage of dematerialization by deflating the distinction between objective and subjective observability. Actual physics practice, however, was not affected by Berkeley's move, noted Hanson. Scientific practice was profoundly affected, on the other hand, as a consequence of the next phase.

The second phase, quantum mechanical dematerialization of the concept of matter, argued Hanson, rocked mechanics to its foundations. This occurred because, according to the orthodox Copenhagen account, observation is inexact and probabilistic and therefore unable to reveal any intrinsic, absolute properties of matter. Rather the detector imposes these properties upon the object being studied, said Hanson. Thus if one accepts the validity of the Copenhagen interpretation, and Hanson's analysis of it, then whoever employs the criterion of observability is beset not only with the practical problem of distinguishing subjective from objective observations (phase one of the dematerialization), but also with (phase two) the in-principle impossibility of avoiding observations of properties which are detector-imposed (Hanson 1963a, pp. 557-58; 1963b, p. 570; McMullin 1978b, pp. 272-73; Feinberg 1977, pp. 77-79, 99).

Although Hanson spoke only of two phases, recent experience with the criterion of observability in the quark case seems to support the existence of a third, and even more radical, step toward the dematerialization of the concept of matter. Since the advent of quantum mechanics, high-energy physicists appear to have moved from the notion that (if the Copenhagen account is correct, and many hope it is not), observation is in-principle inexact and probabilistic, to the notion that (if quark confinement is correct) observation is not merely inexact and probabilistic, but impossible in some key instances. Hence the dematerialization of the concept of matter may have proceeded all the way from (1) the denial of the in-practice possibility of objective observation of particles (first phase), to (2) the affirmation that observation is in-principle inexact and probabilistic (second phase), to (3) the in-principle impossibility of any observation at all of elementary particles (third phase).

4. Noncompositeness as a Criterion for Elementary Particles

Although it sounds revolutionary, if such a deflation of the criterion of observability has taken place (with its apparent effect on the objectivity of our concept of matter), then it is not necessarily an undesirable occurrence. At least in the case of quarks, failure to observe the particles might be a bonus, rather than the problem it is often supposed to be. According to quark theory, the particles are pointlike and structureless; they are said to be the elementary and noncomposite constituents of hadrons (Greenberg 1978, p. 336; Turner and Schramm 1979, p. 42; Marx and Nygren 1978; Drell 1978, p. 27; Glashow 1978, pp. 155-226). If they are such ultimate entities, however, "no detailed description is possible in regions of intense interaction where the characteristic additivity of independent particle contributions ceases to be valid. All that can be done is to compare the state of noninteracting particles after a collision with the state of the generally different number of noninteracting particles prior to the collision." (Schwinger 1970, p. 34). Or, as the point was put, more simply, by Chew, "a truly elementary particle -- completely devoid of internal structure -- could not be subject to any forces that would allow us to detect its existence. The mere knowledge of a particle's existence, that is to say, implies that the particle possesses internal structure." (Cited in Capra 1975, p. 263; see Barut 1967; Pati 1976, pp. 19-23; Frauenfelder and Henley 1974, pp. 133-54; Chew 1966, p. 5; Gale 1971a, pp. 169-72). If this is correct, then the nonobservability of quarks might be a necessary condition of their being elementary. In this sense the problem of their nonobservability might be said to be a bonus for those seeking the ultimate fundamental particle(s).

The claim that observation of a particle implies that it is composite or nonelementary is perhaps best understood in terms of S-matrix theory and QCD, quantum chromodynamics. Nearly everything that is known empirically about subatomic physics has been obtained from particle decays and scattering experiments. According to S-matrix theory, scattering experiments are used to provide a set of amplitudes or reaction lengths (called the S-matrix); the interaction forces manifest themselves as particles whose mass determines the range of the force. If S-matrix theory is correct, its account of interaction forces requires that all "observed" particles must have some internal structure. Only then can they interact with the observer and be detected (Martin and Spearman 1970, pp. 9, 64, 136, 155ff.; Gibson and Pollard 1976, p. 12; Simmons 1975, p. 126; Capra 1975, p. 263).

QCD, another important theory, is by far, the favored candidate fundamental theory for describing strong interactions. It has had numerous qualitative and detailed quantitative successes, and is conceptually simple and symmetric. It is constructed by extending the global $SU(3)_c$ color symmetry to a local gauge symmetry (Applequist *et al.* 1978, pp. 395-96; Turner and Schramm 1979, p. 42; Ramsey 1979; Wilczek 1978, p. 31). From the standpoint of QCD, however, "quarks are composites of quarks and gluons" because the gluon-quark *interaction* accounts for quark binding (Greenberg 1978, pp. 376, 381; Turner and Schramm 1979; Feinberg 1977, pp. 214-15; Glashow 1978, p. 167). This means that if S-matrix theory or the gluon force theory of QCD is correct, or if the quark is subject to forces (yet unknown) allowing us to detect its existence, then there is strong evidence that it is composite.

If fulfilling the criterion of observability is a necessary

condition for terming something a *particle*, and if fulfilling the criterion of noncompositeness is a necessary condition for affirming that it is an *elementary* constituent of of matter, then this suggests that quarks cannot be both observable and elementary particles. In turn, this suggests that the current concept of matter reveals a tangled problematic: do we want a simple, unified, largely-theoretical framework within which we can be said to have grasped the ultimate particles of matter, or do we want experimental evidence for the entities about which we conjecture? (This dilemma is close to the one suggested by an analysis of Leibniz's views. If one finds truly elementary particles, then one can never explain their properties, if they are truly fundamental (Gale 1971a, Chs. 1-5; 1973, p. 35; 1974, p. 344).)

5. Postscript

Although the quark is virtually universally held to be an "elementary particle of matter", small stirrings within the physics community reveal that some scientists are beginning to have doubts about this belief. For one thing, they are troubled by the fact that these allegedly elementary constituents of matter have a great variety of properties; this suggests that they might turn out to be composite (Feinberg 1977, pp. 205, 262). For another thing, the number of quarks has escalated and promises to increase even more the higher the GeV we attain; this proliferation suggests to some that quarks might not provide the simple, unified explanation hoped for from truly elementary constituents of matter (Greenberg 1978, pp. 375-76; see Shrader-Frechette 1979).

What all this comes down to is not an argument that quarks are not elementary particles; that discussion is in the province of physics and it might be presumptuous at this time for anyone to challenge the view of a majority of scientists.[8] Rather the point of this paper is both that observability seems less important to experimental science than it was fifty years ago, and that two intuitively-accepted criteria for elementary particles, observability and noncompositeness, are possibly at odds with each other. If they are, then contemporary scientists might have to revise their notion of *how* elementary particles mean. If we are lucky, consideration of such a revision might also pave the way for a clearer perception of *what* they mean.

Notes

1 This same point is made by McMullin (1978b, p. 276). As he points out, although the theoretical electron (the construct whose properties are defined in terms of a theory) serves as the "matter" for explaining light-emission, the physical referent of this construct, the "real" electron causing cloud tracks and light emission, is not perfectly grasped by means of the theory. Matter in this latter sense "always escapes full statement in any theory."

2 Appropriate answers to the largely scientific "what" question, for example, might be given in terms of Schwinger's operator field theory (1970, 1978) or Gell-Mann's more atomistic "eightfold way" (1964). On the other hand, appropriate responses to the largely epistemological "how" question might be given in terms of criteria for the use of the term, 'matter'. Such criteria might include, for example, observability or predictive power. Although the responses to the "what" question have been characterized as "largely scientific", this should not be taken to suggest that such answers have no philosophical import. Certainly since at least the sixth century BC, humankind has been engaged in both a philosophical and scientific search for the constituents of matter. The early Ionian "metaphysicians" clearly followed a line of cosmological inquiry about "what" matter is. As Sidney Drell, deputy director of the Stanford Linear Accelerator says: Leucippus and Democritus prefigured contemporary atomists, while Anaxagoras had the original bootstrap model of infinitely divisible seeds within seeds. Likewise, he continues, Anaximander, Pythagoras and Plato antedated current theorists who follow abstract models based on mathematical symmetries (Drell 1978, p. 23; Feinberg 1977, p. xiv).

3 Even the precise number of known hadrons (the particles now believed to make up the bulk of matter), for example, depends on how criteria are used to distinguish various manifestations of particulate matter (Feinberg 1977, p. 190). This is because a particle is "defined by the collisions that create it" and is, almost always, itself created by an experimenter "in order to study it" (Schwinger 1970, p. 37).

4 The statement that most physicists currently subscribe to atomistic views is generally not contested and is well-documented (see Treiman, et al. 1972, p. 179; Weinberg 1974, pp. 50-55; Kokkedee 1969, p. 12; Kallén 1964, p. 3). This statement should not be interpreted in too simple a fashion, however; clearly it ought not be taken to mean that there is no place for field-theoretic accounts of matter. For one thing, the Leibniz (field-theoretic) vs. Newton (atomistic) debate has not been clearly resolved in contemporary science; both particle and field concepts are unified in quantum field theory, for example, and several major accounts of particle behavior are expressed in terms of field-theoretic notions, e.g., gauge theories.

5 "Virtual" particles are unobservable in principle because energy conservation prevents their escape and detection, and therefore their existence as "real" (Cline, et al. 1976, pp. 48-51; Davis and Sacton 1967, pp. 366-68; 417-23; Martin and Spearman 1970, pp. 11-12; Frauenfelder and Henley 1974, pp. 16-17, 20).

6 Newton rejected the ether as a material interplanetary medium because he believed that any medium would eventually cause observable slowing of the motions of the planets and no such slowing had been

observed. Einstein, on the other hand, apparently held the view that the ether and its properties were unobservable in principle (Hesse 1967, pp. 67-68).

7 That is, the exclusion principle (Fermi statistics) and half-integer spin for an odd number of particles; Bose statistics and integer spin for an even number of particles.

8 There are, of course, several notable exceptions to the dominance of theory over experiment, even though they are just that, exceptions. Julian Schwinger, co-recipient of the Nobel Prize in 1965, has specifically and sharply criticized the dominant views of the high-energy community. These views, he says, are wrongly founded on "speculative hypotheses about inner structure" of particles. Instead, he proposes that physicists follow source theory, a field theory based only on "the properties of the observed particles." (1978, pp. 367, 348; Schwinger 1970).

9 My own view is that quarks have done too much in explaining hadron multiplets ever to be forgotten by physicists. Likewise they have been made a successful part of nonabelian gauge theories, so that we are now moving toward a quark dynamics. For both these reasons, even if they are discovered to be nonelementary, I do not believe they will disappear, any more than the atom has disappeared now that it is known to be composite.

References

Applequist, T., Barnett, R., and Lane, K. (1978). "Charm and Beyond." In Jackson *et al.* (1978). Pages 397-500.

Avre, C., Gustafson, H., and Jones, L. (1976). "Search for New Massive Long-Lived Neutral Particles." *Physical Review Letters* 37: 474-77.

Barut, A. (1967). *The Theory of the Scattering Matrix*. New York: MacMillan.

Bowman, P. (1970). "Comment." In Stuewer (1970). Pages 364-73.

Brown, L. (1978). "The Idea of the Neutrino." *Physics Today* 31 (No. 9) 23-28.

Capra, F. (1975). *The Tao of Physics*. New York: Bantam.

Cence, R., Dobson, P., Pakvasa, S., and Tuan, S. (eds.). (1978). *Proceedings of the Seventh Hawaii Topical Conference in Particle Physics*. Honolulu: University Press of Hawaii.

Chandrasekhar, S. (1979). "Beauty and the Quest for Beauty in Science." *Physics Today* 32 (No. 7): 25-30.

Chew, G. (1966). "Crisis for the Elementary Particle Concept." *University of California Radiation Laboratory Reprint*. No. 17137: 1-12.

Chupka, W., Stevens, C., and Schiffer, J. (1976). "Search for Fractionally Charged Particles in Lunar Soil." *Physical Review D* 14: 716-27.

Ciardi, J. (1959). "How Does a Poem Mean?" In *Introduction to Literature*. Edited by H. Barrows, H. Heffner, J. Ciardi, and W. Douglas. Boston: Houghton Mifflin. Pages 663-77.

Cline, D., Mann, A., and Rubbia, C. (1976). "The Search for New Families of Elementary Particles." *Scientific American* 234 (No. 1): 44

Cox, P., and Yildiz, H. (1977). "Heavy Leptons in Trimuon Events." *Physical Review D* 16: 2897-99.

Davis, D., and Sacton, J. (1967). "Hypernuclear Physics." In *High Energy Physics*, 2. Edited by E. Burhop. New York: Academic. Pages 365-455.

Drell, S. (1978). "When is a Particle?" *Physics Today* 31 (No. 6): 23-32

Ellis., S., and Kislinger, M. (1974). "Implications of Parton-Model Concepts are Large-Tranverse-Momentum Production of Hadrons." *Physical Review D* 9: 2027-51.

Feinberg, G. (1977). What is the World Made of? Garden City, New York: Doubleday.

Flanagan, D. (1977). "Proliferating Quarks." Scientific American 237 (No. 4): 74.

Frauenfelder, H., and Henley, E. (1974). Subatomic Physics. Englewood Cliffs: Prentice-Hall.

Gale, G. (1971a). Leibniz: The Physicist as Philosopher. Unpublished Ph.D. Dissertation, University of California, Davis.

-----. (1971b). "The Physical System of Leibniz." Studia Leibnitiana 2: 114-27.

-----. (1973). "Forces and Particles: Concepts Again in Conflict." Journal of College Science Teaching 3: 29-35.

-----. (1974). "Chew's Monadology." Journal of the History of Ideas 35: 339-48.

Galik, R., Jordan, C., Richter, B., Seppi, E., Siemann, R., and Ecklund, S. (1974). "Search for Fractionally-Charged Particles Photoproduced from Copper." Physical Review D: 1856-63.

Gaston, J. (1973). Originality and Competition in Science. Chicago: University of Chicago Press.

Gell-Mann, M. and Ne'eman, Y. (1964). The Eightfold Way. New York: W.A. Benjamin.

Gibson, S. and Pollard, B. (1976). Symmetry Principles in Elementary Particle Physics. Cambridge: Cambridge University Press.

Glashow, S. (1975). "Quarks With Color and Flavor." Scientific American 233 (No. 4): 38-50.

------. (1978). "Theoretical Ideas About Charm and the Theory of Flavor Mixing." In Cence et al. (1978). Pages 155-226.

Goldhaber, A. and Heckman, H. (1978). "High Energy Interactions of Nuclei." In Jackson et al. (1978). Pages 161-205.

Greenberg, O. (1978). "Quarks." In Jackson et al. (1978). Pages 327-86.

Hanson, N. (1963a). "The Dematerialization of Matter." In McMullin (1963). Pages 549-61.

--------. (1963b). "Discussion." In McMullin (1963). Pages 570-73.

Heisenberg, W. (1976). "The Nature of Elementary Particles." Physics Today 29 (No. 3): 32-39.

Hesse, M. (1967). "Ether." Encyclopedia of Philosophy, Vol. 3. Edited by P. Edwards. New York: Macmillan. Pages 66-69.

Holton, G. (1978). The Scientific Imagination: Case Studies. London: Cambridge University Press.

Jackson, J., Gove, H., and Schwitters, R. (eds.). (1978). Annual Review of Nuclear and Particle Science, 28. Palo Alto: Annual Reviews.

Kallén, G. (1964). Elementary Particle Physics. London: Addison-Wesley.

Kantorovich, A. (1973). "Structure of Hadron Matter: Hierarchy, Democracy, or Potentiality?" Foundations of Physics 3: 335-402

Kokkedee, J. (1969). The Quark Model. New York: Benjamin.

Lannutti, J., and Williams, P. (eds.). (1978). Current Trends in the Theory of Fields: AIP Conference Proceedings. New York: American Institute of Physics.

Leon, M. (1973). Particle Physics. New York: Academic Press.

Lubkin, G. (1974). "Colliding-Beam Results Upset Quark Advocates." Physics Today 27 (No. 3): 17,20.

_________. (1975). "Mixed Reception for Magnetic Monopole Announcement." Physics Today 28 (No. 10): 17-20.

---------. (1977a). "Evidence Grows for Charged Heavy Lepton at 1.8-2.0 GeV." Physics Today 30 (No. 11): 17,19,20.

---------. (1977b). "Upsilon Particles...Suggest New Quark." Physics Today 30 (No. 10): 17-20.

Martin, A., and Spearman, T. (1970). Elementary Particle Theory. New York: Wiley Interscience.

Marx, J., and Nygren D. (1978). "The Time Projection Chamber." Physics Today 31 (No. 10): 46-53.

McMullin, E. (ed.). (1963). The Concept of Matter. Notre Dame: University Press.

----------. (ed.). (1978a). The Concept of Matter in Modern Philosophy. Notre Dame: University Press.

----------. (1978b). "Epilogue: Eight Problems." In McMullin (1978a). Pages 271-98.

----------. (1978c). "Introduction." In McMullin (1978a). Pages 1-55.

Misner, C. "Mass as a Form of Vacuum." In McMullin (1963). Pages 596-608.

Nambu, Y. (1976). "The Confinement of Quarks." Scientific American 235 (No. 5): 48-60.

Nash, T., and Yamanouchi, T. (1974). "Search for Fractionally-Charged Quarks Produced by 200-and-300 GeV Proton-Nuclear Interactions. Physical Review Letters 32: 858-62.

Novozhilov, Yu. (1975). Elementary Particle Theory. First edition. Translation of Vvedenie v teoriiu͡ elementarnykh chastit͡s, by J. Rosner. Oxford: Pergamon.

Pati, J. (1976). "The World of Basic Attributes: Valency and Color." In Gauge Theories and Modern Field Theory. Edited by R. Arnowitt and P. Nath. Cambridge: MIT Press. Pages 27-28.

Price, P., et al. (1975). "Evidence for the Detection of a Moving Magnetic Monopole." Physical Review Letters 35: 487-90.

Ramsey, N. (1979). "The State of the APS and of Physics in 1978." Physics Today 32 (No. 4): 25-30.

Sandorfi, A., Kilius, L., and Litherland, A. (1978). "Search for Quarklei--A New Form of Matter?" In Van Oers et al. (1978). Pages 756-58.

Schaffner, K. (1970). "Outlines of a Logic of Comparative Theory Evaluation." In Stuewer (1970). Pages 311-54.

Schwinger, J. (1970). Particles, Sources, and Fields. Reading, Massachusetts: Addison Wesley.

-------------. (1978). "Introduction to Source Theory with Applications to High Energy Physics." In Cence et al. (1978). Pages 341-81.

Schwitters, R. (1977). "Fundamental Particles with Charm." Scientific American 237 (No. 4): 56-70.

Shrader-Frechette, K. (1977). "Atomism in Crisis." Philosophy of Science 44: 409-40.

-----------------. (1979). "High-Energy Models and the Ontological Status of the Quark." Synthese 42: 173-89.

Simmons, J. (1975). "Some Topics Concerning N-N and N-D Experiments at Medium Energy." In High-Energy Physics and Nuclear Structure-

1975:AIP Conference Proceedings. Edited by D.E. Nagle, A. Goldhaber, C. Hargrove, R. Burman, and B. Storms. New York: American Institute of Physics. Pages 103-27.

Stuewer, R. (ed.). (1970). Historical and Philosophical Perspectives of Science. (Minnesota Studies in Philosophy of Science, Volume V.) Minneapolis: University of Minnesota Press.

't Hooft, G. (1978). "Non-Abelian Gauge Theories and Quark Confinement." In Lannuti and Williams (1978). Pages 1-8.

Treiman, S. Feynman, R. Jackson, J. Lee, T., and Low, F. (1972). "Panel Discussion." In AIP Conference Proceedings: Particle Physics. Edited by M. Bander, G. Shaw, and D. Wong. New York: American Institute of Physics. Pages 164-185.

Turner, M., and Schramm, D. (1979). "Cosmology and Elementary Particle Physics." Physics Today 32 (No. 9): 42-48.

Van Oers, W., Svenne, J., McKee, J. and Falk, W. (eds.). (1978). Clustering Aspects of Nuclear Structure and Nuclear Reactions: AIP Conference Proceedings. New York: American Institute of Physics.

Weinberg, S. (1974). "Unified Theories of Elementary Particle Interaction." Scientific American 231 (No. 1): 50-59.

Wilczek, F. (1978). "Steps Toward the Heavy Quark Potential." In Lannuti and Williams (1978). Pages 30-37.

Woodruff, A. (1963). "The Elementary Particles of Matter." In McMullin (1963). Pages 578-84.

PART XI

EXPLANATION

What Should We Expect of a Theory of Explanation?[1]

Barbara V. E. Klein

Yale University

Few topics in the philosophy of science have received as much discussion as the topic of explanation.[2] Yet as Hanna's (1979, pp. 291-316) survey of recent research on scientific explanation makes abundantly clear, the field is riddled with competing "paradigms", "models", and "formulas" with little sign of any consensus.

It is useful in this sort of situation to engage in metatheoretic ascent. What do we or ought we to expect of a theory of explanation? Unless some measure of agreement can be reached about that question, how can we expect agreement (or even rationally adjudicable disagreement) on the adequacy of particular candidate theories? The purpose of this paper is to sketch a meta-theoretic characterization of what a theory of explanation ought to be like. The view I will put forth is surely not the only one, but it is a view which I believe to be implicit in the work of a number of influential writers on the subject, including Hempel (1965). For lack of a better name, I shall call it "the logico-normative view" of a theory of explanation. My aim is not to defend the view (although I am quite sympathetic to it), but rather to articulate it clearly enough so that those in agreement with it will be in a better position to assess candidate theories of explanation and those not in agreement will have a clear target for their disagreements. I hope thus to promote, if not "progress" in the field, at least more rational "cross-paradigmatic" (Hanna 1979, p. 292) discussion.

My discussion shall fall into three parts. First, I shall lay out a number of pre-theoretic assumptions which underlie the logico-normative view. Second, I shall present that aspect of the logico-normative view which concerns what a general theory of explanation should be like. Finally, I shall distinguish such a general theory of explanation from both a theory of scientific explanation and a theory of some explanation kind.

PSA 1980, Volume 1, pp. 319-328

1. Some Pre-theoretic Assumptions

The following pre-theoretic assumptions underlie the logico-normative view of a theory of explanation in the sense that they give rise to the questions which, on the logico-normative view, a theory of explanation should answer.

(PA 1) The term 'explanation' is used, ordinarily, to refer both to a certain type of communicative linguistic activity and to what is conveyed by that communicative activity. An explanation of the first sort consists of an episode in which, typically, one person is engaged in explaining something, usually, to someone else, by means of uttering (or writing) a series of sentences. It is the sort of explanation we have in mind when we say things like "President Carter's explanation lasted five minutes." An explanation of the second sort is what is expressed by the sentences which are uttered or written in an explanation of the first sort (or, on some views, these sentences themselves). It is the sort of explanation we have in mind when we say things like "Freud's theory of dream formation provides an explanation of why there is distortion in dreams." To avoid confusion, we shall henceforth refer to explanations of the first sort as "explaining episodes", following Bromberger (1962, pp. 72-105). The term 'explanation' shall be reserved for the second sort.

(PA 2) Associated with any explanation are two logical components: what is explained and what is doing the explaining. Following Hempel (1965, p. 247) we shall call what is explained the "explanandum" and what does the explaining the "explanans". We will leave open for the moment what kind of thing each is. In particular, by using Hempel's terminology we will _not_ be committed to the particular characterization of explanandum and explanans which he gives. Two views can be taken on the relationship between any given explanation and its associated explanans and explanandum. On one view, that taken by Hempel, the explanans and explanandum are the "major constituents" of an explanation. That is, they are what an explanation consists of. On the other, seemingly more in tune with ordinary practice, the relationship between an explanation and explanans is one of identity. The explanans _is_ the explanation. The explanandum gets into the act not by being a part of an explanation but by bearing the right relation to the explanans--the relation of being explained by the explanans. In other words, on this view, explanations are like fathers in the sense that someone is a father only if he is father _of_ someone. Similarly, something is an explanation only if it is an explanation of something. That something is the explanandum.

(PA 3) Relative to any explanandum, we can distinguish a possible or potential explanans from an actual or correct explanans. A potential explanans for why John went to the store is that he wanted to buy some paperclips. The correct explanans is that he promised Harry he would meet him there at 3:00 p.m. What distinguishes the two is that a potential explanans for some explanandum has all the features of a correct explanans except that it may or may not be true. A correct explanans, by virtue of being correct, is always true.

In addition to distinguishing between potential and correct explanans relative to some explanandum, we will also have occasion to distinguish between potential and correct explanations. If an explanation is taken to consist of an explanans plus an explanandum, then a potential explanation is simply a potential explanans for some explanandum plus that explanandum. If an explanation is identical to an explanans, then a potential explanation is simply identical to a potential explanans for some explanandum.

2. The First Concern of a Theory: The Nature of Explanation

These assumptions naturally give rise to a host of questions about explanation which fall roughly into two areas: questions concerning the nature of explanation and questions concerning its assessment. According to the logico-normative view, the project of providing a general theory of explanation is the project of constructing some sort of systematic, useful account of explanation which responds to these various questions, either by answering them directly or by showing that they are misguided in some way.

The central questions pertaining to the nature of explanation are as follows:

(1) What is an explanation?
(2) What is an explaining episode?
(3) What is the relationship between explanations and explaining episodes?
(4) Are there kinds of explanation? If so, what are they?

Questions (1) and (2) are basically requests for ontological classification, classically answered by providing a genus and differentia of the Aristotelian sort. Thus, in answer to (1), one might begin by claiming that an explanation is a set of propositions of some sort. That provides the genus. The remaining task would be to say what distinguishes those sets of propositions which count as explanations from those that don't. Or, in answer to (2), one might decide that an explaining episode is a speech act of some sort, and then the task would be to distinguish those speech acts that are explaining episodes from those that aren't. Things might not be that simple, of course. For example, explanations might turn out to be context-dependent in the sense that something would only count as an explanation relative to a certain context. It might even be that explanations are not the sort of thing whose ontology can be specified in an Aristotelian way. Then, a non-classical response would have to be provided.

One of the issues that arises in connection with question (3) concerns the interdefinability of 'explanation' and 'explaining episode'. In particular, the following questions can be raised: Is it possible to provide a satisfactory answer to (1) (or (2)) without reference to explaining episodes (or explanations)? If the answer is 'yes' to both questions, then, we can say that, in some sense, the two notions are logically independent. If, on the other hand, one can be "defined" independently but the other can't, then the second is logically dependent on the first. Finally, if both require reference to the other, then a relationship of mutual logical interdependence exists.

Since the set of all possible explanations can surely be cross-classified in hundreds of different ways, the important issue underlying question (4) is not whether there are kinds of explanation, but whether there are any theoretically interesting kinds. I take kind differentiation to be of interest primarily where such differentiation is motivated by normative considerations. I shall say more about this shortly.

3. Second Concern: The Normative Problem

The second basic concern of a general theory of explanation, according to the meta-theoretic view we are presenting, is the normative one of assessing putative explanations. (Note that a putative explanation is not the same as a potential one. Something is a putative explanation if it is offered or considered as a candidate for a potential (or correct) explanation. Thus to be a putative explanation, it need not even meet the conditions for being a potential one.) This concern is, in fact, the central one. It is here that the spirit, if not the letter, of the logical positivists is reflected. For, according to Bromberger (1971, p. 47), the logical positivists "were committed to the task of making explicit the types of considerations that govern--or that ought to govern--the acceptance and the assessment of any alleged contribution to scientific knowledge."

Ideally, a solution to the normative problem vis-à-vis explanation will provide a set of principles which (a) can be applied to any putative explanation, and (b) are useful for purposes of evaluation. The standard approach to the normative problem has considered the crucial feature of explanations to be the relationship of explanans to explananda. Hence, the problem has been reformulated as follows: to articulate a set of conditions relevant to the evaluation of any given putative explanans as either a potential or correct explanans for a given explanandum. And since it is assumed that a correct explanans for a given explanandum differs from a potential one only in having to be true, the problem boils down to the search for conditions relevant to a putative explanans being a potential one.

The problem can be made precise in the following way. Let 'E' be a variable ranging over particular explananda (for example, why John Jones went to the store on Tuesday, November 6, 1979); let 'N' range over putative explanans, and let 'O' range over particular occasions. Then let us assume that associated with any explanandum expressed or uttered (depending on whether explananda are regarded as abstract objects or as linguistic entities) on a particular occasion, there exists a set of potential explanans for that explanandum on that occasion. For example, associated with the explanandum why John Jones went to the store on Tuesday, November 6, 1979, expressed on some given occasion O_1 is a set of potential explanans including such things as that he wanted to buy an ice cream cone, that he promised to meet Freddy there, that his mother told him to, etc. An ideally successful general theory of explanation should be able to generate a statement of the following form for every possible explanandum E for every possible occasion O:

(5) Evaluation Schema I:

(N)(Pe(N,E,O) just in case N satisfies evaluation conditions_____)

where 'Pe(N,E,O)' means 'N is a potential explanans for E on occasion O' and the blank is to be filled in by a statement of the evaluation conditions relevant for that explanandum on that occasion.[3] The result would be a very long list of statements like this:

(6) (N)(Pe(N, why John went to the store, O_1) just in case N satisfies evaluation conditions C_1)
(N)(Pe(N, why John went to the store, O_2) just in case N satisfies evaluation conditions C_2)
⋮
(N)(Pe(N, how children learn their first natural language, O_1) just in case N satisfies evaluation conditions C_{24})
(N)(Pe(N, how children learn their first natural language, O_2) just in case N satisfies evaluation conditions C_{25})
etc.

where 'O_1', 'O_2' refer to particular occasions and 'C_1', 'C_2', 'C_{24}', and 'C_{25}' refer to specific sets of evaluation conditions.

What makes the problem difficult is that there are indefinitely many, if not infinitely many, possible explananda as well as indefinitely many occasions, and hence, there will be indefinitely many evaluation rules required. I am not speaking here of kinds of explananda (we will get to that shortly) but of particular explananda, individual explananda. Why John went to the store counts as a different particular explanandum from why Mary went to church, although, in some sense, they may share the same logical form. It is assumed, of course, that not all the ways that particular explananda and particular occasions differ from one another are relevant to the evaluation problem; just as not all the ways that particular sentences, or even propositions, differ from one another is relevant to whether any given sentence entails or is entailed by some set of sentences. It is therefore necessary to abstract away from their relevant differences and fix upon the relevant ones. What this amounts to is figuring out a way to group individual explananda and individual occasions into a manageable number of classes such that the explanandum class and the occasion class to which any particular explanandum expressed on a given occasion belongs will fix the relevant set of evaluation conditions. The result will be a new set of statements of the form:

(7) Evaluation Schema II:

(E)(N)(O)(If E belongs to E-kind ____ and O belongs to O-kind ____, then Pe(N,E,O) just in case N satisfies evaluation conditions __________)

where the first blank is to be filled in by a description of a kind of explananda, the second with a description of a kind of occasion, and the third with a statement of the relevant set of evaluation conditions.

Although this move solves the problem of the multiplicity of explananda and occasions, it falls short of providing a completely satisfactory solution to the normative problem. For unless we have a way of sorting particular explananda expressed on particular occasions into the relevant classes our new set of evaluation rules will not be optimally useful. Thus we must determine what members of the posited explanandum-classes have in common, and, if possible, we must find some property or set of properties associated with each that can actually be used to assign individual explananda expressed on particular occasions to the relevant classes. And, likewise for the posited occasion classes. In other words, what is required are criteria, in the sense of sufficient conditions, for assigning an explanandum expressed on a particular occasion to both an explanandum class and an occasion class. Again the result will be a new kind of evaluation schema:

(8) Evaluation Schema III:

(E)(N)(O)(If E satisfies E-kind criterion _____ and O satisfies O-kind criterion _____, then Pe(N,E,O) just in case N satisfies evaluation conditions _______)

It is at this point that philosophers committed to the logico-normative view typically invoke the apparatus of a canonical system of representation. The idea is to utilize a formal notational system to represent explanans, explananda, and possibly even occasions so as to make salient those properties relevant for the formulation of formal and explicit evaluation rules. The classical case of this sort of attempt is, of course, Hempel and Oppenheim's (1948) statement of the evaluation principle for the explanation of particular events in terms of "a model language L." (Hempel 1965, pp. 270-278).

Evaluation Schema III encompasses two very different kinds of evaluation rules, each of which captures a set of suggestions and intuitions to be found in the literature. The first is what we might call a context-free rule. A context-free rule is one according to which what counts as a potential explanans for a given explanandum on a given occasion is determined completely by intrinsic features of the explanandum. That is, the occasion on which the explanandum is expressed is completely irrelevant to determining the class of potential explanans.

In contrast, a context-dependent rule is one in which intrinsic features of the explanandum underdetermine the set of potential explanans for a given explanandum expressed on a particular occasion. Features of the occasion must be invoked in addition to secure the determination. In particular, features of the explaining episode are usually taken to be the most relevant--features such as the beliefs, interests, desires, intentions, and degree of understanding of the agents involved.

A close reading of Hempel indicates that his theory of explanation includes several such context-dependent rules. This fact is obvious in the case of statistical explanation for particular events which, according to Hempel, is "essentially relative to a given knowledge situation as represented by a class K of accepted statements." (1965, p. 402). Not so obvious is the fact that it is true of all explanations of particular fact on Hempel's theory. For to determine whether a putative explanans counts as a potential explanans for some statement of particular fact (explananda are sentences for Hempel), it is not enough to take into account intrinsic features of that statement. We must also take into account the desired mode of explanation, that is, whether we wish to account for the phenomenon at hand in terms of universal or statistical laws. And this desire can only be a feature of the explaining episode in which the explanandum plays a role.

4. How the Two Problems Interact

We are now in a position to say something about the ways in which the normative problem interacts with the problem of the nature of explanation. First, ascertaining the nature of the relationship between explanations and explaining episodes may shed light on whether the evaluation of explanations is context-free or context-dependent, an issue with an important bearing on precisely how we define the normative problem (that is, on what sorts of evaluation rules we seek). Second, the most theoretically central way to answer the question of kind differentiation is in terms of a solution to the normative problem. The question of kind differentiation is the question whether there are any theoretically interesting distinctions into kinds within the class of explanations. Since the central focus of a general theory of explanation is the normative problem, the taxonomy of kinds which is theoretically most basic, on the logico-normative view, will be one in which sameness and difference of kinds correspond to sameness and difference of the relevant sets of evaluation conditions. (There can, of course, be other kinds of explanation-kinds such as kinds based on subject matter or domain, but these will not be the theoretically basic kinds.)

Finally, how we answer the ontological questions 'What is an explanation?' and 'What is an explaining episode?' will depend on how we solve the normative problem. For answering the question 'What is an explanation?' will require not only distinguishing explanations from closely related objects such as justifications and criticisms and descriptions, but also distinguishing putative explanations (which aren't really explanations at all) from potential and correct ones. And the latter task will involve recourse to a solution to the normative problem.

5. Different Sorts of Theories of Explanation

Thus far I have been talking about what a general theory of explanation would be like on the logico-normative view. I would like now to distinguish this sort of general theory from both a theory of scientific explanation and a theory of some particular kind of explanation. As should be clear by now, according to the logico-normative view, a general theory of explanation is a set of systematically related claims about explanation which responds in some way to our four questions about the nature of explanation and provides a putative solution to the normative problem. The domain of such a theory includes anything that counts as an explanation or an explaining episode, whatever the context in which it might occur. In contrast, a theory of scientific explanation deals with the more restricted domain of explanations offered in the context of scientific inquiry. It is plausible to assume that explaining episodes in the scientific context do not differ from those in a non-scientific context except insofar as there are differences in the explanations conveyed by those episodes. If so, then a theory of scientific explanation need not address itself to any version of questions (2) or (3). What remains for such a theory to do, then, is provide an answer to a modified version of (1), viz., What is a scientific explanation?; provide an answer to a modified version of (4), viz., Are there kinds of scientific explanation? If so, what are they?; and provide a solution to the normative problem for the domain of scientific explanations.

The third kind of theory, a theory of a kind of explanation, will consist of that part of either a general theory of explanation or a theory of scientific explanation which addresses the normative problem for a specific explanation kind. It will include a specification of the criteria for belonging to that kind as well as a description of the relevant set of evaluation conditions. Much of the discussion in the philosophical literature on explanation is relevant to this third kind of theory rather than to that part of a general theory of explanation which deals with features of explanations (or scientific explanations) in general.

Notes

[1]I am grateful to Scott Soames and Jeffrey Poland for comments on an earlier version of this paper.

[2]A recent computer search with DIALOG of the Philosopher's Index for articles on scientific explanation published between 1966 and the present in English language journals yielded 232 items.

[3]The inclusion of reference to occasions in this and subsequent schemata does not mean I am endorsing the view that what counts as a potential explanans for a given explanandum can or does vary from occasion to occasion; rather, the schemata are formulated in this way to allow for the possibility of stating an evaluation rule which embodies such a view.

References

Bromberger, S. (1962). "An Approach to Explanation." In Analytical Philosophy. Edited by R. J. Butler. Oxford, England: Basil Blackwell. Pages 72-105.

-------------. (1971). "Science and the Forms of Ignorance." In Observation and Theory in Science. Edited by M. Mandelbaum. Baltimore: Johns Hopkins Press. Pages 45-67.

Hanna, Joseph F. (1979). "An Interpretive Survey of Recent Research on Scientific Explanation." In Current Research in Philosophy of Science. Edited by P. D. Asquith and H. E. Kyburg, Jr. East Lansing, Michigan: Philosophy of Science Association. Pages 291-316.

Hempel, C. G. (1965). Aspects of Scientific Explanation. New York: The Free Press.

------------- and Oppenheim, P. (1948). "Studies in the Logic of Explanation." Philosophy of Science 15: 135-75. (As Reprinted in Hempel (1965), pp. 245-295.).

Scientific Explanation and Norms in Science

David Gruender

Florida State University

For some thirty years, now, discussions of explanation in science have begun with the theory most clearly articulated by Hempel and Oppenheim (1948). It has been modified, extended, and defended by Hempel with perserverance and patience since then (1965). As a result, it has served as the standard against which competitors strived, and the clarity with which it has been set forth and defended has played a large role in the fecundity of the debate it has aroused.

It is not my intention here to attack or defend the theory in any of its versions, but to examine some aspects of the debate as they bear on the development of our concepts of what science is about. For the theory has been taken by its friends and its enemies to play a dual role: it is, on the one hand, intended to be descriptive of what science does; while, at the same time, it is to serve as a model or ideal of what science ought to strive to be. This latter role, the normative one, is often taken in a stronger sense than I have put it here. In this stronger sense, the theory is not so much an ideal as a minimum standard for science. The difference is more important than it may seem at first.

The ideal physician, for example, is a skilled diagnostician, has a warm and supportive bedside manner, is ingenious, persistent, and effective in planning individual therapeutic strategies, and never duns his patients. A physician may occasionally fall slightly short of this ideal, but no one would suggest he was any the less a physician for it. However, the minimum standard is set as graduation from a course of medical training and the passing of a licensing examination. The candidate who does not meet this standard is not a physician. Similarly, on this interpretation, the inquiry which fails to meet the minimum standard of scientific explanation is not merely science which does not measure up to the highest ideals--it is not science at all! We would deny it that honorific title with the same firmness which we deny the title of physician to one who fails his medical examination.

With either of these approaches, the theory of explanation plays a key role in our very concept of what constitutes scientific inquiry; on the latter, it is

---

PSA 1980, Volume 1, pp. 329-335

decisive. In what follows, I will try to show that both the stronger and weaker of these normative claims go too far. I shall contend that, in the form they are stated above, they are incompatible with what we know of science. I shall offer a hypothesis to explain this, and a modest remedy.

Let us first look at the content of the theory. Explanation in science, on this view, is essentially deductive in character. If we confine our attention at the outset to individual events, and ask how, in science, they are to be explained, the answer is that we must somehow identify a number of universal generalizations or law-like statements, some statements about particular matters of fact, and then logically deduce from all these the sentence describing the puzzling individual event we had, at the outset, wanted explained. Or in place of the universal generalizations, we may use sentences stating tendencies, likelihoods, or probabilities of classes of events, and then proceed as before; except that the outcome in this case will be a probability for the sentence to be explained.

Of course, in addition, we shall require those things that are necessary to make deduction possible and clearly demonstrable: formation rules, transformation rules, definitions, consistency, and so forth; with corresponding standards for probabilistic inference. Hempel and Oppenheim also specified some further, now familiar, requirements: that the sentences be testable, that the universal generalizations be true, and that they be actually required for the deduction of the sentences describing the phenomena to be explained. Each of these requirements has been spelled out in some detail, and in the ensuing discussion, a consideration of the exact form they ought to take has resulted in the recognition of a variety of further philosophical problems, for each of which there remain disagreements on what the best course of action might be. For now, I wish to pass over these problems.

The way I have formulated the theory is neutral as to whether the phenomenon to be explained is a single event, a class of similar events, or a law-like generalization itself. In the former case, at least one particular or nongeneral sentence is needed in the explanation to make the deduction possible. Thus, if one is to explain the Great Chicago Fire, one will need, in addition to the laws of combustion and a description of the lamentable wooden building practices of the time, a statement about Mrs. O'Leary's cow knocking over a lantern in the hay. That such a sentence is continually offered as part of the explanation of an event even in cases in which we cannot be confident that it *was* Mrs. O'Leary's cow that fateful night, may be taken as testimony of the power and attractiveness of this theory of explanation. In any case, I have chosen this formulation because what I have to say is similarly neutral to the level of abstraction of an explanation.

Now the attractiveness of this theory is that it fits classic cases of modern science, and to the extent that classic cases serve as the paradigms or models for further scientific work, the theory both explains much of what goes on in contemporary science, and stands as the methodological norm in judging it (Compare Feyerabend 1977). In addition, from a historical point of view, it provides after-the-fact insight into what went wrong when earlier societies used familiarity, verisimilitude with respect to already accepted knowledge, authority, or other alternatives as standards for judging explanations of natural events. Reflection on this point adds further urgency to our use of Hempel and Oppenheim's theory normatively, as we may notice the recurring seductiveness of the alternatives and renew our resolve to avoid their dangers.

Now, how is it that the Hempel and Oppenheim theory fits classic science? By classic science I mean that development from Pythagoras to Plato, Aristotle, Euclid, Archimedes, Pappus, Galileo, Descartes, Leibniz, and Newton--and many others--which resulted in the discovery of abstract logical or mathematical laws or systems of laws and their application to puzzling phenomena in nature. One can identify a tangible beginning of the process to Pythagoras' discovery that the relationship of string length to pitch can be expressed by ratios of simple integers, and a mature stage to Newton's three laws of mechanics. The process of development was long and complex, and even today not fully understood. The participants in it held a wide variety of philosophical views about knowledge and their own aims and methods, most of which are different from the variety of views we are inclined to take today. The important point for our purposes is that abstract laws for natural phenomena came to be identified, along with methods for helping us to decide when we had the right laws by a study of the phenomena themselves.

These laws have taken many forms, but in what became the ideal case, they constitute an axiom system which, together with a body of methods that serve to apply the system to a range of natural processes, constitutes our knowledge of that area. Experimental methods test the further applicability of the system to a broader area, and from time to time systems are modified or replaced (Kuhn 1970). But within the period that a system is in use, it defines what counts as knowledge of that area. Its postulates serve as the laws for the field, and its logical structure supports extensive and ramified deductions which are, so to speak, an exploration of the logical space of the theory. And yet, since the system is interpreted in terms of natural phenomena, these deductive consequences signal new tests or further applications of the theory, while the recognition of new facts in the field stimulates attempts to account for them by deductive extensions of the existing theory. Such a theory is not come upon all at once, nor without painful difficulties that take years--or generations--to surmount. But once developed, the uniting of deductive power with empirical interpretation provides a potentiality for the understanding, further exploration, and control of natural phenomena that itself explains why we should take it as a model of what scientific knowledge should be. Work of Euclid, interpreted as physical geometry, and Newton's mechanics both stand as exemplars of such scientific achievements, and both have, for the same reason, stood as ideals of what science can be.

The Hempel and Oppenheim theory serves largely to explicate this ideal methodologically, adding some ideas on verifiability that grew out of the positivist movement. Aside from the latter, which it is not my purpose to explore here, it is remarkable that discussion of the theory, also called the deductive-nomological theory (for obvious reasons), should be conducted with such heat throughout the some thirty years since its first articulation. It is remarkable, for no one seriously wishes to denigrate the achievements of Euclid or Newton, nor the many other powerful abstract systems that have been developed since then to illuminate puzzling phenomena, from genetics to relativity. While most of these fall short of being fully worked out axiom systems, they provide just that basis of deductive exploration coupled with the observation and measurement of phenomena that the Hempel and Oppenheim theory attempts to characterize. Hence, the normative appeal of this methodological or second-order theory is precisely that of the Euclidean or Newtonian model of science, whose success has been such that it would seem to be beyond criticism.

In addition, the Hempel and Oppenheim theory has been extended to cover areas in which our knowledge as yet provides no laws or generalizations of great rigor. Where we have a rational basis for estimates of the likelihood of the phenomena we are interested in, the theory, as mentioned initially, provides for the use of probabilistic methods, treating the ensuing reasoning rather as a weak form of deduction. And where that general knowledge we have is of very limited scope and highly idiosyncratic, as our knowledge of humanity might fairly be characterized at this stage, the theory allows us to offer explanation sketches, so to speak, to be filled out by such generalizations as most are wont to think appropriate in the circumstances described. In this fashion we can continue to use the knowledge we have while, at the same time, looking forward to a future in which knowledge of more rigorous laws would form the basis for deductive and nomological explanation in the fullest sense.

One might suppose that the graciousness and generosity of these extensions of the theory would have covered all possibilities and stilled all opposition. Unfortunately, they have served neither to increase its acceptability nor to reduce the heat of the debate. This heat suggests that perhaps we are facing issues that are deeper than methodological: metaphysical. Nor does one have to look far to find them. A deductive nomological explanation requires laws for the behavior of the phenomena it explains. And if we insist that *all* explanations are to be of this kind, it appears that we have, like Laplace, opted for universal determinism, but without actually quite saying so. Probabilistic explanations and explanations of human events are not required to meet these standards yet, but the standards are there, and the implication is a double one: 1) we or our posterity will meet them some day, and 2) truly scientific method demands no less. Thus does a methodological reflection become converted to a metaphysical creed. The process is subtle; the result especially ironic for having grown out of a tradition that rejected metaphysics.

A similar pattern appears when we look at arguments commonly offered against the theory. Some object to the subsidiary role played by probabilistic reasoning in the Hempel and Oppenheim theory. It is not deductive relationships that are central to scientific theories, Salmon tells us, for example, but statistical ones (1966). Many reasons are offered for this view, including the widespread use of probabilistic reasoning in science, from quantum mechanics to psychology. On such a view, we may replace deductive with probabilistic reasoning, and dispense with universal laws; the few cases in which they seem important may be covered by the extreme limiting cases in which the probabilities of the events we are interested in go to one or to zero. If we adopt *this* pattern as the ideal for scientific explanation, we will not look for universal laws. Instead, the norm for scientific knowledge will be the statistical relationship. Through this methodological path we have chosen the metaphysics of indeterminism.

Or consider the problems raised by Donagan and Dray. In papers that have become classic, they point out that, whatever advantages it may accrue to us to pursue universal laws in the natural sciences, where people are the subject of study this does not work. Donagan argues that the history of human action supports no universal law statements that are not question-begging (1957). Dray maintains that accounts of rational human decision-making must be framed of entirely different materials (1957). I make no comment on the intrinsic merits of their points. But the metaphysical lesson is plain here also: responding to the odor of universal determinism, they are saying its scope must be cut down to allow some freedom for human action.

I do not wish to mislead anyone. I regard these metaphysical issues as serious in their own right. They are complex, ramified, and puzzling at many levels. They have engaged the attention of philosophers and others in all periods and from many points of view, and have even survived being dubbed nonsense. But, whatever one's views on this head, we cannot solve such problems by fiat or indirection. Yet that is the effect of converting a norm of scientific method to a minimum standard for scientific inquiry, and it is in just this sense that the normative claim of theories of explanation goes too far.

Let us remind ourselves that theories of explanation, like the rest of the philosophy of science, are a second-order activity with respect to science itself. Of course this second-order activity can be engaged in by scientists as well as by philosophers, mathematicians and others. And it is frequently interactive with primary work in science in that its results may affect scientific inquiry or the other way about, or both may occur so closely we have the impression they happen at the same time. All of this is healthy, but we must not fool ourselves about what we know: namely, at any one time, the results of scientific inquiry as they stand at that time, and the burden of whatever philosophical reflection these may bear. That we can be quite wrong at either level has been demonstrated amply throughout history, and we have no right to expect this to change in the future.

The deductive theory of explanation points to classic achievements of science as the very models of scientific inquiry, and models, indeed, they must be! But such reflections do not license the claim that nothing else is science, nor that universal determinism is true. It is easy to fall into the same trap with opposing probabilistic theories, which point to statistical mechanics, quantum indeterminism, and the use of statistics in the social sciences as the very models of scientific inquiry. And these, too, must stand as models. But, as in the first case, our adopting them as models does not serve to support the claim that there is no other way to conduct science, much less that indeterminism is a brute characteristic of the universe.

In neither case can the proferred model be taken as exhaustive of the possibilities, for we do not really know that all possible competing models will someday be shown to be special cases of our favorite theory. If we could know that now, it would not be by means of science or philosophy, but prognostication. And if that were a tool we reliably possessed, we could abandon science and its philosophy altogether. My hypothesis is that we are sometimes led to prognostication through an excess of zeal in defense of our vision of science, and thus, unwittingly, serve it ill. The modest remedy I promised must now be evident: it is modesty about our knowledge of nature and about the science through which we come to know it.

I have mentioned two sorts of models of scientific achievements. Others could perhaps be developed and made the basis for theories of scientific explanation, and nothing in the brief history of science on earth rules out such future developments. To recognize such possibilities is not to prognosticate their occurrence, much less their content, but merely to remind us of our limitations at this point in history. We do have the scientific knowledge that has been achieved thus far, and the results of reflection and philosophical analysis of it to date. Both can be pushed forward, and in the meantime we can use the models of science we have, while at the same time stepping back from the claim that the exclusive adoption of any one of them is imperative, or that, through its exclusive licensing, we can legitimately settle a major metaphysical problem covertly. In that way we might make some problems easier for our posterity to resolve.

References

Donagan, Alan. (1957). "Explanation in History." Mind. 66: 145-64.

Dray, Wiliam H. (1957). Laws and Explanation in History. Oxford: Oxford University.

Feyerabend, P.K. (1977). Against Method. London: New Left Books.

Hempel, Carl, and Oppenheim, Paul. (1948). "Studies in the Logic of Explanation." Philosophy of Science 15: 135-75.

------------. (1965). Aspects of Scientific Explanation; and Other Essays in the Philosophy of Science. New York: Free Press.

Kuhn, Thomas. (1970). The Structure of Scientific Revolutions. 2nd ed. Chicago: University of Chicago Press.

Salmon, Wesley. (1966). The Foundations of Scientific Inference. Pittsburgh: University of Pittsburgh Press.

Idealization, Explanation, and Confirmation

Ronald Laymon

The Ohio State University

Scriven, with his well-known bridge example, showed that explanations need not carry with them accurate or useful predictions. A bridge has collapsed, but there is no history of overload or natural shock. Metal samples taken near the break show severely reduced elastic capabilities of a sort usually associated with natural aging. Therefore, given that the bridge did fall, it fell because of this sort of metal fatigue. However, on the basis of current theoretical and measurement capabilities, it could not have been predicted that it would fall. (Scriven 1962, pp. 181-185).

The existence of explanations which lack true or even usefully approximate predictions is not restricted to complex situations in engineering. Nor are such explanations the exclusive property of the non-physical sciences. The existence of such explanations, I want to claim, is common in the physical sciences as well. For example, the usual explanation of the Michelson-Morley experiment is that a non-null result would violate the principle of relativity (as expressed in Special Theory). The theory predicts, we are told, the null result. Yet as Miller, a later associate of Michelson and Morley, correctly pointed out, there is in fact a shift of anticipated frequency and phase, though not of amplitude. A similar example is Newton's *experimentum crucis*. If uniformly colored light from a prism generated spectrum is isolated, by means of an aperture, and projected into another prism, there will be, according to Newton, no further color generation or separation. Newton explained this difference of behavior (among the prisms) in terms of his ray composition theory of white light. But as Lucas, again correctly, pointed out, there is color variation after the second prism. As a final example, consider that the French experimentalist Perrin received the Nobel prize in large part for his confirmation of the new Einstein-Smoluchowski statistical mechanics. Yet, as admitted by Perrin, the principal prediction of the new theory, about root-mean-square displacement, exceeds the calculated limits of experimental error by a factor of no

---

PSA 1980, Volume 1, pp. 336-350

less than ten![1]

There is a facile and, I believe, common response to these claimed difficulties: it is to assert that the predictions of my examples are approximately true. Hence, they pose no problem for the Hempel-Oppenheim or Deductive-Nomological account of explanation.[2] Furthermore, since the scientific explanations of my examples carry with them these approximately true predictions, they differ radically from Scriven's bridge example.

Such a response is in difficulty on at least three counts. First, it seems that the concept of approximately true must be stretched beyond ordinary limits to cover our cases. In each of the cases, there is a prediction (the null result, no color modification) which is either qualitatively false (there was a shift, there was color modification) or which considerably exceeded the limits of known experimental error. Second, Scriven's bridge example is perhaps not really all that different from our scientific cases. The bridge was constructed according to standard engineering practice and theory. Furthermore, there was a prediction, perhaps only implicit, that it would not collapse. Otherwise, the bridge would have been closed. And this prediction, like those of my scientific cases, is false. Finally, just to say that these scientific predictions are approximately true is really just to name the problem. It remains to give some analysis or account of this concept. I shall attempt such an analysis in this paper. More specifically, I shall begin with this problem: if explanations in the physical sciences typically do not lead to true predictions, then of what value are their false predictions for confirmation or disconfirmation? Expressed less tendentiously, how do we construct a theory of confirmation that uses the concept of approximately true instead of simply being true? There is, too, the prior problem: if explanations are not glorified predictions with a latinized taxonomy, what are they? After suggesting some answers to these questions, I shall reappraise the similarities and differences between Scriven's bridge example and the above scientific cases.

1. Idealized Sketches and Modal Auxiliaries: Newton's Optical Work

One elegant approach to the problem of predictions being only approximately true is due to Kuhn, who argues that since there is no universally agreed standard of "acceptable experimental fit", even within a scientific speciality, it must be the case that acceptable experimental fit, i.e., being true enough for the purposes of confirmation, is a relative notion. That is, theory T has acceptable experimental fit with respect to experimental result e if T's prediction is closer to e than the predictions of T's competitors. I shall argue that this approach, however seductive, is incorrect.

There is obvious truth to claims that explanations are best understood as being relative to particular human needs and problems, and furthermore, that they can be expected to exemplify a bewildering variety of forms. Despite my sympathy with such approaches, I want to claim that there is a single sense or form of explanation that is of central importance--if you allow me the metaphor--for understanding the confirmation and disconfirmation of theories in the physical sciences. My theses in stark unqualified form are these. First, explanations (of the above sort) consist of two parts: (1) an idealized or simplified deductive-nomological sketch (henceforth the idealized sketch) and (2) an auxiliary argument or set of arguments showing that if the idealizations of the sketch are improved, i.e., made more exact and realistic, then the prediction of the idealized sketch will be correspondingly more exact and realistic. I shall refer to this second component as the modal auxiliary, since the argument purports to show that an improved idealized sketch is possible. My second thesis is that confirmation and disconfirmation occur in the realm of the modal auxiliaries. That is, a theory is confirmed if it can be shown that it is possible to improve its idealized sketches. Similarly, a theory is disconfirmed if it can be shown that the associated idealized sketches cannot be improved.

A striking example of a disconfirmation obtained by a well-designed negative modal argument is provided by Newton's use of the spectrum to refute the theory that all light is characterized by a single refractive index, the then received view. In his famous 1671 optics paper, Newton calculated, on the basis of the received view, that the vertical angle of dispersion for the exiting rays of his prism should be 31 minutes. His experimentally determined value, however, was $2^{o}49'$. That is, the spectrum was oblong and not circular. On a traditional view of disconfirmation by modus tollens, we would expect Newton to announce, on the basis of this disparity, that a successful refutation of the received view had been obtained. But this is not what Newton does!

> But because this computation was founded on the Hypothesis of proportionality of the sines of Incidence, and Refraction, which through by my own & others Experience I could not imagine to be so erroneous, as to make that Angle but 31', which in reality was 2 deg.49'; yet my curiosity caused me again to take my Prisme. And having placed it at my window, as before, I observed that by turning it a little about its axis to and fro, so as to vary its obliquity to the light, more then by an angle of 4 or 5 degrees, the Colours were not thereby sensibly translated from their place on the wall, and consequently by that variation of Incidence, the quantity of Refraction was not sensibly varied. (Corr. I, pp. 93-94).

The point of the rotation is this: it shows that even if Newton were seriously in error in his determination of the key initial condition of prism orientation, different values here would not help the received theory. In other words, the prediction of the idealized sketch based on the received view cannot be improved by a more careful quantitative determination of the initial conditions. The explanation afforded by the received view is not acceptable because it has no corresponding modal auxiliary. Newton has shown, at least with respect to the possibility of better initial data, that an improved idealized sketch is not forthcoming. It is only after his presentation of this negative modal argument that Newton announces the refutation of the received view.

Newton's treatment of his experimentum crucis provides a supporting example from the point of view of confirmation. In the experiment, a beam of sunlight was projected on a prism and the spectrum formed. A narrow cylinder of uniformly colored light was then separated, via an aperture, and projected onto another prism. Unlike the first prism, there was no diffusion or separation of colors after the second prism. Newton's explanation was that white light consists of independent light "rays" which are characterized by specific combinations of refractability and color. I should note that in the 1671 paper, where the experiment was first announced, Newton gives only what I have called an idealized sketch. There is no modal auxiliary. There were, however, many criticisms of the experiment, and these provoked modal auxiliaries from Newton. Let me give just one. Lucas, a continental scientist, asserted that on repetition of the experiment, unlike Newton, he observed color change and separation after the second prism. Such an observation clearly differed from the prediction of Newton's idealized sketch. Newton's response was to accept Lucas' description as being more accurate than his own. Newton, though, went on to assert that if aperture size were taken into account, then such an improved idealized sketch would predict color separation at the second prism. Newton did not actually construct such an improved idealized sketch; he merely argued for its possibility.

> But why the image is in one case circular, and in others a little oblong, and how the diffusion of light lengthwise may in any case be diminished at pleasure, I leave to be determined by geometricians, and compared with experiment. (Corr. I, p. 167).

There is one more feature of Newton's treatment of the experimentum crucis that needs, for present purposes, to be highlighted. And that is his initial misdescription of the result of the experiment. At first glance, Newton seems simply to have deliberately falsified his report so as to save his own theory. This accusation gains strength because the alternative theories to Newton's, namely the medium disturbance theories of Hooke, Grimaldi, and others, predicted color change after the second prism. This line of attack is based on too simple an approach. For as eloquently noted by Duhem, no description can capture with precision the complexity and irregularity of natural objects. Roughly, the choice is between true but vague descriptions or precise but false ones.[3] But as Duhem also noted, if descriptions of natural phenomena are to logically attach to our theories, we must opt for the latter alternative. The color separation and diffusion after the second prism were both small and irregular. Hence, in the interests of providing a tidy deduction with a conclusion compatible with his theory, Newton opted for the particular idealized description that he did. What provides prima facie justification for a description being an adequate idealization or useful approximation is that it readily attach logically to a theory and that it be compatible with the theory. More than prima facie justification is provided by the modal auxiliary, since it shows that the idealizations can be improved.[4] Furthermore, it is in the modal auxiliary that one can expect to find true but vague descriptions. Hence, Newton, in the discussion of his idealized sketch for the experimentum crucis, notes that "because of the refraction of the second prism, the colored light is much less diffused and less divergent, then when it is quite white, so that the image...is nearly circular." (My italics, Corr. I, p. 167).

2. The Michelson-Morley Experiment as a Supporting Example

The problem of the selection and justification of the appropriate idealized or approximate description of a phenomenon is to be distinguished from the problem of auxiliary assumptions. There are two ways to save a theory in the face of conflicting evidence. First, one can adjust the auxiliary assumptions so that the prediction of the idealized sketch is changed. Second, one can ignore the existence of more accurate descriptions of reality, describe the phenomenon so as to achieve an idealized sketch compatible with theory, and then try to construct a saving modal auxiliary. Newton, as we have seen, chose the second route for his experimentum crucis. Lorentz and Fitzgerald, with their contraction hypothesis, chose the first route for the Michelson-Morley experiment. This, of course, is well-known. What probably is not so well-known is that there were also considerable efforts along the second approach, i.e., to justify false but tidy misdescriptions of the phenomenon. Shankland's response to Miller's criticism is a good example of such an approach.

Miller began his criticism of the Relativistic explanation of the interferometer by noting the falsity of the null description.

> ...this fact must be emphasized, the indicated effect was not zero; the sensitivity of the apparatus was such that the conclusion, published in 1887, stated that the observed relative motion of the earth and ether did not exceed one-fourth of the earth's orbital velocity. This is quite different from a null effect now so frequently imputed to this experiment by writers on Relativity. (Miller 1933, p. 206).

After a very extensive analysis of the available data from the several repetitions of the experiment, Miller was able to show, as later verified by Shankland, that the fringes shifted with the frequency and phase, though not the amplitude, to be expected given the uncontracted aether theory. In terms of the concepts of this paper, Miller substituted for the null description his more accurate (but still idealized because of residual deviations) shift description. Furthermore, except for amplitude, this was the prediction of the idealized aether sketch.

Miller's next move was to show that it was not possible to improve the Relativistic idealized sketch by appealing to more realistic initial conditions.[5] The list of proposed candidates included variations due to "radiant heat, centrifugal and gyrostatic action, irregular gravitation effects, yielding of the foundation, magnetic polarization and magnetrostriction," (Miller 1933, p. 218). Miller's procedure was to vary these proposed causes and to note the absence of the hoped for, by the Relativists, corresponding effects. With respect to temperature variations, the leading candidate, Miller concluded that his "experiments proved that under the conditions of actual observation, the periodic displacements could not possibly be produced by temperature effects." (Miller 1933, p. 212).

It is fortunate for the history of physics that Miller was late for his calling, as defender of the aether theory, by a few decades. By the time Miller announced his comprehensive analysis in 1933, the battle for aetherial views had long since been lost and Relativity had won the day. Nevertheless, Shankland considered it of some value to carefully appraise Miller's arguments. One curious result of Shankland's appraisal is his flat contradiction of Miller's claims to have shown that a variation of heat sources produced no corresponding change (of the appropriate sort) in fringe shift.

> It must be emphasized that the foregoing analysis of these tests reveals small but certain temperature effects, in contrast to Miller's statement that he had shown the absence of periodic effects caused by artificial heating when the light path was thermally insulated as previously described. (Shankland 1955, p. 174).

Shankland gives no explanation of this discrepancy, nor has anyone else. In any case, Shankland provides a saving modal auxiliary for Relativity Theory: "an interpretation of the systematic effects in terms of the radiation field...is not in quantitative contradiction with the physical conditions of the experiment." (Ibid., p. 175). This episode, and in particular Shankland's defense of Relativity Theory, is very similar to Newton's defense of his ray theory. For in both cases, the predictions of the original idealized sketches were shown, by the opposition, to be replaceable by more accurate idealizations that were, at least in certain respects, straightforward deductions from alternative theories. And in both cases the original idealized sketch was defended by providing a saving modal auxiliary. There is an important difference, however, between Newton's defense and that of Shankland; it is discussed below.

There was, in addition to Miller's efforts, another class of attempts to show that the idealized sketch of the aether theory could be improved by making some of the idealizations more realistic. These attempts began with the observation that Michelson's single ray account, his idealized sketch, is really by itself (or as I would say: in the absence of a supporting modal auxiliary) no account at all, since it does not entail a finitely sized shifting fringe pattern. The surprising result of many of the more complete and realistic accounts of the interferometer was a null prediction! Since these idealized sketches were extremely complex and varied greatly in terms of approximation and mathematical techniques used, it was difficult to appraise them. Lorentz cut through the maze with a very elegant proof, utilizing Huygen's principle and the calculus of variations, to show that a null prediction could not be derived from more complete descriptions. That is, Lorentz gave a possibility argument: it was not possible by improving the idealizations used in the idealization sketch of the uncontracted aether theory to improve on the original prediction of a fringe shift.

3. A Taxonomy for Modal Auxiliaries

Kuhn and Lakatos have focused attention on the actual succession of increasingly superior idealized sketches. Lakatos (1970) speaks of "theories constructed in accordance with a program," while Kuhn (1970) describes the process as one of "paradigm articulation." For both philosophers, such series consist of actually constructed theories or articulations. As a historical phenomenon, there can be no doubt that such series can be reconstructed from the scientific record. What I have been drawing attention to is somewhat different: the role and importance of arguments whose conclusions concern the possibility of an improved idealized sketch or theory or articulation. Furthermore, if I am right in my claim that explanations include modal auxiliaries, then series of the sort discussed by Kuhn and Lakatos will contain series of associated modal auxiliaries. Finally, some of the work to be done by the proposed series of actually existing idealized sketches can be done, and historically was done, by the modal auxiliaries alone. I shall not pursue in this paper the connections between my views and those of Kuhn and Lakatos, but shall instead[6] opt for a further articulation of the concept of a modal auxiliary.

Since to be possible is always with respect to some set of constraints, the question here becomes, what sorts of constraints are presupposed by what I have called modal auxiliaries. This is, I believe, a very difficult question and I shall only make a cautious approach to an answer. The simplest sense of possible is what in ordinary parlance is usually called possible in fact. To show that x is possible in fact, one need merely give a recipe for the production of x that utilizes materials and procedures already in existence. The obvious analogue is the metaproof in constructivist logic: if we have but enough patience and symbols, the proof is constructible by means of the recipe. Newton's modal auxiliary for the *experimentum crucis* is of this type. There existed an aperture description, namely, "about 1/4 inch", that was more accurate than "infinitesimally small". The mathematics required to take account of a finite aperture was only simple geometry. Newton gave no formal proof to show that the utilization of a finite aperture would yield the desired effect; it could easily be "seen", perhaps on the basis of past experience with similar geometrical constructions.

Newton's comparison of the acceleration of the Moon with that of falling bodies is another case of the same sense of possible. Again, superior data and the required mathematical techniques were both available. There are two interesting differences though: (1) Newton gives, in the Moon case, a formal proof, and (2) what is shown is not that the prediction will improve, but that the prediction will remain the same in the face of more realistic descriptions of the initial conditions. In Proposition IV, Book III, _Principia_, Newton demonstrates this counterfactual: _if_ (1) the Sun is stationary (in absolute space), (2) the distance between Earth and Sun is 60 earth radii, (3) the period of the Moon's orbit is $27^d7^h43^m$, and (4) the gravitational force exerted by the earth varies as the inverse square, _then_ the acceleration of falling bodies at the surface of the earth _will_ be "15 and 1/12 Paris feet." The first item was thought by Newton to be false. In fact, its falsity is demanded by universal gravitation. The second item, as admitted by Newton, was not consistent with the best estimates of the time. Despite these false assumptions, Newton went on to argue on the basis of the counterfactual and the claimed correctness of its prediction that there was a universal and reciprocal gravitation. However, Newton _also_ gave a modal auxiliary to show that if more realistic assumptions were used, namely a moving Sun and a distance of 60 and 1/2 radii, the predicted acceleration of terrestrial bodies would remain unchanged. (The modal auxiliary is briefly discussed at the end of the Proposition, is given in more detail in section 11 of _De mundi systemate_, and utilizes Proposition LX of Book I, _Principia_.)

Lorentz had a similar aim for his possibility proof about the Michelson-Morley experiment: to show that the positive shift prediction would remain unchanged if more complete and realistic descriptions of the interferometer were substituted for Michelson's original single ray account. However, here the sense of possibility is not so clear since Lorentz's proof does not contain, in any straightforward way, a recipe for the construction of the proposed initial conditions.

The case of planetary perturbations differs from the above cases since neither the necessary data nor the required mathematical techniques actually existed (i.e., in the seventeenth century). Early perturbation predictions could be improved only if better planetary data were available as well as improved techniques of mathematical approximation. Since there was no guarantee that such data and approximation techniques were forthcoming, the possibility was, in ordinary parlance, open or unknown.

Another example of the "open possibility" is provided by the Einstein-Smoluchowski explanation of Brownian-motion. The idealized sketches used by Einstein and Smoluchowski made predictions that to an outsider must seem blatantly off the mark. Svedberg's experiments indicated displacements six to seven times larger than those predicted. The experiments of Henri brought the discrepancy down to only four times larger. Chaudesaigues, a student of Perrin, was able to get results significantly better, but still not exact or even within the range of calculated experimental errors. Nevertheless, Feyerabend and others have spoken of the "predictive success" of statistical mechanics and have seen these experiments as confirming the theory. While I agree with this appraisal, it is not for the reason that Feyerabend suggests, namely, the predictive advantage enjoyed by statistical mechanics over classical thermodynamics.[7] My contention is that there were at the time perceived open possibilities for improving the prediction of statistical mechanics. It is these possibilities that made the experiments confirming. For example, Einstein assumed that each increment of the path of the Brownian particle is independent of previous increments. But given the theoretical framework of statistical mechanics, there is reason to think that such an assumption is false. There was also the fact that Einstein, in the absence of a theory of statistical fluid mechanics, used Stokes' law for resistance in a continuous fluid. While there were open possibilities for Brownian motion, there were no possibilities for improving the (classical) statistical or kinetic predictions for specific heats. It had been shown analytically that all available means of describing gas molecules more realistically (because of the increased number of degrees of freedom of motion) would lead to predictions that diverged from experimental values.

I come now to my last category of possible, which can be somewhat misleadingly described as possible in principle only. Brownian motion provided an interesting puzzle for classical thermodynamicists, since the motion continued unabated despite all attempts to more perfectly isolate the fluid environment of the particle from outside influences. The standard idealized sketch had the fluid isolated from outside disturbances and hence in thermal equilibrium. There should therefore have been no particle motion. Given the assumed initial conditions, Brownian-motion is, as noted by Gouy and Poincaré, a perpetual motion machine of the second kind. Clearly some modal auxiliary was called for and it was this: since there is motion, it must be due to undetected outside influences. Furthermore, given the magnitude of the effect, it was obvious that measurement of the postulated deviations would be beyond the realm of instrumental capability. And even if such data were available, the necessary fluid dynamics was not available to assimilate it. This case is almost exactly analogous to Scriven's bridge example. In the thermodynamics case (as well as in the bridge example), the modal auxiliary makes no pretense of showing that improved predictions are possible. They are not and never will be. Despite this non-possibility result, such modal auxiliaries are not disconfirming, as one might expect given my view. Neither, or so I contend, are they confirming; they are neutral. They prevent disconfirmation by appeal to what, given the theory in question, must be the case. (I shall argue for this claim below.)

Given the special character of this sort of modal auxiliary, and given its analogy to Scriven's bridge example, I propose to give it a special name: the S-defense. Modal auxiliaries are of value because they point the way to improved idealized sketches. Of what value is the S-defense? Can such defenses ever be rationally rejected?[8] If not, then perhaps my claims about their effectiveness in preventing disconfirmation should be changed to: if it cannot be shown that improvement of idealized sketches is possible, then the theory stands refuted. S-defenses should count as blocking refutation only if they can be overthrown.

Historically, Gouy, Poincare, and Smoluchowski all held that the S-defense of thermodynamics in the face of Brownian-motion was insufficient. Several experimentalists had shown that when the claimed causes were varied, there was an absence of the appropriate variation in Brownian-motion. Hence, the S-defense was shown to be, in a sense, incoherent by the method of concomitant variation. S-defenses therefore are not immune to criticism. This sort of critical attack of an S-defense, you will recall, was also used by Miller against Special Theory: the fringe shift did not vary, or so he claimed, in the appropriate way given variations in the proposed causes.[9] Shankland's response was to restore the S-defense. It should also be noted that Shankland went on to determine if what I have called the S-defense could be strengthened so as to be a possible-in-fact modal auxiliary.[10] His conclusion was that the complexity of the causal interactions precludes, at least given current capabilities, calculation. Furthermore, Shankland went on to note that it is certainly the case that no idealized sketch could be made to approach the historically obtained fringe data since "hardly any of the data for such calculations exist." Therefore, the best that can be done with respect to explaining the historically performed experiments is the S-defense.[11]

4. Final Remarks

I shall now stop my taxonomic endeavors and end this paper with a corollary, a qualification, and a regret.

(a) A simple corollary of the above discussion is that what counts as a residual error or discrepancy to be accommodated by a modal auxiliary is a function of assumed theory. Color change after the second prism was a residual discrepancy for Newton; it was not for Hooke's diffusion theory. Similarly, fringe shift and Brownian-motion were residual discrepancies for respectively Special Theory and thermodynamics; they were not for aether theory and statistical mechanics. This situation is sometimes described by saying that what one theory explains another must explain away. A related point: the bewildering variety of remaining experimental discrepancies suggests that we look to alternative theories and explanations as providing the rationale for focusing on particular residual discrepancies. Brownian-motion became important for thermodynamics precisely because it followed (in some idealized description) from statistical mechanics.[12]

(b) I do not expect that in actual practice the expression 'explanation' will be used to pick out that combination of idealized sketch, modal auxiliary, and S-defense, that I have labeled an "explanation". Since the function of explanations is to fill in gaps in human knowledge and understanding, it is unlikely that the above combination will have to be given completely from one person to another. What I do expect to find is that when the question is the confirmation or disconfirmation of a theory in the physical sciences, what counts as an explanation will be some subset of the above items or of components of these items.[13]

(c) I have not said anything about the so-called statistical explanations. In part this has been because, with but one exception, the cases discussed have all involved idealized sketches that were exclusively deductive. The one exception, of course, is statistical mechanics. I am inclined to view the problems here, as well as with quantum mechanics, as rather different from explaining little Billy's recovery (or for that matter, his demise) from pneumonia in terms of a 90% survival rate for eight year olds. Regardless of the correctness of this appraisal, the basic taxonomy suggested here can accommodate, I believe, statistical idealized sketches. Of more immediate interest to me, but not carefully discussed in the paper, is the fact that statistical forms of argumentation do appear in the modal auxiliaries, as witnessed by the Brownian-motion and interferometer cases. Statistical arguments are of great importance, I would suggest, for the purpose of explaining away untoward variations between one's favorite idealized description and more accurate ones.

Notes

[1]In the interests of reasonable size and clarity, I shall avoid historical qualifications in this paper. I give more complete and accurate descriptions, as well as bibliographical details, of the principal historical cases of this paper in my (1977a), (1977b), (1978a), (1978b), and (1980).

[2]See, for example, Barr (1971), (1974), and Schwartz (1978).

[3](Duhem 1906, pp. 132-138). See Schwartz's distinction between intrinsically and extrinsically approximately true (1978, pp. 602-603). The distinction he has in mind is illustrated by the pair: there are nine marbles in the bag versus the bag weighs ten pounds. See also Scriven's interesting discussion about the gap between phenomena and description (1962, p. 195). The concept of a more accurate (yet no less precise) description is taken to be a primitive notion in this paper. This concept is theory relative and there may be disagreement among theorists over which competing (idealized) descriptions are more accurate, or in some cases, if the descriptions are even comparable. One cause of such disagreement is the fact that very often the theory of experimental instruments is the very theory to be tested. This means that estimates of experimental error will be a varying function of competing theories. For some examples of such disagreements and their rational adjudication, see my (1977a) and (1980). For an example o a disagreement beyond the realm of (direct) experimental determination and an argument form used to decide such disputes see my (1977b, pp. 231-233). Another cause of disagreement is a difference of theory-based opinion about relevant descriptive features; for examples, see my (1978b) and (1980).

[4]I do not mean to be arguing here for a thesis that claims the idealized sketch is psychologically developed first to be followed

by the modal auxiliary. Another qualification: my talk of a deductive-nomological sketch is not meant to indicate a commitment to a positive view about either the desirability or possibility of reconstructing such sketches in terms of first-order predicate calculus. I do believe, however, that there are some additional contraints the idealized sketch must satisfy if the modal auxiliaries are to be confirmationally relevant; these are given by Glymour's instantiation bootstrap theory, 1980. Finally, perceived similarity to past accepted cases (paradigms) can be taken as a reason for thinking that positive modal auxiliaries are forthcoming. It is such similarities in conjunction with compatible theory attachment that provides *prima facie* justification.

[5]Miller also attempted to provide a modal auxiliary for the aether theory, although he and others were greatly troubled by the fact that the phase relations were coherent for only three out of four observational epochs *and* that there was no apparent way of relieving this discrepancy. Appeal was also made to Planck's aether theory. Here, however, it looks as if an idealized sketch based on a new but similar theory is being given rather than a modal auxiliary for the old idealized sketch of the Fresnel aether theory. This shows that there are borderline cases. For similar problems with Lakatos' concept of a theory program see my (1978c).

[6]One sort of negative modal argument does play a central role in Kuhn's discussion of scientific change: if the best minds cannot construct an improved version of some idealized sketch then that sketch cannot be improved. In more Kuhnian terms: if the best efforts cannot further improve the articulation of a paradigm, then that paradigm is to be replaced (See,e.g.,Kuhn 1970, pp. 79-82). This is an important and actually used argument. My point is that it does not exhaust the forms of modal auxiliaries given by scientists.

[7](Feyerabend 1965, pp. 173-176). For a detailed analysis see my (1977b).

[8]A case can be made to show that Kuhn, and more clearly Feyerabend, give this sort of argument: there is always a discrepancy between the predictions of an idealized sketch and maximally accurate descriptions of phenomena; S-defenses are always possible; therefore, there is no refutation of a theory unless there is a predictively superior alternative. The argument of this paper seeks to deny this conclusion by showing that S-defenses can be rejected. (Cf.,my 1977b).

[9]In his refutation of the received laws of refraction (by means of the spectrum experiment) Newton also considered the S-defense of the received view (See *Corr*. I, pp. 92-93), and my(1978a, fn.9, pp. 255-256).

[10]One way of looking at Newton's famous bucket experiment is this: he gives a justification of his S-defense by noting that the effect in question (shape of water surface) does vary directly as the proposed

cause; and he attacks the S-defense of the Cartesians by showing that the effects vary improperly according to their proposed cause. The justification is an S-defense and not an idealized sketch because Newton does not draw a quantitative prediction on the basis of an idealized description of the bucket. It should also be noted that the usual description of the experiment, by, for example, Nagel and Reichenbach, is quite mistaken! Newton did not suddenly stop the bucket, as is usually reported. He did not need to give the nature of his analysis of S-defenses. For more details see my (1979).

[11]There are important distinctions to be made here between explaining particular experimental runs, particular versions of an experiment, and the experiment as ideal type. The idealized sketch is the explanation of the ideal type. Confirmation, however, comes from explaining runs of real versions. For some of the difficulties in analyzing the concepts of experimental types and sub-varieties, see my (1980).

[12]Alternatives also can be used to justify changes in explanatory standards. For example, the failure of theory T to make quantitative predictions about some aspect of a phenomenon of interest may be taken as a reason (if T is successful elsewhere) for thinking it unreasonable to demand such predictions. However, if some alternative T' comes along and does provide the desired predictions (and allows for the improvement of these predictions), then this success of T' is an argument for imposing new explanatory standards. Newton's optics was extremely important in this regard (See my 1978b, p. 70).

[13]An interesting example of a place where only a subset of my explanatory components is given is the scientific textbook. Here one typically finds only idealized sketches presented as the explanations of various phenomena. But as Kuhn has noted in 1961, experiments and data are not presented in textbooks for the purposes of confirmation and disconfirmation. See my (1977a) and (1977b) for an analysis of the changes between Newton's published papers and his quasi-textbook Opticks.

References

Barr, W.F. (1971). "A Syntactic and Semantic Analysis of Idealizations in Science." Philosophy of Science 38: 258-272.

---------. (1974). "A Pragmatic Analysis of Idealizations in Physics." Philosophy of Science 41: 48-64.

Duhem, Pierre. (1906). La theorie physique: son objet et sa structure. Paris: Chevalier & Riviere. (Translated by Philip P. Wiener as The Aim and Structure of Physical Theory. Princeton: Princeton University Press, 1954.)

Feyerabend, P. (1965). "Problems of Empiricism." In Beyond the Edge of Certainty. (University of Pittsburgh Series in Philosophy of Science, Vol. 2.) Edited by Robert Colodny. Pittsburgh: University of Pittsburgh Press. Pages 145-260.

Glymour, Clark. (1980). Theory and Evidence. Princeton: Princeton University Press.

Kuhn, Thomas. (1961). "The Function of Measurement in Modern Physical Science." In Quantification. Edited by Harry Woolf. New York: Bobbs-Merrill. Pages 31-63.

---------. (1970). The Structure of Scientific Revolutions. 2nd ed. Chicago: University of Chicago Press.

Lakatos, Imre. (1970). "Falsification and the Methodology of Scientific Research Programmes." In Criticism and the Growth of Knowledge. Edited by Imre Lakatos and Alan Musgrave. Cambridge: Cambridge University Press. Pages 91-195.

Laymon, Ronald. (1977a). "The Michelson-Morley Experiment: Description Dependence on To-be-tested Theories." In Motion and Time, Space and Matter. Edited by P. Machamer and R. Turnbull. Columbus: Ohio State University Press. Pages 436-464.

---------. (1977b). "Feyerabend, Brownian Motion and the Hiddenness of Refuting Facts." Philosophy of Science 44: 225-247.

---------. (1978a). "Newton's Advertised Precision and His Refutation of the Received Laws of Refraction." In Studies in Perception. Edited by P. Machamer and R. Turnbull. Columbus: Ohio State University Press. Pages 231-258.

---------. (1978b). "Newton's Experimentum Crucis and the Logic of Idealization and Theory Refutation." Studies in the History and Philosophy of Science 9: 51-77.

---------. (1978c). "Colin Howson (ed.) Method and Appraisal in the Physical Sciences." (Review) Philosophy of Science 45: 318-322.

---------. (1979). "Newton's Bucket Experiment." History of Philosophy 16: 399-413.

---------. (1980). "Independent Testability: The Michelson-Morley and Kennedy-Thorndike Experiments." Philosophy of Science 47: 1-37.

Miller, D.C. (1933). "The Ether-Drift Experiment and the Determination of the Absolute Motion of the Earth." Reviews of Modern Physics 5: 204-242.

Newton, Issac. (1959). The Correspondence, Volume I. Edited by H.W. Turnbull. Cambridge: Cambridge University Press.

Schwartz, R.J. (1978). "Idealization and Approximations in Physics." Philosophy of Science 45: 595-603.

Scriven, Michael. (1962). "Explanations, Predictions, and Laws." In Scientific Explanation, Space, and Time, (Minnesota Studies in the Philosophy of Science, Volume III). Edited by Herbert Feigl and Grover Maxwell. Minneapolis: University of Minnesota Press. Pages 170-230.

Shankland, R.S., McCuskey, S.W., Leone, S.W. and Kuerti, G. (1955). "New Analysis of the Interferometer Observations of Dayton C. Miller." Reviews of Modern Physics 27: 167-178.

Part XII

Economics and Sociobiology

How to Do Philosophy of Economics[1]

Daniel M. Hausman

University of Maryland/College Park

Although philosophers of science have always been interested in the actual work of scientists, there has been a strong turn in the last generation away from prescribing how science ought ideally to proceed and toward studying more carefully how science has proceeded. In part this turn has been a reaction to previous work in philosophy of science, which to many seemed misguided and largely irrelevant to the sciences. In part this change reflects a general scepticism about the possibility of doing traditional foundationalist epistemology. Such scepticism is itself a reaction to the failure of the foundationalist program of the logical empiricists. The contemporary turn toward careful empirical study of the sciences constitutes a new program for the philosophy of science, which I shall call 'empirical philosophy of science' or 'the empirical approach to the philosophy of science'.

1. Empirical Philosophy of Science

The credo of the empirical approach may be stated trenchantly and simplistically as follows:

> The philosophy of science is itself an empirical science.

All conclusions about the scientific enterprise that the philosopher of science draws are, or should be, scientific conclusions and must be defended in the same way or ways that the results of the sciences are defended. When the philosopher of science makes pronouncements about the goals of science or the basis or bases upon which scientists accept various theories or about any other feature of science, we should regard these pronouncements as scientific claims and assess them as we would assess the various assertions the sciences make.

The empirical approach to the philosophy of science denies that epistemology can be distinct from empirical study of the human cognitive faculties, the history of the human search for knowledge and the general

---

PSA 1980, Volume 1, pp. 353-362

progress of the sciences. In Quine's terminology (1969), epistemology is "naturalized". It aims no longer to justify kinds of knowledge claims in terms of an epistemologically prior (self-evident or indubitable) foundation. In justification we always take for granted much of our scientific and every-day knowledge. "Justifying" an assertion consists solely of showing that it is supported by evidence in the way or ways that scientific assertions generally are. When our claims to know are challenged, the best we can do is to explain scientifically how we know (or can know) what we do.[2] As empiricists, we accept these explanations ultimately because they help us to organize our experience and are supported by our experience. We have no other ultimate warrant. Our goal is to construct empirical theories of human knowing which are consistent with theories of other subject matters and which explain how we can know all these theories.

The empirical approach to philosophy of science is not purely "descriptive". Although philosophers' claims about sciences should be defended in part by showing their consistency with scientific practice, empirical philosophers of science can still assess the work of scientists and offer advice and instruction. Philosophers of science can sometimes contribute directly to the scientific disciplines they study. What we learn about scientific knowledge acquisition provides the basis for such assessment and advice. Empirical philosophy of science thus does not reduce to history of science. Not all of the history of science is relevant to the questions with which the philosophy of science is concerned. The precise details of how scientific results are reached are only important to philosophers of science when they help them understand how we come to know. On the other hand, there are other sources of evidence (for example, from psychology) about how humans acquire knowledge.

2. The Epistemological Circle

In attempting to study science as an empirical philosopher of science, one falls into a logical circle with at least four forms or manifestations. Such an "epistemological circle" is, in fact, common to every theory of knowledge. Hegel states the predicament well:

> We ought, says Kant, to become acquainted with the instrument [of cognition], before we undertake the work for which it is to be employed; for if the instrument be insufficient, all our trouble will be spent in vain. The plausibility of this suggestion has won for it general assent and admiration;... . In the case of other instruments, we can try and criticize them in other ways than by setting about the special work for which they are destined. But the examination of knowledge can only be carried out by an act of knowledge. To examine this so-called instrument is the same thing as to know it. But to seek to know before we know is as absurd as the wise resolution of Scholasticus, not to venture into the water until he had learned to swim. (1817, p. 14).

Some theories of knowledge find their way through these difficulties easily. If one maintains that there are self-warranting truths, for example, then one can easily meet the demand that we know some of the results of epistemology in order to do epistemology. The empirical philosopher of science, on the other hand, has some serious problems.

The first form of the epistemological circle is perhaps most striking. Empirical philosophy of science is itself a science. In doing philosophy of science empirically, one should thus follow scientific method or scientific methods. But one of the goals of the empirical philosophy of science is to find out what scientific methods are. It thus seems that one must already know at least tacitly what one is supposed to find out.[3] If we do not already know how to do science, how can we find out (scientifically) how to do science?

The circularity is not vicious. Empirical philosophers of science disavow seeking any justification for scientific knowledge other than the broadest possible coherence among our theories, including our theories of knowledge acquisition and our perceptual beliefs. There is thus nothing improper in beginning empirical philosophy of science as it were mid-stream, believing already that we know something tacitly or consciously about how to acquire knowledge. Justification, although philosophically interesting, is not the immediate task. Investigating scientific knowledge in accord with our initial conception of scientific investigation, we improve and articulate this conception (and revise our procedures for carrying out this improving and articulating) as we proceed. We are not guaranteed that we will not be forced to change our minds and our procedures. Although we cannot start learning about the sciences from scratch, we can learn about the sciences. This circle remains disturbing, since many philosophers find it difficult whole-heartedly to eschew searching for justification for our knowledge that goes beyond such broad coherence. Contemporary philosophers show, however, little enthusiasm for any alternative. The talk of "coherence" here should not be misconstrued. Since perceptual claims are for the most part knowledge claims, "coherence" here incorporates a sort of correspondence with "reality".

When one questions the philosophical theses upon which the empirical approach is based, the epistemological circle manifests itself again. Suppose some traditional philosopher maintains, as many have, that there is knowledge to be gained in epistemology which is different in kind from the empirical knowledge the sciences provide. Such a philosopher would accuse the empirical philosopher of science of avoiding the real epistemological tasks of assessing and justifying (not merely explaining) our scientific knowledge. In answer to such a challenge, the empirical philosopher of science must either deny that there are any such justificatory tasks or deny that there is any way to tackle them. But on what basis is either of these denials to be made? The grounds must themselves be the results of empirical philosophy of science (or of naturalized epistemology) or an anticipation of those results. But the traditional philosopher of science denies that philosophers ought to rely on (or ought to rely _only_ on) such grounds. All empirical philosophers of science can do is to repeat their (scientific) reasons for surrendering the ambitions of traditional foundationalist epistemology. They can, of course, also criticize in detail epistemologies which attempt to do more.

The third way in which the epistemological circle manifests itself is somewhat different. Much of the evidence upon which empirical philosophy of science bases its conclusions comes from the history of science. Unless, however, empirical philosophers of science are content only to describe all cognitive enterprises whatsoever, they must add to the presuppositions of their investigations discriminations between good and bad science, between science and pseudo-science, and between knowledge and conjecture. These initial discriminations are revisable as the inquiry proceeds, but they are indispensable. If an investigation of, for example, an economic theory is to contribute to understanding how humans acquire scientific knowledge, the philosopher must be able to assess that theory. An informed assessment demands that one *do* economics--that one finds out what there is to be learned at present about economies. The philosopher of economics must be a competent economic theorist. Standards to assess scientific work are also needed. Yet the standards of assessment and the methods to be employed in learning about economies can only be anticipated now. In trying to learn more, philosophers need to rely on all the knowledge they think they have, even if some of it is not well founded and turns out not to have been knowledge at all.[4] There are, however, special difficulties when one's subject matter is a discipline like economics, whose conclusions are disputed and important to people's material interests. I will return to these last difficulties below.

The fourth form of the epistemological circle concerns the relations between empirical philosophy of science and empirical philosophical investigations of particular sciences. The conclusions of empirical philosophy of science rest largely on investigation of the history of actual sciences. To that extent empirical philosophy of science in general depends on empirical investigations of particular theories, disciplines, incidents, etc. General conclusions in the philosophy of science must rest on particular inquiries into particular sciences. Yet in order to investigate some limited area in science, one needs a great deal of philosophical apparatus. One has no choice except hesitantly and critically to rely on philosophical models of theories, explanations, laws, confirmation, objectivity and the like. Once again the philosopher must anticipate the answers to his or her questions. If the conclusions of current philosophy of science were already well supported and already merited the esteem and confidence of philosophers and scientists, these anticipations would not be troubling. But a great deal of "established" philosophy of science is poorly confirmed and has been cast into doubt. In this last form the epistemological circle presents a pressing practical problem. In my own work concerning theories of capital and interest, I have made use of whatever philosophical wisdom I could; but the limitations in that wisdom have been palpable. Yet there is no way to increase our knowledge in the philosophy of economics or the philosophy of science in general except to rely on (while attempting to improve upon) conclusions of the past.

Empirical philosophers of science are caught in at least these four ways in the epistemological circle. Does this fact make dubious an empirical approach to the philosophy of science? Should we worry about whether the conclusions of empirical philosophical investigations of particular sciences are prejudiced by the presuppositions with which they begin? Note that many of these presuppositions come from less self-conscious "investigations" of just the same data (from the history of science and from experience of human learning) that the philosophy of science now examines more systematically. We already know a good deal about the world and about how to get knowledge about the world. Without that knowledge, we could not inquire into the nature of our knowledge and the means of its acquisition--but then we would lack not only the means to carry out such an investigation, but also an object to investigate. If we really lacked even tacit knowledge about how to acquire knowledge, we would be unable to find out how to learn by investigating scientifically the knowledge we had. Not only would we not know how to inquire, but we would have little or no knowledge to inquire about. The possibility of doing epistemology arises with the possibility of having serious epistemological questions. One may, of course, have good reasons to suspect bias in particular cases. General doubt about whether we can achieve any knowledge through an empirical approach to the philosophy of science on the other hand merely expresses scepticism about the possibility of human knowledge in general. It may turn out, of course, that we are unable in doing philosophy of science to come up with any interesting general results.

3. Philosophy of Science as a Social Science

The empirical philosophy of science, if itself a science at all, is a social science (where 'social science' is understood to include history and psychology). Thus it may be that the structure, methods, etc.,of philosophy of science will be unlike those of the natural sciences. Social scientific naturalists argue that, in crucial respects (goals, methods of justification, logical structure, fundamental ontology, or whatever), the social sciences are or should be identical to the natural sciences. Anti-naturalists argue for an essential difference in one or more of these respects. The debate over naturalism, as the last two sentences suggest, is exceedingly messy. It seems to me that the empirical approach to the philosophy of science ought not itself to prejudge this debate. Both naturalists and anti-naturalists ought to be able to adopt empirical approaches to the philosophy of science. Otherwise it is hard to see how the philosophy of science can contribute to clarifying and resolving the many disputes between them. Individual empirical philosophers of science may anticipate the resolution of the debate over naturalism.

The empirical approach to the philosophy of science does not presuppose that the structure, methods, etc. of the social sciences (and thus of the philosophy of science itself) are the same as those of the natural sciences. In fact, philosophers of science rarely study the sciences the way physicists study motion or matter. The actual practice of empirical philosophy of science is diverse. Much of it will remain for the foreseeable future more like intellectual history than like physics. While the object of the philosophy of science is all of science, its structure(s), methodology(ies) and the like should be that of (some of) the social sciences. The worst social scientific naturalists can say of this distinction is that it is empty.

Notice that the question of social scientific naturalism is only a special form of the question of whether the methods, structure, goals and the like are, at a suitable level of generality, one and the same for all sciences in all historical periods. Although we may sometimes have to beg this general question, we should not forget that it is there. It should not be a condition of doing the philosophy of science empirically that this question have only one answer. Otherwise we could not learn its answer in doing philosophy.

4. How to Do Philosophy of Economics

If the above general view of the philosophy of science is correct, how is one to do the philosophy of economics? How is one to answer a much debated question like:"Is microeconomic theory a good scientific theory despite the fact that its basic lawlike claims appear to be false?"

The general technique--to study the works of economists and philosophers which develop, apply and discuss the theory--is certainly not novel. In the actual course of such study the philosopher of economics will have to rely heavily on the tentative results of contemporary philosophy of science and on initial judgments concerning the nature and worth of economic theory and of economics as a discipline. Merely to classify and to order what one finds when reading economics books, one must have some idea of what a science is, what a theory is, what count as laws and so forth. The richness of philosophical work on the natural sciences and the extent of its influence makes it tempting to suppose that a moderate naturalism is correct. Economists talk about their own work in many ways. They write, for example, about "principles", "models", "theories", "assumptions", and "definitions" and make use of previous work by epistemologists and philosophers of science. To interpret their comments and to describe accurately what they have done, one needs to know a great deal of philosophy of science. How else is one to decide whether microeconomic theory is even a theory?

Some of those most critical of traditional philosophy of science and most insistent on the need for a new empirical philosophy of science might object that we do not know enough philosophy now to understand the structure or methods of microeconomics. There is some merit in this objection, although I believe that it is overstated. It would help if we could begin with solid and well-confirmed philosophical theses. But no philosopher of science can now begin with these, since they are unavailable. A philosopher of economics studying economic theory is in the same philosophical position as any empirical philosopher of science seeking knowledge about the sciences. The only important difference is that philosophers of physics, for example, can begin with fewer doubts about the worth of the physics they study. Philosophers of physics are unlikely ever to conclude that Newton was a mediocre physicist. They can safely begin by regarding a large body of physics as "good physics". Revisions may of course be needed later. Philosophers of physics have, however, comparatively few practical problems deciding what to do when conventional philosophical wisdom does not fit the "good physics" studied. The difficulties facing a philosopher of economics are much greater.

Yet I do not think that this contrast with philosophy of physics shows that we should postpone philosophical examination of more dubious sciences like economics. What we learn about knowledge acquisition in physics may not apply to economics. Even if it does, philosophers of economics will probably have to find this out through their investigations of economics. Furthermore, although the practical differences between the tasks of philosophers of economics and philosophers of physics are considerable, they are differences in degree, not in kind. Philosophers of physics can hardly assume that Newton or Einstein never blundered.

How are philosophers of economics to proceed, if they cannot simply import categories and theses concerning theories, laws and so forth upon which philosophers agree? When microeconomics fails to fit current philosophical conceptions, philosophers cannot automatically conclude that something is wrong with microeconomics. Philosophers of economics will have to trim, revise and even invent philosophical categories and theses in trying to make sense of economic theory. They can neither start from scratch nor rely on authoritative philosophical dicta. Cautiously and critically, the philosopher of economics must make use of the most plausible among current philosophical views of the sciences, as ill-founded and wrong-headed as they may be. There is no alternative.

To make sense of, for example, microeconomic theory, one needs not only philosophical apparatus to systematize what one finds, but also an idea of the sort of sense to make of the theory. Histories full of rational decision-making and debate certainly make sense, but so do histories full of stupidity, stubbornness, dishonesty and ideological distortion. When the economist's practice conflicts with the philosopher's dicta, which should be criticized? The question arises frequently. Controversies concerning the merits of economic theories often, for example, do not resemble philosophical models of how scientists assess competing theories. Should one "make sense" of such controversies as a different sort of rational debate or should one "make sense" of them by concluding that they are shot-full of confusion, misunderstanding and ideological distortion? Obvious answers may be deceptive. Perhaps there are unfamiliar methods or structures of sciences which are of great value. Yet we must make such assessments.

The difficulties are aggravated because we know that discussions of economic issues are often biased and distorted because of their importance to interests of individuals of various social groups. As Marx luridly put it: "In the domain of Political Economy, free scientific inquiry meets not merely the same enemies as in all other domains. The peculiar nature of the material it deals with summons as foes into the field of battle the most violent, mean and malignant passions of the human breast, the Furies of private interest." (1867, p. 10). Although I am sceptical of the possibility of finding a completely neutral starting point and of avoiding commitments, the philosopher of economics can address a broader audience and a wider spectrum of issues if he or she does not start by taking neo-classical (or Marxian, or institutionalist or monetarist) economics as the paradigm for what economics should be. The philosophy of economics must struggle to avoid becoming apologetics for any school of economics.

My own work leads me to believe that the task of the philosopher of economics should be to show that the state and development of economics manifest imperfect rationality. The standard of scientific rationality comes, as it must, from existing philosophy of science as inadequate as it may be. One should expect to find deviations because of the Furies' influence, but I believe that these will be important exceptions and complications, not the center of the story. If one succeeds in providing a compelling account that is in accordance with these expectations, one thereby provides evidence (not proof) that these expectations are correct.

Seeking to find imperfect rationality in economics comes down to looking for good reasons for whatever one finds unless there are specific grounds to expect or to substantiate bias. Given the dubiousness of many of the conclusions of economics, it is crucial to distinguish carefully between judging the enterprise to be rational and judging its results to be correct. When, according to the standards of accepted philosophy of science, some feature of, for example, microeconomics appears irrational, one should look both for ways of improving the philosophical model and for evidence of the influence of ideology or of simple error. I know of no precise rules to decide such cases.

5. Conclusion

The methodology of the philosophy of economics is thus vague and imprecise. It hardly evidences a dramatic new approach to the philosophy of science, such as the empirical approach might initially appear to be. What the empirical approach implies in practice are the following: (1) Philosophers should demand historical and psychological evidence for their conclusions and, in so far as that evidence is scanty (which it has been), should be hesitant about accepted philosophical "wisdom" concerning the sciences. (2) Philosophers of science should be more willing to study and to learn from particular sciences than they sometimes have been. Much can, I think, be learned by employing this homey advice.

Notes

[1]I am indebted to Philip Ehrlich, Michael Gardner, Jonathan Lieberson, Stephen Stich and Paul Thagard for criticism of earlier versions and to unpublished work of Dudley Shapere.

[2]One might argue that a scientific explanation of knowledge acquisition is inconceivable because empirical scientists cannot explain why some methods of acquiring beliefs justify beliefs while some do not. Determining the standards for justified belief is not and cannot be the task of any empirical science. I do not find this claim compelling. A psychologist might, by means of sufficiently cunning experiments, be able to show that certain methods of acquiring beliefs are more likely to lead to true beliefs in certain circumstances than are others. If the psychologist could moreover show us why some grounds for believing lead to more reliable beliefs, then he or she would be in a position to explain how in certain circumstances people acquire knowledge. There are obviously many circularities here, but I argue in the body of the paper that they are benign. I thus see no reason to believe it impossible that we can explain how we know what we do.

[3]I doubt that the distinction between knowing how and knowing that is sharp and significant enough to provide a way out of such circularity, although I cannot argue the case here.

[4]Since empirical philosophers of science must begin by discriminating knowledge from superstition and science from pseudo-science, is not the way open for astrologers, for example, to begin by regarding astrology as the paradigm of a science? Might they not then come up with an empirical philosophy of science which shows how we can acquire such astrological knowledge? After all, a crucial test, among astrologers, for any philosophical account of science will be whether it successfully shows us how we can know astrology. But if astrologers can invent an empirical philosophy of science that "justifies" the claims of astrology, what does our empirical philosophy of science, which criticizes astrology, accomplish? The argument is deceptive. Alternative philosophies of science are, I suspect, not easily created. Astrologers who attempt to come up with a naturalized epistemology which coheres with both their purported knowledge of astrology and their non-astrological knowledge of the everyday world will face a difficult task. If they find that epistemology is to be done as a non-astrological science is done, they will discover that their attachment to astrology is irrational. If they develop some other sort of epistemology, they might (in some sense) be able to come up with a coherent body of knowledge. Yet this body of knowledge would have to be so different from ours, that our inability to show the astrologers that they are in error is not so disturbing. Once we deny that there is any certain or self-evident foundation for human knowledge, the possibility of consistent and incommensurable knowledge systems cannot be denied. The fact that the astrologer (or theologian or paranoid) begins with different beliefs does not, however, itself show that such incommensurable knowledge systems can be constructed.

References

Hausman, D. M. (Forthcoming), A Philosophical Inquiry into Capital Theory. New York: Columbia University Press.

Hegel, G.W.F. (1817). Encyklopädie der philosophischen Wissenschaften in Grundrisse. Erster Teil. Heidelberg: A. Oswald. (Translated by William Wallace as Hegel's Logic. (Part One of the Encyclopedia of the Philosophical Sciences.) Oxford: Clarendon Press, 1975.)

Marx, Karl. (1867). Das Kapital, Vol. 1. Hamburg: O. Meissner. (Translated by Samuel Moore and Edward Aveling as Capital, Vol. 1. New York: International Publishers, 1967.)

Quine, W. V. O. (1969), "Epistemology Naturalized." In Ontological Relativity and Other Essays. New York: Columbia University Press. Pages 69-90.

Is Sociobiology A Pseudoscience?

R. Paul Thompson

Scarborough College
University of Toronto

Among the numerous criticisms of sociobiology is the criticism that it is not genuine science. The attempts to support this charge are based on the beliefs that sociobiology is in some sense untestable (Allen et al 1976 and 1977; Sahlins 1977; Lewontin 1977) and untestable hypotheses and theories are pseudoscientific. There are, however, a number of difficulties with both of these beliefs. First, the testability criterion of genuine science is far from clear or well accepted. It is now obvious to most philosophers of science that hypotheses and theories do not occur in isolation. Numerous assumptions and hypotheses together form part of a general framework within which any hypothesis is formed, applied and tested (see, Quine 1951; Quine and Ullian 1978, pp. 96-107; and Kuhn 1970). The refusal to give up a hypothesis in the face of a negative test result is not only not grounds for considering it pseudoscientific but is also a common occurrence in scientific research. Scientists can and do employ many techniques to save a hypothesis. The less isolated the hypothesis the more acceptable are these various moves. This is in large part due to the often drastic implications of the rejection of the hypothesis for the framework of which it is a part.

In the second place, even if the criterion were acceptable it would not mark sociobiology as pseudoscience since it is in important respects testable. Sociobiological models permit the prediction of certain behaviour and gene ratios, and the failure of these to obtain in a number of interesting cases would clearly provide a basis for rejecting a model. And rejections of models on this ground by sociobiologists have occurred (see for example, Alexander and Sherman 1977).

Despite the problems with the testability criterion of pseudoscience and despite the criterion's inapplicability to sociobiology, there is little doubt that sociobiology will continue to be viewed by many as metaphysical or a caricature of genuine science. While testability was, for the critics, a convenient expression of the underlying belief that sociobiology is not genuine science, it does not appear to be the sole or

---

PSA 1980, Volume 1, pp. 363-370

most important ground for the belief. The critics might well argue that the failure of this criterion does not mean that other criteria will not work and if other criteria also involve difficulties, that would not establish that sociobiology was not a pseudoscience any more than similar failures with respect to astrology or biorhythms would vindicate them. Despite the lack of an unproblematic criterion of pseudoscience there are identifiable pseudosciences and sociobiology is one of them. What this line of argument makes clear is that any attempt to claim that sociobiology is genuine science will have to do more than show that the testability criterion is unacceptable or inapplicable.

In an attempt to meet the implicit challenge of this position and also to define sociobiology's status as genuine science, I take two complementary approaches. First, I argue that a reasonable criterion of pseudoscience developed by Thagard (1978) makes possible the application of the term 'pseudoscience' to astrology but not to sociobiology. Second, I argue that there are positive grounds for accepting sociobiology as genuine science.

Paul Thagard has argued that a demarcation criterion for pseudoscience, "requires a matrix of three elements: [theory, community, historical context]." (p. 227). The theory element is comprised of the traditional concerns of explanation, prediction, etc. The community element is comprised of concerns about: agreement of the practitioners on the concepts, principles, and methodology; the degree of commitment of the practitioners to explaining anomalies and to comparing their success with others; and the extent to which the practitioners are, "actively involved in attempts at confirming and disconfirming their theory." The historical element is comprised of concerns about, "the record of a theory over time in explaining new facts and dealing with anomalies, and the availability of alternative theories." On the basis of this matrix of elements Thagard proposes the following principle of demarcation:

> A theory or discipline which purports to be scientific is <u>pseudoscientific</u> if and only if:
>
> 1) it has been less progressive than alternative theories over a long period of time, and faces many unsolved problems; but
>
> 2) the community of practitioners makes little attempt to develop the theory towards solutions of the problems, shows no concern for attempts to evaluate the theory in relation to others, and is selective in considering confirmations and disconfirmations.
>
> Progressiveness is a matter of the success of the theory in adding to its set of facts explained and problems solved. (pp. 227-228).

Thagard then argues that on this principle of demarcation astrology is marked as a pseudoscience.

> This principle captures, I believe, what is most importantly unscientific about astrology. First, astrology is dramatically unprogressive, in that it has changed little and has added nothing to its explanatory power since the time of Ptolemy. Second, problems such as the precession of equinoxes are outstanding. Third, there are alternative theories of personality and behaviour available: one need not be an uncritical advocate of behaviourist, Freudian, or Gestalt theories to see that since the nineteenth century psychological theories have been expanding to deal with many of the phenomena which astrology explains in terms of heavenly influences. The important point is not that any of these psychological theories is established or true, only that they are growing alternatives to a long-static astrology. Fourth and finally, the community of astrologers is generally unconcerned with advancing astrology to deal with outstanding problems or with evaluating the theory in relation to others. For these reasons, my criterion marks astrology as pseudoscientific. (p. 228).

The criterion, however, does not mark sociobiology as a pseudoscience. There is a trivial reason why this is the case and more substantive reason. In a trivial sense it fails to mark sociobiology as a pseudoscience because sociobiology is a very recent development and cannot, by virture of this, "have been less progressive than alternative theories over a long period of time." (p. 228). Were this the only reason for asserting that sociobiology is not a pseudoscience on this criterion, the case would be exceptionally weak. There are, however, two other considerations that strengthen the case considerably. First, with respect to all the other elements in the criterion, sociobiology passes the test of genuine science. Second, the absence of alternative theories or the lack of sufficient time, while saving a theory from being clearly branded a pseudoscience on this criterion, does not make a discarding of it impossible. Thagard considers this situation in connection with the history of astrology.

> But there remains a challenging historical problem. According to my criterion, astrology only became pseudoscientific with the rise of modern psychology in the nineteenth century. But astrology was already virtually excised from scientific circles by the beginning of the eighteenth. How could this be? The simple answer is that a theory can take on the appearance of an unpromising project well before it deserves the label of pseudoscience. The Copernican revolution and the mechanism of Newton, Descartes and Hobbes undermined the plausibility of astrology. Lynn Thorndike has described how the Newtonian theory pushed aside what had been accepted as a universal natural law, that inferiors such as inhabitants of earth are ruled and governed by superiors such as the stars and the planets. William Stahlman has described how the immense growth of science in the

> seventeenth century contrasted with stagnation of astrology. These developments provided good reasons for discarding astrology as a promising pursuit, but they were not yet enough to brand it as pseudoscientific, or even to refute it. (p. 229).

In the case of sociobiology, there is no reason at this point to consider it an unpromising pursuit. Indeed, it appears to be a very promising pursuit. Certainly in the case of insect research it has shown exceptional promise. Again, however, it must be admitted that if showing promise was the sole ground for claiming that sociobiology was, on Thagard's criterion, not a pseudoscience it would be a weak case. Hence, the substantive case rests with the first point, i.e., with respect to all the other elements in the criterion sociobiology passes the test of genuine science. This in conjunction with its potential promise is sufficient, on Thagard's criterion, to mark sociobiology as a genuine science.

The second condition of Thagard's criterion deals with the theory and community elements of his matrix. There is little doubt that sociobiologists have made considerable attempts, "to develop the theory toward solutions of the problems." These attempts have not always been successful or readily accepted by the community of practitioners but there can be no doubt that the practitioners have made such attempts. Wilson's book (1975) alone demonstrates this fact. Also, sociobiologists show considerable concern for attempts to evaluate the theory in relation to others. As just one example, the very first chapter of Barish's book (1977) is devoted to discussions of past theories of animal behaviour, the limitations of current theories, how sociobiology provides a framework (paradigm) for the study and interpretation of animal behaviour, and how it integrates other areas of study. It is hard to see how in light of this sociobiologists could be charged with a lack of concern for attempts to evaluate the theory in relation to others. Finally, sociobiologists are not selective in considering confirmations and disconfirmations. This element is, of course, the falsifiability (testability) issue in different garb and Thagard is quite right to include it. The value of Thagard's criterion is its comprehensiveness. It makes clear that the falsifiability element is not alone sufficient to mark a purported science as pseudoscience but it is an element in the assessment of the purported science. As indicated earlier in the discussion, sociobiology is open to refutation by evidence.

One can conclude from the above discussion that on Thagard's criterion of pseudoscience sociobiology is not a pseudoscience. Failure of sociobiology to satisfy even one of the elements would be grounds for declaring that it was not a pseudoscience. Sociobiology, however, fails to satisfy any of the elements. Thus, it passes the test. There remains only one question: "Why should Thagard's criterion be given such importance in this issue?" The response to this question is threefold: first, Thagard's criterion has the undeniably important virtue of marking recognized pseudosciences as pseudoscientific. As Thagard himself argues in his article, other criterion such as those implicit in Lakatos and Kuhn and explicit in Popper, employ central notions but are not by themselves sufficient to mark genuine science from pseudoscience. Second, Thagard's criterion takes seriously the wide range of elements that causes doubt about a branch of science. The criterion represent the now obvious fact that genuine sciences do from time to time satisfy some of these elements. Only when they overstep the mark by satisfying all the elements in the criterion can they be considered pseudoscientific. Third, because Thagard's criterion is so comprehensive it is unlikely that any purported science that satisfies none of the elements of his criterion will satisfy any elements of a more restricted criterion. Thus, by measuring sociobiology against Thagard's criterion, sociobiology is measured against a number of more restricted criteria at the same time since the more restricted criteria have been worked into Thagard's comprehensive criterion.

Since suspicion will no doubt linger despite the fact that there appears to be no adequate grounds for considering sociobiology a pseudoscience, consideration of a possible argument in favour of its acceptance as genuine science is appropriate. The place to begin is with the views of sociobiologists themselves. Why do sociobiologists think that they are doing genuine science? The answer seems to be that sociobiologists consider sociobiology to be "the application of evolutionary biology to social behaviour." (Barish 1977, p. ix). Wilson, for example, in Sociobiology: The New Synthesis claims, "This book makes an attempt to codify sociobiology into a branch of evolutionary biology and particularly of modern population biology." (p. 4). The contention seems to be that sociobiology involves the application directly and via models, of the concepts and principals of evolutionary theory to the explanation of social behaviour. This position gives rise to two issues. First, there is the issue of whether sociobiology does involve the application of the concepts and models of evolutionary theory to the explanation of social behaviour. Second, there is the issue of whether a branch of science that involves the application of the concepts and principles of a genuine science is, by virtue of this, a genuine science. Let me discuss these separately.

Typical of sociobiological model-explanations is the explanation of altruistic behaviour given by W.D. Hamilton (1964a and 1964b). This model-explanation is interesting and appropriate for the purpose at hand because it attempts to provide an evolutionary explanation of a social behaviour that on the surface appears to reduce an organism's chances of reproductive success. Hamilton's model makes use of Wright's coefficient of relationship as a measure of the proportion of genes of common descent in individuals with particular genetic relationships. Hamilton's central thesis is succinctly stated by J.M. Smith (1978) as follows:

> Consider two individuals, a "donor" and a "recipient." The donor performs an act that reduces its own Darwinian fitness, or expected number of surviving offspring, by a cost C but increases the recipient's fitness by a benefit B. Suppose that there is a pair of allelic, or alternative, genes A and a and that the presence of A makes an individual

> more likely to perform the act. Hamilton showed that the change in the frequency of gene *A* in the population after the act depends on the coefficient of relationship *r* between the donor and the recipient, that is, on the average fraction of genes of common descent individuals with the genetic relationship of the donor and the recipient. More precisely, he showed that (with certain approximations) the frequency of gene *A* will increase because of the donor's act if the coefficient or relationship *r* is greater than *C/B*.
>
> For example, if an individual has two sets of genes, one from a father with two sets and one from a mother with two sets, then on the average the probability is 1/2 that any particular gene in the individual is present in a full sibling. Hence the coefficient or relationship between the individual and its full sibling is 1/2. Therefore, according to Hamilton's argument, if a gene in the individual causes it to sacrifice its life to save the lives of more than two siblings, then the number of replicas of the gene present after the sacrifice is made is greater than the number present would be if the sacrifice had not been made. The sacrifice is selectively advantageous. (In this instance the cost *C* is equal to 1 and the benefit *B* is equal to more than 2, so that the coefficient of relationship 1/2 is indeed greater than *C/B*.) (p. 178).

Hamilton's model involves the concepts of 'gene frequency', 'Darwinian fitness' and 'genetic relationships of members of a population'. Gene frequency is understood and calculated in the same manner as in population genetics. Darwinian fitness is still understood in terms of the expected number of surviving offspring. The important concept is that of 'genetic relationship' understood in terms of Wright's coefficient of relationship. This can hardly be considered a departure from population genetical principles. This concept does not change existing concepts or change fundamental principles, but rather is a concept that captures an already existing presupposition of population genetics, i.e., that individuals in a population are related to other individuals genetically. Could anyone have believed otherwise? What Wright provided was a way to quantify the concept of genetic relationship and this method of quantifying was clearly within the framework of population genetics. It was a method of assessing the ratio of share genes. Hamilton's demonstration of the role of the coefficient of relationship was a simple application of existing population genetical algebra. In light of this is it hard to see how the sociobiologist's claim that sociobiology involves an application of evolutionary theory (i.e., population genetics) to social behaviour could be doubted? It should be noted that after Hamilton formulated his model, considerable confirmation from the study of the social behaviour of insects was achieved. Hamilton's model was capable of explaining the evolution of the highly developed social systems of such insects as those of the hymenoptera order. Also, the model in recent years has been applied to the explanation of the evolution of certain behaviours of birds and mammals. The model, then, not only is firmly entrenched in the framework of population genetics but has also been applied to phenomena of a far wider scope than was conceived at its inception.

Insofar as Hamilton's model is representative of sociobiology, and it is, there can be little question that sociobiology is indeed best understood as the application of the concepts and principles of evolutionary theory to social behaviour. The remaining question is whether that understanding of sociobiology is, in itself, support for the claim that sociobiology is genuine science. I must confess that I find it hard to imagine how it could fail to constitute support. There is, of course, the obvious problem that some applications of the concepts and principles of a genuine science can so alter those concepts and principles that only the words remain the same. In such cases there would be clear grounds for rejecting the claim that the application of the concepts and principles supports the use of the title genuine science for the applying discipline. This, however, is not, in general, the case with sociobiology. It does not, in its application of the concept 'gene', change that concept. It continues to use such concepts in the same sense that they have always been used in population genetics. Clearly sociobiology has added concepts to those available in population genetics but these new concepts have been characterized in terms that are consistent with the existing concepts and principles of population genetics. The only criticism of sociobiology that attempts to show that it has changed a fundamental concept of evolutionary theory is that of Sahlins (1977) in his chapter, 'Ideological Transformations of Natural Selection' (pp. 71-91). The criticism, however, is confused. As Ruse has pointed out, "it is clear that Sahlins exhibits a fundamental misconception of evolutionary theory: that is, of evolutionary theory of the most orthodox kind." (p. 110). The misconception concerns the meaning of differential reproduction. Sahlins (p. 87) characterizes differential reproduction in terms of differential advantage and much of his argument at this point rests on this characterization. This, however, is too narrow a characterization. Differential reproduction also involves differential disadvantages. Those with disadvantages will be disfavoured. When this is introduced into discussion Sahlin's case is substantially weakened if not destroyed (see Ruse 1979, pp. 106-111).

Sociobiology, then, passes the test of Thagard's criterion and it seems reasonable to assert that there is a positive and impressive ground for claiming that it is a genuine science. While I have no doubt that this will not convince many of the critics, the ball is now in their park. Their previous case against sociobiology as genuine science has not only been undermined but in its place a case in favour of sociobiology's acceptance as genuine science has been offered.

References

Alexander, R.D. and Sherman, P.W. (1977). "Local Mate Competition and Parental Investment Patterns in the Social Insects." Science 196: 494-500.

Allen, E. et al. (1976). "Sociobiology: Another Biological Determinism." Bioscience 26: 182-186.

--------------. (1977). "Sociobiology: A New Biological Determinism." In Biology as a Social Weapon. Edited by The Ann Arbor Science for the People Editorial Collective. Minneapolis: Burgess Publishing Company. Pages 133-149.

Barish, D.P. (1977). Sociobiology and Behaviour. New York: Elsevier.

Hamilton, W.D. (1964a). "The Genetical Theory of Social Behaviour I." Journal of Theoretical Biology 7: 1-16.

-------------. (1964b). "The Genetical Theory of Social Behaviour II." Journal of Theoretical Biology 7: 17-32.

Kuhn, T.S. (1970). The Structure of Scientific Revolutions. 2nd ed. Chicago: The University of Chicago Press.

Lewontin, R.C. (1977). "Sociobiology - A Caricature of Darwinism." In PSA 1976, Volume Two. Edited by F. Suppe and P.D. Asquith. East Lansing: Philosophy of Science Association. Pages 22-31.

Quine, W.V. (1951). "Two Dogmas of Empiricism." Philosophical Review 60: 20-43.(Reprinted in From a Logical Point of View. 2nd ed. Cambridge: Harvard University Press, 1961. Pages 20-46.)

---------- and Ullian, J.S. (1978). The Web of Belief. 2nd ed. New York: Random House.

Ruse, M. (1977). "Sociobiology: Sound Science or Muddled Metaphysics?" In PSA 1976, Volume Two. Edited by F. Suppe and P.D. Asquith. East Lansing: Philosophy of Science Association. Pages 48-73.

-------. (1979). Sociobiology: Sense or Nonsense? Boston: D. Reidel.

Sahlins, M. (1977). The Uses and Abuses of Biology. London: Tavistock.

Smith, J.M. (1978). "The Evolution of Behaviour." Scientific American 239 (September): 176-192.

Thagard, P.R. (1978). "Why is Astrology a Pseudoscience?" In PSA 1978, Volume One. Edited by P.D. Asquith and I. Hacking. East Lansing: Philosophy of Science Association. Pages 223-234.

Wilson, E.O. (1975). Sociobiology: The New Synthesis. Cambridge: Belknap Press.